Infectious Disease Ecology and Conservation

Infectious Disease Ecology and Conservation

Johannes Foufopoulos

Associate Professor, School for Environment and Sustainability,
University of Michigan, USA

Gary A. Wobeser

Professor Emeritus, Department of Veterinary Pathology,
University of Saskatchewan, Canada

Hamish McCallum

Professor and Director, Centre for Planetary Health and Food Security, Griffith University
and Environmental Futures Research Institute, Australia

OXFORD
UNIVERSITY PRESS

Great Clarendon Street, Oxford, OX2 6DP,
United Kingdom

Oxford University Press is a department of the University of Oxford.
It furthers the University's objective of excellence in research, scholarship,
and education by publishing worldwide. Oxford is a registered trade mark of
Oxford University Press in the UK and in certain other countries

Published in the United States of America by Oxford University Press
198 Madison Avenue, New York, NY 10016, United States of America

British Library Cataloguing in Publication Data
Data available

Library of Congress Control Number: 2021942639

ISBN 978–0–19–958350–8 (hbk)
ISBN 978–0–19–958351–5 (pbk)

DOI: 10.1093/oso/9780199583508.001.0001

Printed and bound by
CPI Group (UK) Ltd, Croydon, CR0 4YY

Links to third party websites are provided by Oxford in good faith and
for information only. Oxford disclaims any responsibility for the materials
contained in any third party website referenced in this work.

This book is dedicated to the countless field conservation biologists and wildlife veterinarians who struggle daily, often at great personal cost, for the protection of the planet's biodiversity. It is also dedicated to the present generation of students whose idealism and energy make us hopeful that a better world is within reach.

Contents

Acknowledgments

This book is the product of the work of countless individuals who have contributed to it directly, or indirectly. First, we would like to acknowledge the institutional support from our academic homes: the School for Environment and Sustainability at the University of Michigan, the School of Environment and Science and Centre for Planetary Health and Food Security at Griffith University, and the Western College of Veterinary Medicine at the University of Saskatchewan, without which this book would not have been possible. JF would also like to acknowledge the support of Griffith University and the Max Planck Institute for Animal Movement (Konstanz, Germany), which provided intellectual homes during two sabbatical stints of writing this book. HM would like to acknowledge the support of All Souls College Oxford, through a Visiting Fellowship in Michaelmas term 2015, during which a substantial amount of time was spent working on the book.

We are greatly indebted to multiple students, and especially David Faulker, Kesiree Thiamkeelakul, Kristine Meader, and Hal Terry, for reading earlier versions of the manuscript and improving the text through numerous suggestions, corrections, and edits. Chrysi Beltsou and Laura McNeil were instrumental in compiling, organizing, and formatting the extensive bibliography. In addition, we are thankful to the members of the UM Ecosystem Health class who read individual chapters and provided comments and insights. Last, but not least, this book reflects the indirect inputs of many colleagues who, through innumerable conversations at conferences and professional meetings, helped shape the ideas in this book. We thank our colleagues Rebecca Hardin and Joe Eisenberg, who acted as sounding boards for many of the ideas in this book. We also would like to express our gratitude to the many colleagues who so generously provided us with images of their work that are included in this book.

We are deeply indebted to our editors at OUP, Lucy Nash, Bethany Kershaw and Charles Bath for their patience, support, and attention to detail in improving the manuscript. However, our deepest thanks go to Ian Sherman, who with unfailing poise, patience, and persistence kept us focused, and helped us navigate this project through the many years to completion.

Most of all, we are thankful for our friends and especially our spouses (JF: Jennifer; HM: Barbara; Gary: Amy as well as our children (JF: Alex; HM: Alasdair and Maddy) and parents (JF: Christel and Manoli) for putting up with us while we spent endless hours on writing this book.

List of Acronyms and Abbreviations

ACE-2   angiotensin-converting enzyme 2
AIC   Akaike information criterion
ANOVA   analysis of variance
bTB   bovine tuberculosis
CM   capture myopathy
CSF   classical swine fever
CWD   chronic wasting disease
CyHV-3   cyprinid herpesvirus-3
DRA   disease risk analysis
ELISA   enzyme-linked immunosorbent assay
ESU   Evolutionarily Significant Unit
FIV   feline immunodeficiency virus
FPV   feline panleukopenia virus
GC   glucocorticoid
GLMM   generalized linear mixed model
GPS   Global Positioning System
HA   hunted animals
IA   indicator animals
IPM   integral projection model
IUCN   International Union for Conservation of Nature
MCMV   murine cytomegalovirus
NIAID   National Institute of Allergy and Infectious Diseases
OIE   World Organisation for Animal Health
PAA   pre-analytical artifact
PCR   polymerase chain reaction
PVA   population viability analysis
RCV-A1   rabbit calicivirus Australia 1
RML   Rocky Mountain Laboratories
RHDV   rabbit hemorrhagic disease virus
SARS   severe acute respiratory syndrome
TTP   total testing process
TST   tuberculin skin test

Glossary

"When I use a word" Humpty Dumpty said, in rather a scornful tone, "it means just what I choose it to mean— neither more nor less."

Lewis Carroll, Through the Looking-Glass

In the spirit of Humpty Dumpty and Lewis Carroll, the definitions we give in this glossary are the way in which we have used the terms in this book. Many of these terms may have slightly differing definitions throughout the biological and epidemiological literature.

Terms adapted from Grenfell and Dobson (1997), Meffe and Carroll (2006), Loker and Hofkin (2015), Allaby (2005), and various internet resources.

Acaricide A chemical agent used to kill mites or ticks.

Accuracy The degree to which an individual measurement or estimate represents the true value of the attribute being measured; the proportion of all tests, both positive and negative, that are correct. (cf. **Precision**)

Adaptive management A management style in which the strategy is altered as additional information becomes available as management proceeds. Passive adaptive management uses additional information as it becomes available. Active adaptive management deliberately alters the management strategy in order to gain information to aid management.

Aetiological agent The causative agent of a disease or condition, often a microorganism or a toxin.

Aggregation Organisms show an aggregated distribution when the numbers per sampling unit (often a quadrat or, in the case of parasites, a host) are more variable than would be expected from a random (**Poisson**) distribution. The variance in counts per sampling unit will then be larger than the mean count per sampling unit. **Macroparasites** are almost invariably aggregated in their host population, the majority of hosts harboring a few or no parasites and a few hosts harboring large parasite burdens. Aggregation generally arises from some source **heterogeneity** in the host or parasite population. Clustering and overdispersion are synonymous terms. Aggregated distributions are often described by the **negative binomial distribution**.

Agroecosystem Land used for crops, pasture, and livestock; the adjacent uncultivated land that supports other vegetation and wildlife; and the associated atmosphere, the underlying soils, groundwater, and drainage networks.

Amplification host A host in which infectious agents multiply to high levels, providing an important source of infection for **vectors** or other species.

Anagenesis Evolutionary change, especially along a single, unbranched lineage.

Antibody A protein produced in the blood of vertebrates in response to an **antigen**. The antibody produced is able to bind specifically to that antigen and plays a role in its inactivation or removal by the immune system.

Antigen A substance, generally foreign, capable of inducing **antibody** formation.

Antihelminthic A drug used specifically to control helminth (i.e., parasitic worm) infections.

Arbovirus A member of a diverse group of viruses that use arthropods as **vectors** and are transmitted in their saliva to the **definitive host**.

Background extinction rate Historical rates of extinction owing to environmental causes not influenced by human activities. Distinct and much lower than **mass extinction events**.

Barrier culling A disease elimination approach in which populations of susceptible hosts are removed from a specific region ahead of an epizootic in order to prevent entry of the disease into an area where it does not occur, or from where it has been eliminated in the past.

Basic reproductive number (basic reproduction number, basic reproductive ratio R_0) A theoretical value representing the average number of new infections that arise during the period of infectiousness of a single infectious individual who has entered a population of completely susceptible hosts. (cf. **effective reproduction number**)

Bayesian statistical methods Statistical methods derived from Thomas Bayes' theorem of conditional probability. Bayesian methods are characterized by their incorporation of prior knowledge about the quantity to be estimated, which is then modified by the observed data to generate a posterior distribution.

Binomial distribution A statistical distribution that describes the probability of observing the number of "successes" x, from a series of n independent trials in which the probability of success p remains constant. In disease ecology, if the **prevalence** of infection in a population is p, the number of infected individuals x in a sample size of n individuals is likely to follow a binomial distribution.

Biocontrol or biological control The addition of a species to an ecological community with the intention of controlling an invasive species.

Biomagnification The accumulation of toxic substances in species occupying higher trophic levels.

Bridge host A host (in most usages, other than a **vector**) that transmits infection from a **maintenance** or **reservoir host** to a **target host**. (cf. **amplification host** and **reservoir host**)

Burden of disease Mortality, morbidity, and disability in a population caused by a disease.

Bushmeat Meat from animals (usually terrestrial) that is harvested in the wild. Often a cause of wildlife endangerment and a portal for the introduction of zoonotic pathogens into human populations.

Carrier (asymptomatic) An individual infected with a parasite that may transmit infection, but which does not display symptoms of disease.

Case definition A standard set of criteria for determining whether an individual has a particular infection, disease, or syndrome. Use of an agreed-upon standard case definition ensures that every case is equivalent, regardless of when or where it occurred. Furthermore, it allows for rigorous comparison of case numbers or rate of disease, identified in one time or place against the number or rate from another time or place (Centers for Disease Control and Prevention).

Chemoprophylaxis The use of chemicals to prevent infection or disease.

Chemotherapy The treatment of infection by means of chemicals (drugs) that have a specific toxic effect on the parasite or pathogen.

Coccidia An order of parasitic protozoa.

Cohort (observational) study A particular form of longitudinal study that samples the performance of a cohort (a group of individuals who share a defining characteristic, typically those who experienced a common event, such as birth during a selected period), at various intervals throughout time. Cohort studies represent one of the fundamental designs of epidemiology and are used on "difficult to reach" answers, for instance on how risk factors affect the incidence of diseases.

Commensalism A form of interspecific association in which two species live in close association with each other, with one deriving a benefit, and the other being neither positively nor negatively affected. (cf. **Mutualism** and **Parasitism**)

Complex life cycle A life cycle in which a parasite needs to be transmitted through at least two different host species in order to complete its development. (also **Indirect life cycle**)

Confidence interval (xx%) A range of values within which the true value of a **parameter** will fall xx% of the time if a given estimation procedure is repeated a large number of times. Loosely, it is often said that there is an xx% chance that the true value of the parameter will fall within the confidence interval. This definition is, however, closer to the **credible interval** used in **Bayesian statistics**.

Contact rate   The average frequency per unit time with which susceptible individuals contact (or are sufficiently close to) infected individuals or infective stages of a parasite, such that they can potentially acquire infection.

Credible interval   In **Bayesian statistics**, an interval in which an unknown parameter falls with a given probability.

Cross-sectional observation study   A study that examines the infection or disease status of a host population (generally subdivided by age or sex) at a moment in time.

Definitive host   see **Primary host**.

Degrees of freedom   (of an estimate or test statistic) The number of independent observations used to calculate the statistic or estimate, minus the number of quantities calculated from the same dataset that were required to calculate the statistic or estimate.

Density-dependent transmission   Transmission of a pathogen/parasite at a rate that is dependent (typically in a positive manner) on the population density of the host.

Digenean   A parasitic helminth belonging to the platyhelminthic fluke subgroup Digenea.

Dilution effect   The hypothesis that pathogen or parasite transmission to a **target host** decreases with increasing diversity of the ecological community.

Direct life cycle   (or **Simple life cycle**) A life cycle in which a parasite is transmitted directly from one host to the next without an intermediate host or **vector** of another species.

Disease   An abnormal condition affecting the functioning of an organism, not caused by external injury. Often used incorrectly or as shorthand to refer to a **parasite** or **pathogen** that causes disease.

Disease triangle   The concept that infectious disease (both at a population and individual level) is a result of an interaction between the host, a disease-causing organism (parasite or pathogen), and the environment in which both host and pathogen occur.

Edge effect   The altered biological and environmental conditions at the perimeter of a fragmented habitat; see **Habitat fragmentation**.

Effective reproduction number (R_e)   The average number of secondary cases resulting from each infected individual at any point after disease has been introduced into a population, taking into account factors such as an acquired immunity or vaccination. (cf. **Basic reproduction number R_0**)

Efficacy   An index of the potency of a drug or treatment, usually estimated as the average proportion of parasites in any host killed by a single dose or a short-term course of the treatment.

Eigenvalue   A fundamental mathematical parameter associated with square matrices that has many important applications in population biology and epidemiology. For example, the largest (dominant) eigenvalue of a matrix describing an age-structured population represents the logarithmic growth rate of that population.

Elimination   of a pathogen. Reduction to zero of the incidence of infection caused by a specific agent, in a defined geographical area, as a result of deliberate efforts. (cf. **Eradication**)

Environmental reservoir   Part of the environment, either other organisms (see **Reservoir population**), or abiotic parts of the environment, in which a parasite or pathogen can be maintained in the long term, and from which it can infect a **target population**.

Enzootic transmission   Relatively stable transmission of an infectious agent in an animal population.

Epizootic transmission   Transmission of an infectious agent in an animal population in excess of what is normally observed in a region in a given period. The resulting wave of infection is called an *epizootic*.

Eradication   of a pathogen The reduction to zero of the worldwide incidence of infection caused by a specific agent, as a result of deliberate effort. (cf. **Elimination**)

Eutrophication   The overabundance of algal life (an algal bloom) in a body of water, typically as the result of human-caused nutrient (N, P) inflow. Following the eventual death of these algae, their decomposition leads to pronounced removal of oxygen from the water column, therefore creating conditions conducive to pathogen spread.

Exotic species Species introduced to a different ecosystem than their native one, often through anthropogenic processes; such nonnative species however are not necessarily invasive.

Exposure Contact with an infectious agent required, but not sufficient for infection. Not the same as dose.

Ex situ preservation Managing animals in zoos, or otherwise away from their natural habitat in order to protect them. (Contrast with *In situ conservation*)

Extinction cascade A series of linked extinctions whereby the extinction of one species leads to the extinction of one or more different species.

Extinction vortex A set of self-reinforcing internal processes (such as inbreeding depression or demographic stochasticity) that can lead to the eventual extinction of small populations even in the absence of external stressors. The tendency of small populations to become progressively smaller in each generation, eventually going extinct.

Extinction rate (background) "Normal" low-level rates of extinction owing to background environmental causes, not influenced by human activities or other catastrophic events (see **Mass extinction event**).

Fitness The extent to which an organism is adapted to its environment, measured by the number of viable offspring it leaves behind relative to other individuals in the population.

Focal culling ("reactive culling," "point infection control") An approach used to eliminate disease in animals. The entire group on an "infected" farm or a herd is euthanized, and culling may extend to adjacent premises, to remove potential sources of infection and reduce availability of susceptible animals. The technique requires rapid identification of cases, prompt intervention, and restriction of movement of animals into and out of the cull site and is best used in situations where good access exists to the animal population.

Fomite An inanimate object or substance that transfers a pathogen from one host to another.

Force of infection For **microparasites**, the rate (per unit of time) at which susceptible individuals in a given population acquire infection. For **macroparasites**, the rate at which hosts acquire additional parasites.

Frequency-dependent transmission A term used to describe disease transmission that does *not* depend on the density of infected and susceptible hosts but instead on the frequency with which a transmitting event occurs. Spread of vector-transmitted pathogens and sexually transmitted diseases is often frequency dependent.

Genetic drift Genetic changes and losses that occur by chance, especially evident in small populations.

Glochidium/ia The larval stage typical of many types of freshwater bivalves (mussels) that attaches to the gill filaments of fish where it lives as an ectoparasite. Ultimately it drops off and settles on the bottom of the water body to continue its development into an adult mussel.

Habitat The physical environment in which a species is found.

Habitat corridors Strips of land, typically covered by natural vegetation, which connect patches of natural habitat.

Habitat fragmentation The process by which a large, continuous area of habitat is both reduced in area and divided into two or more sections (fragments).

Helminth A member of one the five classes of parasitic worms: monogeneans, digeneans, cestodes, nematodes, and acanthocephalans.

Herd immunity Reached when a sufficiently high proportion of a population is immune (either as a result of vaccination or acquired immunity) that the prevalence of infection no longer increases in the population.

Herd immunity threshold The minimum level of herd immunity that prevents sustained transmission of an infectious agent in a population. Depends on characteristics of the infectious agent, the host population, and the environmental conditions.

Heteroxenous A parasitic organism that utilizes more than one host during its life cycle.

Hematozoan A parasitic organism residing primarily in the blood of the host.

Host Any animal or plant that under natural conditions provides sustenance or shelter to a parasitic organism.

Horizontal transmission The most common type of pathogen/parasite **transmission**, in which an infected individual infects another susceptible individual of the same species.

Hypothesis testing A statistical approach in which a **null hypothesis** continues to be accepted following an experiment, unless the data collected indicate that the null hypothesis is sufficiently unlikely to be true. In that case an alternative hypothesis is accepted.

Immunity the ability to combat **infection** or **disease** owing to the presence of **antibodies** or activated cells. Typically divided into three types: **acquired immunity** is conferred on an individual following recovery from a disease; **natural** or **innate immunity** is inherited from parents, or in some cases antibodies may be passed across the placenta and therefore are present in the blood at birth; and **artificial immunity** may be induced by the injection of a vaccine or antiserum and confers protection of variable duration to a pathogen.

Immunosuppression Suppression of the immune response by drugs, parasites, or the host's own immune regulatory mechanisms.

Inbreeding The mating of individuals who are more closely related than by chance alone.

Inbreeding depression Lowered reproductive rates, or production of offspring with lowered survival and reproduction, following mating among close relatives or self-fertilization.

Incidence rate The ratio of the number of new events (e.g., infections) during a defined time period to the population at risk of experiencing the event.

Incubation period The time interval between the acquisition of infection and the onset of clinical signs. (cf. **latent period** and **serial interval**)

Indirect Life cycle (or **Complex life cycle**) A life cycle that requires one or more **intermediate hosts**, or **vectors** before the **definitive host** is reinfected.

Infection The colonization and replication of a parasitic organism that gains sustenance or shelter from the body of a host, whether or not that host experiences ill health.

Infectious (or contagious) disease Disease caused by infection with a parasite that can be transmitted from one individual to another either directly or, indirectly, through a **vector**.

In situ conservation Managing species in their natural environment. (cf. *Ex situ* **conservation**)

Intensity of infection Used inconsistently in the literature: either the mean number of parasites within infected members of the host population or the mean parasite burden of the entire population. Unless **prevalence** is 100%, the latter will be smaller than the former.

Intermediate host (also **secondary host**) A host required for a parasite's life cycle, but not the host inside which parasite sexual reproduction occurs.

Invasive species A species occurring and expanding its distribution outside its natural range, typically introduced directly or indirectly by to human activities (see **exotic species**).

Keystone species Select species that have a disproportionate effect on the organization of a biological community and the loss of which may have far-reaching consequences for community structure.

Koch's postulates A set of four criteria used to determine whether a causative relationship exists between a microorganism and a disease. First expressed formally by the German physician Robert Koch in the nineteenth century.

Latent infection An **infection** that is causing no disease.

Latent period Interval between acquisition of infection by a host and its ability to transmit infection. (cf. **Incubation period** and **Serial interval**)

Likelihood In statistics, the probability of observing values of one or more model parameters, given a particular set of data. The set of parameters at which the likelihood is maximized are called the **maximum likelihood estimates**.

Macroparasites   Parasites that in general do not multiply within their definitive hosts but instead produce transmission states (eggs and larvae) that pass into the external environment or to vectors (e.g., the parasitic helminths and arthropods). Typically multicellular and visible with the naked eye.

Maintenance host   A host species, in which a parasite or pathogen can be maintained indefinitely and from which infection can be transmitted to another **target** species. See **Reservoir population**, which is a slightly more general term.

Mass extinction event   The terminal extinction of a very large number of taxa that occurs over a relatively short geologic time. Mass extinction events are thought to be caused by catastrophic insults of global reach such as disruptions of the planetary biogeochemical cycles following extensive volcanic eruptions or asteroid impacts. Five major mass extinctions have been identified in the fossil record, one each at the end of the Ordovician, Devonian, Permian, Triassic, and Cretaceous geologic periods. Earth is presently at the beginning of a sixth mass extinction triggered by human activities.

Maximum likelihood estimates   For a given statistical model and set of data, the parameter estimates at which the **likelihood** is maximized.

Microparasites   Parasites that undergo direct multiplication within their definitive hosts. Microparasites are characterized by small size, short generation time, and a tendency to induce immunity to reinfection in those hosts that survive the primary infection. Duration of infection is usually short in relation to the expected life span of the host.

Microparasitism   Infection with a microparasite.

Metapopulation   A shifting mosaic of frequently transient populations, linked by some degree of migration; a population of populations.

Morbidity   Any departure, subjective or objective, from a state of physiological or psychological well-being.

Mortality rate   The rate per unit time at which individuals die; or alternatively the ratio of the number of deaths during a defined time period to the number of hosts at risk of dying during that period.

Mutualism   A form of interspecific association in which two species live in close association with each other, with both deriving a benefit. (cf. **Commensalism** and **Parasitism**)

Negative binomial distribution   A discrete probability distribution for counts, giving the probability that a sampling unit has x items in it, in situations where items are **aggregated**. A key property of the negative binomial distribution is that the variance in the number of items per sampling unit is greater than the mean number of items per sampling unit. Negative binomial distributions are described by two **parameters**: the mean, and a parameter usually denoted k, which describes the degree of aggregation. Small values of k represent highly aggregated distributions, and as k approaches infinity, the negative binomial distribution approaches a **Poisson** distribution. Distributions of parasites among hosts are often well described by negative binomial distributions.

Nidus   Specific location of a given disease; the result of a unique combination of ecological factors that favors the maintenance and transmission of the disease organism.

Nonselective culling, area-wide   ("proactive culling," "host population reduction") A disease management approach that aims to stop the spread of a density-dependently transmitted pathogen by reducing wildlife host population densities through indiscriminate culling. See **Selective culling**.

Null hypothesis   In hypothesis testing, what will continue to be believed in the absence of sufficient evidence to the contrary. Usually, it is that "nothing is going on": in an experiment it might be that the response does not differ depending on whether experimental subjects are exposed to a treatment or a control.

Ontogenetic   Pertaining to the development of an individual from fertilization of an egg to adulthood and death.

Oocyst   The intermediate stage in the Apicomplexan life cycle following the union of the micro- and macrogametocyte in which sporozoites are produced.

Parameter   A constant number that forms part of the specification of a mathematical model (e.g., the transmission rate of a pathogen, or the variance in a normal distribution). Parameters are usually theoretical, unknown quantities that need to be estimated from data, frequently derived from a sample taken from a wider population.

Parasite An organism exhibiting a varying but obligatory dependence on another organism, its **host**; typically detrimental to the survival and/or **fecundity** of the host.

Parasitemia The presence of a parasite in the blood of the host.

Parasitism A form of interspecific association in which two species live in close association with each other, with one (the **parasite**) living in or on the other and deriving a benefit, and the other (the **host**) being negatively affected. (cf. **Mutualism** and **Commensalism**)

Pathogen A parasitic organism, typically unicellular or a virus (hence also a **microparasite**), that causes disease or morbidity in its host.

Pathogenicity The degree to which a parasite tends to cause disease in its host, and the severity of the disease caused.

Poisson distribution A discrete probability distribution for counts, giving the probability that a sampling unit has x items in it, given that the overall mean number of items per unit is known. It applies if items are distributed independently between sampling units. A key property of the Poisson distribution is that the variance in the number of items per sampling unit is equal to the mean number of items per sampling unit. In the context of parasite distributions among hosts, the number of parasites per host would follow a Poisson distribution if the fact that a host has acquired one parasite provides no information on whether it is more or less likely to acquire another parasite, compared with other members of the host population. This is not a common situation: parasites are usually **aggregated** in their distribution among hosts and the distribution is often better described by a **negative binomial distribution**.

Power Statistical power in a hypothesis test is the probability that the **null hypothesis** will be rejected in favor of the **alternative hypothesis**, given that the alternative hypothesis is true.

Therefore, the higher the power, the more likely the test can detect a true effect. A variety of factors affect the power of a test including the sample size, the effect size, and the inherent variability in the data.

Power analysis An analysis used to determine the minimum necessary sample size so that a test or experiment can detect an effect at the desired level of significance.

Precision The closeness to each other of repeated measurements or estimates of the same quantity. (cf. **Accuracy**)

Prepatent period The time from infection (generally first infection) until the host shows symptoms (synonymous with **incubation period**).

Prevalence (or **Prevalence Rate**) The proportion of the host population with infection or disease, often expressed as a percentage. A measure of how widespread an infection or disease is.

Primary host The host in which a parasite's sexual reproduction normally occurs (also termed **definitive host**).

Random sample A sample is a subset of the population and in statistics it is used to provide information about the whole population. As a result, it needs to be an unbiased representation of the entire population. Drawing a random sample is a common method for achieving this unbiased representation. In a simple random sample, each member of the population has an equal probability of being included in the sample. The number of units in the sample is called the sample size, often denoted n.

Recrudescence Reappearance of disease in a host whose infection has been quiescent (i.e., without symptoms).

Reintroduction An attempt to establish a species in an area that was once part of its historical range, but from which it was been extirpated. Reintroduction is prime occasion during which attendant parasites/pathogens may also get accidentally introduced. (cf. **translocation**)

Replicate A random subset of the entire available sample (i.e., sampling pool) that has been drawn for a particular survey. Sample replicates help measure variation in an experiment, so that any differences between treatment groups can be better evaluated.

Representativeness In a population, the extent to which a selected sample represents the overall population in terms of specific biological characteristics; in a reserve system, the quality of a set of sites that together include all or most existing biodiversity elements.

Reservoir population A host population, or group of populations, in which a parasite or pathogen can be maintained indefinitely and from which infection can be transmitted to another **target population**. Reservoirs are often (but not necessarily) different species from the target population and will frequently be more **tolerant** of infection than the target population. See **maintenance host**.

Resistance (as applied to parasites) The reduction in susceptibility to chemotherapy or to chemical vector control.

Resistance (as applied to hosts) The ability to limit parasite burden if exposed to infection. One host is more resistant than another if it develops a lower parasite burden when exposed to the same infective dose. (cf. **Tolerance**)

Response variable The variable that investigators are trying to explain or predict and typically the focus of a question in a study. Also known as the dependent or outcome variable, its value is predicted, or its variation is explained by explanatory (independent) variable(s). In an experimental study, it is the outcome that is measured following manipulation of the explanatory variable(s).

Risk assessment The process of identifying, evaluating, and managing the risks that may arise from a given action or activity.

Sample A group of individuals or items that are selected from a larger population for measurement. The sample should be representative of the population to ensure that one can generalize the findings from the sample to the population as a whole.

Sampling frame A sampling frame is a list or other device used to define a researcher's population of interest. Hence a sampling frame defines a set of elements from which a researcher can select a sample of the target population.

Saprophytic (or **saprotroph**) An organism that absorbs soluble organic nutrients from inanimate sources (e.g., dead plant or animal matter or dung).

Secondary host See **Intermediate host**.

Selective culling ("test-and-cull") A disease eradication approach in which animals are tested for infection, and those that test positive are killed to prevent further disease transmission. It is the least controversial form of culling, because only infected individuals are removed; however it is also challenging because it requires both the identification of infected individuals under field conditions, and the ability to test a substantial proportion of the population. See **Nonselective culling**.

Sensitivity analysis A method used to determine the robustness of an assessment by investigating the degree to which the results and conclusions of the assessment are shaped by the methods, models, or assumptions that went into the analysis. It often involves modifying these methods/models/assumptions and then determining to what extent the results change.

Serial interval The time between the acquisition of infection by a host and transmission onto another host. The serial interval is usually longer than the **latent period** because onward transmission does not necessarily occur as soon as a host becomes infectious.

Significance level A measure of the strength of the evidence that must be present in a sample before one rejects the **null hypothesis** (of no difference) and accepts the **alternative hypothesis** that an effect is statistically significant. In classical **hypothesis testing**, the researcher selects the significance level before conducting the experiment. Typically denoted as alpha or α, it is the probability of rejecting the null hypothesis when it is true (making a **Type I error**). For example, a significance level of 0.05 indicates a 5% risk of concluding that a difference exists when there is no actual difference.

Simple life cycle See **Direct life cycle**.

Species–area relationship The positive relationship between the size of natural areas and the number of species there. One of the fundamental patterns in ecology.

Stochastic Random; a random process.

Stratified random sampling A method of sampling that involves the division of a population into smaller subgroups known as **strata**. In stratified random sampling, or stratification, the strata are formed based on individuals' common characteristics such as sex or age. Within each stratum individuals are then selected randomly.

Stratum In statistics, a stratum (plural strata) refers to a subset of the population that is being sampled. Stratification therefore consists of dividing the population into strata within each of which an independent sample can be chosen. Stratification may be conducted based on geographical criteria, e.g., by dividing up the sampled area into subareas; or by referring to some other quality of an individual, e.g., by dividing the individuals into strata according to sex. It ensures that all subsets of a population are equally sampled.

Superspreaders A small proportion of the infected members of a population that transmit the parasite or pathogen at a rate substantially higher than most infected members of the population.

Susceptibility Accessible to, or liable to infection by a particular parasite.

Susceptible individual An individual that can get infected by a pathogen either because it has never developed immunity (i.e., it is immunologically naive) or because it has lost past immunity.

Synergism An interaction in the way two factors affect another variable. This may be either positive or negative (in which case the final effect is bigger/smaller than the sum of the individual factor effects).

Systematic samples Samples collected using a systematic sampling format. In this format, an investigator chooses individuals from a target population by choosing a random starting point and then selecting sample members after a fixed "sampling interval," e.g., every 10th individual. If the population order is random or random-like (e.g., alphabetical), then this method will provide a representative sample that can be used to draw conclusions about the population. Systematic sampling is popular because it is simpler and more straightforward than random sampling and more conducive to covering a large area.

Target host species or population A host species or population of particular interest in a given situation. In the context of conservation biology, the target population will often be an endangered species or a species highly susceptible to the parasite.

Test accuracy This refers to the ability of a test to differentiate between the patient (infected) and healthy cases correctly. To estimate the accuracy of a test, one needs to calculate the proportion of true positive and true negative in all evaluated cases.

Test reliability This refers to how dependably or consistently a test can measure a characteristic. This consistency of a measure relates to whether the test results can be reproduced repeatedly under the same conditions.

Test sensitivity Sensitivity and specificity are measures of a test's ability to correctly classify an individual as having a disease or not having a disease. Sensitivity refers to a test's ability to designate an individual with disease as positive. A highly sensitive test means that there are few false negative results, and thus fewer cases of disease are missed. It is desirable to have a test that is both highly sensitive and highly specific. This is frequently not possible, and typically there exists a trade-off. In many clinical tests, there are some individuals who are clearly normal, some clearly abnormal, and some that fall into the gray area between the two. As a result, careful choices must be made in establishing test criteria for positive and negative results (see **Test specificity**).

Test specificity Specificity and sensitivity are measures of a test's ability to correctly classify an individual as having a disease or not having a disease. The specificity of a test is its ability to designate an individual who does not have a disease as negative. A highly specific test means that there are few false positive results. It may not be feasible to use a test with low specificity for screening, since many individuals without the disease will screen positive, and potentially receive unnecessary diagnostic procedures (see **Test sensitivity**).

Test validity This is the extent to which a screening test accurately identifies diseased and nondiseased individuals. A test that has high validity produces results that correspond to the "real" infection status of an individual, as determined by some "gold standard" test.

Tolerance   The ability to limit the harm caused by a given parasite burden. One host is more tolerant of infection than another if, given the same parasite burden, its fitness is higher. (cf. **Resistance**)

Translocation   A management technique often used in mitigation for endangered species protection whereby an individual, population, or species is removed from its habitat to be established in another area of similar or identical habitat. An occasion during which attendant parasites/pathogens may also get accidentally introduced.

Transmission   The process by which a **parasite** passes from a source of **infection** to a new **host**. Multiple types occur including **horizontal**, **vertical**, sexual, trophic, vector-borne, waterborne etc., with the first two being the most common.

Trematode   Any of the parasitic flatworms of the class Trematoda; characterized by a thick outer cuticle and one or more suckers to attach to the host.

Trophic transmission   A mode of transmission of an infectious agent that depends on the consumption of the one host by another.

t-statistic   In statistics, the t-statistic is the difference between the estimated value of a parameter and a hypothesized value (which is often zero), divided by the standard error of the estimate. It is often used to test whether there is evidence that two means differ. Another common use is to test the hypothesis that the estimated value of a parameter differs significantly from zero.

Two-tailed   In statistics, a two-tailed test is a method in which the critical area of a distribution is two-sided, and tests whether a sample is greater than, or less than, a certain range of values. It is used in null hypothesis testing and testing for statistical significance. If the sample being tested falls into either of the lateral critical areas, the alternative hypothesis is accepted instead of the null hypothesis. A test of a statistical hypothesis, where the region of rejection is on both sides of the sampling distribution, is called a two-tailed test.

Type I error   In a hypothesis test, a type I error occurs when a null hypothesis that is actually correct is rejected. The probability of committing a type I error equals the **significance level** one sets for the hypothesis test. For example, a significance level of 0.05 indicates that one is willing to accept a 5% chance of wrongly rejecting the null hypothesis. For a given set of data and statistical test, decreasing the probability of making a type I error increases the probability of making a **Type II error**.

Type II error   In a hypothesis test, a type II error occurs when one fails to reject a null hypothesis that is actually false. In other words, one obtains a nonsignificant test result even though a true population effect actually exists. Some combination of a small sample size, inherent variability in the data, and bad luck with random sample error might have obscured the population effect. One can decrease the probability of a type II error by increasing the **power** of the test.

Vector   An organism that carries and transmits a pathogen between hosts. Vectors are typically small-bodied and mobile and are often insects or other invertebrates.

Vector-borne transmission   A mode of transmission of an infectious agent that depends on infection of a relatively small-bodied mobile host (termed **vector**), typically an invertebrate such as an insect or a tick.

Vertical transmission   A mode of transmission in which an infectious agent passes from a parent host to its offspring.

Virulence   Broadly defined the ability of a pathogen to cause damage to the host.

Zoogeography   The study of the geographic distribution of animals at different taxonomic levels.

Zoonosis   Infectious disease of humans caused by an infectious agent that normally circulates among nonhuman animal **reservoirs**.

Introduction

As we write this, the raging COVID-19 pandemic is reminding humanity of a lesson many of us had forgotten over the past century of rapid medical advances: the power of infectious disease to affect not just our health and well-being, but to drastically reshape the world around us. At the same time, this pandemic is also a reminder of the critical role of the environment, including ecological species interactions, in shaping disease outbreaks. While the circumstances of the origin of SARS-Cov-2 will undoubtedly become clearer in the future, its animal origins are undeniable, thus underscoring the role of free-ranging wildlife as a perennial source of new and emerging pathogens.

An extensive body of research over the past 20 years has documented the origins and processes of pathogen emergence, and has also offered important avenues for infectious disease control and prevention. The same research has also revealed that infectious diseases pose a significant danger not just for humans but also for many wildlife species. Indeed, emerging infectious diseases are increasingly recognized as one of the major forces driving species extinction. The same anthropogenic processes that have led to increasing emergence of zoonotic diseases affecting humans, namely habitat modification and destruction, climate change, and increased global connectivity, also have led to increasing emergence of wildlife disease. Addressing the challenges that wildlife faces from new pathogens is particularly difficult for numerous reasons, and no work currently exists that synthetically integrates recent advances in the field. With this book, we aim to fill this gap and help elucidate the role of emerging disease ecology in conservation.

To paraphrase our colleague Bart Kempenaers, books differ from scientific papers because they not only state and establish facts, they also establish arguments, synthesize fields, and propose new directions—in so doing, they have the ability to change perspectives and minds. To that extent, the purpose of this book is threefold: first and foremost, to make the argument that parasites and pathogens play a critical role in species conservation and need to be considered throughout any conservation management effort; second, to serve as a gentle academic introduction to wildlife infectious disease ecology for people not directly involved in the field; and last, to serve as a practical "how-to" guide for setting up a study of pathogenic organisms in natural populations.

Overall, this book is intended to stimulate interest and summarize the current state of the field. It is also our hope that it can serve as an introduction to the often-overwhelming literature for people who are not experts in this rapidly expanding and particularly interdisciplinary field.

Intended audience

This book, focusing largely on the topic of disease ecology, sits thematically at the nexus of at least three more established disciplines: wildlife veterinary science, conservation biology, and theoretical population ecology. Because each of these fields comes with its own priorities, values, biases, and its own specific vocabulary, one of the early challenges while writing the book was to integrate these distinct approaches and to produce a text that would be accessible to readers from each field.

Infectious Disease Ecology and Conservation. Johannes Foufopoulos, Gary A. Wobeser and Hamish McCallum, Oxford University Press.
© Johannes Foufopoulos, Gary A. Wobeser and Hamish McCallum (2022). DOI: 10.1093/oso/9780199583508.001.0001

The intended audience of this book includes students and practitioners from each of these fields. We hope that it can act as a bridge that will allow members of different disciplines to become familiar with basic concepts in disease ecology and conservation. More specifically, this book is aimed at advanced students, practicing conservation biologists, and anyone interested in the study of wildlife diseases in nature. To ensure accessibility of the content, we have tried to limit the amount of information to the most pertinent works and to enrich the book with numerous figures and images. References to technical literature are restricted to the most important works, and each chapter is accompanied by its own reference section to ensure its standalone relevance. Furthermore, we have tried to use approachable language and have included an extensive glossary that handles the more technical terms and concepts discussed in the book.

Background and focal systems

As already pointed out by others before us, the study of infectious diseases has been one of the triumphs of modern ecology. The past 20 years have seen a blossoming of the field of disease ecology with numerous key studies advancing our understanding, not just of the role of parasites and pathogens in natural ecosystems but also of the associated ecological and evolutionary host–pathogen dynamics. At this point, significant progress has been made and it is an excellent time to summarize the present state of the field.

We have intentionally focused this book on terrestrial vertebrate organisms. While occasional examples involving marine, invertebrate, and even plant species, are included, we have limited ourselves to terrestrial vertebrates for two major reasons: first, host–parasite interactions are much better understood in terrestrial vertebrates than in other groups of hosts; and second, whether one agrees with this or not, they are the most common focus of conservation management and restoration projects.

Within this group, we return repeatedly to three study systems that are, by virtue of the known detail and our personal research background, disproportionally important. We have chosen these

because they are particularly prominent in the conservation literature and because they illustrate different aspects of managing infectious diseases of conservation significance. Further, by returning to the same case studies repeatedly, we show how a range of approaches can be applied to solve these difficult problems. More specifically:

Avian malaria and bird pox have devastated the Hawaiian avifauna following the introduction of the mosquito vector *Culex pipiens* in the mid-nineteenth century. Several dozen species of endemic birds have become extinct and several others have contracted in distribution largely because of these diseases. Owing to the disproportionate conservation impact and available funding, extensive research has already been conducted on this system. As a result, we have a comparatively good understanding of the transmission dynamics of these pathogens, their impact on the native bird populations, as well as possible management approaches. Avian malaria and bird pox are important examples of the class of vector-borne pathogens, and this system also illustrates multiple disease-related issues associated with anthropogenic climate change. This case study is first introduced in Chapter 1 and then also referred to in Chapters 2, 3, 6, 7, 12, and 14.

Chytridiomycosis is an emerging fungal disease that has had the greatest biodiversity impact of any known vertebrate pathogen. First identified in 1998, *Batrachochytrium dendrobatidis*, the causative agent of the disease, is directly responsible for over 100 global frog species extinctions, particularly in Latin America and Australia. In addition, this pathogen has precipitated broad-scale population declines on every continent where frogs can be found. More recently, *Batrachochytrium salamandrivorans*, a related pathogen, has also led to mass mortalities of newts and salamanders in Europe. The origin of both waterborne fungi appears to be in Asia, and they have been disseminated worldwide owing to human activities. Supported in each area by a few host species that are relatively tolerant of infection and can act as reservoirs of infection, these pathogens have been transmitted to other highly susceptible species, which can then be driven to extinction. Chytridiomycosis holds important lessons for conservation biologists as an example of the threats posed by persistent, multi-host pathogens. Chytridiomycosis is introduced in Chapter 1 and discussed in Chapters 2, 3, 7, 9, 10, 11, 12, and 14.

Tasmanian devil facial tumor disease (DFTD) is a transmissible cancer, first detected in 1996, which threatens to cause the extinction of the largest marsupial carnivore still surviving. Tumor cells are transmitted between Tasmanian devils by biting during competitive or mating interactions. These tumors are clonal descendants of a single tumor that arose in a nerve cell of a female devil, which must now be long dead. More recently, a second transmissible tumor has appeared in Tasmanian devils, with a male origin. The spread of this cancer has been facilitated by the underlying severe loss of genetic diversity in Tasmanian devil populations. While this is a highly unusual pathogen, it is one of the very few cases in which a novel infectious disease has been extensively researched from its first emergence, so that insights can be gained into its impact and spread throughout almost the entire geographic range of a species. This case study also illustrates the power of the diverse approaches that have been applied to understand the disease and manage its impact on the host population. DFTD is introduced in Chapter 1, and is further discussed in Chapters 9, 10, 11, 14, and 16.

Beyond these key study systems, we have gone to great pains to expand coverage of the pathogens discussed. We have tried to include information from disease and conservation from diverse parts of the world, especially from less developed countries that harbor important biological diversity and underappreciated ecological systems.

Book structure

The book is structured around three sections. First, there are three general overview chapters, starting with an introduction of the general principles of conservation biology for non-conservationists, some principles of infectious disease ecology in natural ecosystems for conservation biologists, and, drawing the two strands together, a chapter on interactions between human activities and wildlife disease. Second, a series of methodology chapters follows, discussing principles of planning and

experimental design for wildlife disease studies, techniques for animal capture and data acquisition, methods for detecting disease agents in the field and laboratory, and development of mathematical models together with their parameterization. The final section of the book deals with combining the theory and methodology to manage emerging disease threats. Following an introductory chapter on planning management strategies, we discuss preventing infectious disease arriving in an ecological community, eliminating disease once it has arrived, and ensuring that populations persist if disease cannot be successfully eliminated. The next chapter considers an entirely different application of management of infectious disease in a conservation context, which is to use a parasite or pathogen to control a pest species. The final chapter considers the ethical and societal issues that may arise when managing infectious diseases in wildlife populations.

In summary, this book combines an introduction to basic principles of wildlife epidemiology with practical veterinary approaches and theoretical ecology insights. It is not a detailed mathematical treatise of theoretical wildlife disease ecology, nor a comprehensive review of veterinary field methods. Readers interested in this should consider the specialized literature available from other authors cited throughout the book.

Disease ecology has expanded so rapidly in the past several decades that it is no longer possible to include all relevant references in this work. One of the hardest choices has been to decide which works to cite in this book. We apologize to our colleagues who may not find their work cited—any such omissions are not intentional.

We tried to ensure the book is as accurate and free of errors as possible. Nevertheless, this highly interdisciplinary subject has become sufficiently complex and broad that it is easy for mistakes and omissions to occur. Any mistakes are ours alone.

We hope that this book will raise awareness of the dangers presented of infectious organisms for threatened biodiversity and encourage further study of this fascinating and rewarding field.

JF, Ann Arbor
GW, Saskatoon
HM, Brisbane

PART I

Epidemiological Background

CHAPTER 1

Conservation Biology and Parasitism

> Conservation biology differs from most other biological sciences in one important way: it is often a crisis discipline.
>
> **Michael Soulé (1985, p. 727)**

Much of what will be discussed in this book lies at the interface between multiple distinct disciplines, including ecology, epidemiology, and veterinary science. Yet, squarely in the center of the book's focus is the science of conservation biology, a relatively new field that may be unfamiliar to some of the readership. As a result, depending on their background, users of this book may encounter unfamiliar conservation biology terminology and concepts. This chapter serves as an informational foundation for readers wanting to familiarize themselves with the basics of conservation biology, especially in the context of disease ecology. The first half of the chapter therefore discusses basic concepts in conservation science, as well as the central question of the role that infectious diseases play in the extinction of species. In contrast, the second half focuses on the problem of parasitic organism endangerment and conservation, and on the ecological implications of parasite extinction. Together the two sections provide a thematic bridge that connects many of the issues that will be discussed in subsequent chapters of the book.

1.1 What is conservation biology?

Conservation biology is a relatively new, synthetic area of study that emerged from the field of ecology in the early 1980s (Soulé 1986). Soulé describes conservation biology as a "crisis discipline"—in contrast to conventional ecology, it is a value-laden, mission-oriented science with the principal aim to manage and protect the biological diversity of the planet. As such, conservation biologists do not have the luxury of choosing their problems the way conventional ecologists do; instead they often have problems thrust upon them and are asked to solve them, frequently with little prior knowledge about the system involved. Conservation biologists have, from the very inception of the field, embraced an interdisciplinary approach. The science combines elements of basic organismal biology (such as population biology, community ecology, population genetics, and ecophysiology) with applied approaches from forestry, wildlife management, and fisheries. Because essentially all of the crises the field deals with are human-made, conservation biologists need to be aware of, and fluent in, basic concepts of social science, environmental planning, law, and education, as well as public communication. As a matter of fact, many of these themes will appear throughout this book when we discuss the various dimensions of wildlife disease problems.

1.2 Biodiversity

"Biodiversity" is a relatively new term, coined in the mid-1980s by Dr. Walter Rosen of the US National

Infectious Disease Ecology and Conservation. Johannes Foufopoulos, Gary A. Wobeser and Hamish McCallum, Oxford University Press.
© Johannes Foufopoulos, Gary A. Wobeser and Hamish McCallum (2022). DOI: 10.1093/oso/9780199583508.003.0001

Research Council (Wilson 1988). It has by now become well established and is an accepted concept in public discourse. In academic circles some discussions still persist on the specifics of the definition of the concept—nonetheless, it is typically used in daily life as a summary term to describe the richness of life on earth. In addition to the total number of species in a region ("species richness"), biodiversity also includes all levels of organization of life, ranging from genes to ecosystems, as well as all of the ecological processes of which they are part (United Nations 1992).

Parasites and pathogens, the focal organisms of this book, are obviously part of the planet's biodiversity (Dobson et al. 2008). Beyond this, infectious disease and biodiversity have a complex two-way relationship: on the one hand, infectious diseases can in some circumstances cause species endangerment and even extinction, therefore undermining biodiversity (de Castro and Bolker 2005). On the other hand, however, it has been shown that high levels of biodiversity have a protective effect against overwhelming epizootics (Schmidt and Ostfeld 2001; Keesing et al. 2006, 2010; Johnson et al. 2013). Much of this book will focus on exploring the relationships and the various aspects between these two concepts.

At the core of the debate about the nature of biodiversity and its conservation lies the question of what constitutes its basic unit, and how it can be best measured. Whereas most biologists would agree that the foundational unit of biodiversity is the species, this also raises the question of how best to define it.

This is not merely an academic debate: without a clear definition of what constitutes a species, it is difficult to know what exactly to protect. The fact that biological diversity exists in discrete units rather than as a continuum has been recognized since antiquity. While Carl Linnaeus, the eighteenth-century Swedish botanist, was the first to provide a formal definition of a species, this effort has been the focus of active scientific debate ever since (Cracraft 1983; Coyne and Orr 2004). Numerous interpretations have been proposed, based on evolutionary, morphological, phylogenetic, and genetic criteria, but despite the diversity of opinions, a consensus appears to be slowly emerging in recent years (de Queiroz 1998; Mallet 2001; de Queiroz 2005,

2007). In practice, different criteria are applied for different groups of organisms; what represents a distinct species of bird is very different from what constitutes a new species of bacterium or parasitic worm. The standard definition involving reproductive isolation taught to high school biology students, derived from Mayr (1963), is problematical in many cases, particularly in the absence of sexual reproduction. Because of this variation, the recommendation often given is that "a species is what a competent taxonomist says it is."

Related to the question of the definition of a species is the issue of which units of biological diversity deserve protection (Fraser and Bernatchez 2001; Coates et al. 2018). Many national conservation laws are based on the unit of a species as the most obvious conservation target. Several, including the US Endangered Species Act, bestow protection not just to species but also to "distinct biological units" such as subspecies within a species (Coates et al. 2018). Many population geneticists suggest that Evolutionarily Significant Units (ESUs), which are clearly delimited evolutionary lineages within species, are an important unit for conservation. However, too strict an adherence to managing ESUs separately may lead to losses of genetic diversity and inbreeding depressions, if ESUs consist of too few individuals (Coates et al. 2018).

1.3 Extinction

While extinction is on some level a fact of life, there are numerous reasons why we should be concerned about the ballooning rates of species extinction, and why it is critically important to act to limit this (Cardinale et al. 2019). There is the moral argument that humans have an ethical duty to maintain biodiversity, and there are aesthetic reasons why natural environments need to be protected because they are beautiful or fulfilling. However, most of the reasons proposed for preserving biodiversity are more utilitarian—biodiversity in general, and individual species in particular, are known now to provide valuable services to human societies (Diaz et al. 2018), and it is typically not possible to predict in advance which particular species may turn out to be of benefit.

Earth's biodiversity is currently facing a severe existential threat originating in human activities (Ceballos et al. 2017). While estimates and assessments of the risks that biodiversity faces may vary, most conservation biologists agree that the web of life is facing the most severe peril to its existence in as long as the past 65 million years. A combination of drivers, ultimately caused by the soaring human population and the advances of technology, is undermining both the species communities and the natural processes that sustain them. Annual rates of global species extinctions are increasing exponentially as human populations and activities expand across the globe. Species extinction, originally confined to rare, large-bodied species with small populations, especially on islands, is now spreading across mainland ecosystems and is progressively affecting whole species communities and ecosystems, which now become threatened in increasingly complex ways (Diaz et al. 2019).

Species extinction has always been a key element of paleontology—after all, most of the earth's species known from the fossil record do not exist today (Lawton and May 2005). The occurrence of these baseline, "normal" extinctions, captured in the concept of a *background extinction rate*, may create the false impression that the present-day extinction crisis is nothing to worry about. Despite any superficial similarities between this past continuous backdrop of extinctions and what is going on today, the two processes are fundamentally different. Extinction of a species in the fossil record generally means that this species has either gradually evolved into a different organism (a process called *anagenesis*) or has been replaced ecologically by another species. In either case, this means there has been a disappearance with a replacement by another species. As a result, despite the fact that 99% of all species that existed over earth's history have gone extinct, the planet has steadily gained species over the past 1 billion years (Rohde and Muller 2005; Alroy 2008). In contrast, present-day extinctions represent the terminal disappearance of taxa without something equivalent replacing them; what we are witnessing is a rapid decline in the total number of species (Dirzo et al. 2014; Ceballos et al. 2015). Often these are unique, irreplaceable organisms that are the result of tens of millions of years of

evolution, and which are now disappearing in the geologic blink of an eye (Ceballos et al. 2017).

Sharply different from background extinctions are *mass extinction events*, well-defined catastrophic losses of life that have occurred at several distinct times over earth's history. While multiple lesser extinction periods are also discernable in the fossil record, five major mega-extinction spasms—termed the "Big Five"—stand out for their severity, the large number of organism groups affected, and their global reach (Raup and Sepkoski 1982). Active debate continues in the scientific community about the causes of most of these events and various hypotheses, ranging from volcanic eruptions to gamma-ray bursts, have been proposed as driving factors. Even diseases, especially for coral reef communities, have been proposed as a cause, although there are several reasons why they are an unlikely explanation (Veron 2008; Elewa and Joseph 2009). Ultimately, all mass extinctions likely involved some combination of disturbances in the earth's carbon cycle, changes in ocean chemistry, and shifts in global sea levels. Each of these traumatic events led to the extinction of a sufficiently large proportion of the earth's communities so as to "scramble the deck" and usher in the establishment of a new, characteristically different communities of species. Consequently, mass extinction events tend to mark the transition points from one geologic period to the next, each recognizable by its own characteristic fossil organism assemblages.

There is little doubt that the current rate of extinction exceeds that which has existed over geological time by several orders of magnitude (De Vos et al. 2015). Mounting evidence suggests that we are presently at the beginning of the sixth such mega-extinction spasm, one that, if continued, will result in the disappearance of a substantial fraction of the earth's biodiversity (Wake and Vredenburg 2008; Barnosky et al. 2011; IPBES 2019).

1.4 Drivers of extinction

The primary threatening processes, originally dubbed the "evil quartet" by Jared Diamond (1984), are conventionally listed as:

1. habitat destruction;
2. overkill;

(a)   (b)

Figure 1.1 (*Left*) The Panamanian golden frog (*Atelopus zeteki*) is one of many species of frogs that have now become extinct in the wild following the spread of the invasive chytrid fungus *Batrachochytrium dendrobatidis*.(*Right*) Scanning electron microphotograph of a frozen intact sporangium and zoospores of *B. dendrobatidis*. (Sources: *A. zeteki*, Brian Gratwicke; *B. dendrobatidis*, Alex Hyatt, Commonwealth Scientific and Industrial Research Organisation.)

3. introduction of exotic species;
4. secondary extinctions.

Most conservation biologists would now also add to this list various forms of ecosystem stress, such as pollution and especially climate change, which is rapidly gaining in importance.

1.4.1 The role of parasites and pathogens in extinction

Infectious disease is not explicitly listed as a member of this "evil quartet," although several notorious extinctions can be attributed to the introduction of either a pathogen itself, or a reservoir host for a pathogen, into an environment from which it was previously absent.

Emergence of infectious disease may also be a consequence of habitat destruction or modification, or a result of extinction or declines in other species in an ecosystem, precipitating in turn secondary extinctions among the remaining taxa. Disease-induced extinctions can therefore be regarded as relating to at least three members of the evil quartet. In Chapter 3 we discuss how pathogens and parasites may exacerbate the impact of these threatening processes.

Relatively few reviews have tried to evaluate the importance of infectious disease as a driver of species extinction. Analyzing the 2004 International

Union for Conservation of Nature (IUCN) Red List, Smith et al. (2006) found that only 31 of the 833 listed extinctions of plants and animals could be attributed to infectious disease. In an analysis of the 2006 Red List, Pedersen et al. (2007) found that parasites and pathogens were listed as threatening processes for only 54 mammal species, with the majority (88%) coming from just two orders: Artiodactyla (13% of 218 species) and Carnivora (5.3% of 281 species). Bacterial infections were especially important in the Artiodactyla, representing 47% of all the infectious disease threats, whereas viruses were dominant among the Carnivora, representing 56% of infectious disease threats. It is important to recognize, however, that many extinctions and/or endangerments are a result of synergies between multiple independent stressors (Brook et al. 2008), and that the impact of infectious diseases may be most important when interacting with other extinction drivers rather than when acting alone.

For example, pathogens may have particularly severe effects in populations that have lost genetic diversity. Hence, small host populations that have declined owing to other anthropogenic stressors and are subject to genetic drift are likely to be more susceptible to epizootics. While this remains an active area of investigation, evidence from island populations seems to point in this direction: genetically impoverished and inbred island wildlife mount different responses than mainland taxa (Lindstrom et al. 2004), and seem less able to control infections (Paterson et al. 1998; Spielman et al. 2004; Wikelski et al. 2004).

The complexity of the interplay between disease and other stressors in causing species disappearance is highlighted by the example of climate change and infectious disease, a topic of intense research interest but few generalizable conclusions. Most of the pathogen responses to changing temperatures are species-specific and nonlinear (Mordecai et al. 2013, 2017), meaning that it is difficult to make broad generalizations about the role of climate change in threats posed by pathogens. Climate change is causing major reassortments of ecological communities as different species respond in distinct ways to climate change and do so at dissimilar rates. This means that species will be increasingly brought into contact with parasites, pathogens, and reservoir species with which they have not coevolved, likely resulting in elevated disease risks to wildlife (for more details, see Chapter 3, Section 3.5).

There are also good reasons to believe that infectious disease has been underrecognized as a threatening process to wild species (MacPhee and Greenwood 2013). The study of disease in wild species is a fairly new field, and many health conditions in natural populations remain simply unrecognized. Except in the most dramatic epizootics, it is uncommon to detect in nature wildlife that is obviously being impacted by infectious disease. On the one hand, animals have likely evolved to hide disease symptoms in order to avoid signaling vulnerability to predators. On the other hand, diseased or dying animals are typically consumed rapidly by predators or scavengers, leaving little physical evidence for investigators (Preece et al. 2017).

The cryptic nature of wildlife disease is exemplified in the case of two endemic rat species (*Rattus macleari* and *R. nativitatis*) that went extinct shortly after the accidental introduction of black rats (*R. rattus*) and their attendant fleas onto Christmas Island in 1900 (Pickering and Norris 1996; Green 2014). Despite contemporary field observations suggesting a role of disease in the declines of the native rat populations, and field autopsy results consistent with spreading trypanosome infections (Durham 1908), there was no conclusive evidence

for an epizootic wiping out these two species until approximately a century later, when Wyatt et al. (2008), using molecular diagnostic tools on museum specimens, were able to demonstrate the spread of *Trypanosoma lewisi* into the native *Rattus* populations from the invading black rats and their fleas. The alternative hypothesis, that this was a pre-existing island disease, was disproved by the fact that the pathogen was absent from preinvasion native rat samples, as well as the lack of any genetic differentiation between mainland and Christmas Island trypanosome strains (Wyatt et al. 2008; Dunlop 2015). Underscoring the often-transient nature of infection, and the challenge of implicating disease in species extinctions, is the fact that trypanosomes can no longer be detected in either black rat or feral cat populations on Christmas Island today (Dybing et al. 2016). In the absence of museum specimen evidence, present-day field investigations would not have been able to implicate this pathogen in the extinctions of the two endemic rats.

In summary, infectious diseases likely play a more important role in the endangerment and extinction of wildlife than has been appreciated until now. Because infectious disease is often transient in its presence, and cryptic in its impacts, it has not been the focus of traditional research investigations. It is only in recent years that scientists, using interdisciplinary approaches, have been able to uncover the hidden yet pervasive effects that infectious pathogens, acting alone or in conjunction with other factors, have had on the survival and fecundity of wild populations.

1.4.2 Possible mechanisms of parasite-induced extinction/endangerment

Epidemiological and evolutionary theory (see Chapter 2) provides some guidance about when parasites and pathogens are likely to constitute a substantial threat to wild species. For example, computer modeling investigations into one of the simplest disease scenarios suggest that host-specific pathogens transmitted through direct contact are unlikely to drive their hosts to extinction. The reason for this is that transmission of a pathogen is, in most cases, a process that depends on host population density: in more dense populations, infected hosts are more likely to encounter and then transmit the pathogen to other, uninfected hosts. Consequently, if the host population declines owing to disease mortality, transmission will also decrease, eventually falling to a rate insufficient to maintain the epizootic. The pathogen should therefore die out in the host population before the host itself becomes extinct. Nevertheless, even such a pathogen can still contribute to its host's extinction if it interacts with another stressor, or if it reduces the population to a level where it becomes susceptible to other problems like inbreeding depression. For some types of pathogens, like those transmitted sexually or via vectors (e.g. mosquitoes), the contact rate between infected and susceptible hosts does not depend strongly on host density (see Chapter 2, Section 2.2). Therefore, such pathogens are more likely to drive their sole host to extinction (de Castro and Bolker 2005). In another complication, many pathogens are able to infect multiple host species and are therefore more likely to represent a conservation threat. In this scenario, a host species with low tolerance or resistance to a pathogen may be driven to extinction if there are other, resilient host species present that are able to maintain the epizootic with limited impact on their own population. Such resilient hosts will continue to shed large amounts of infectious particles and continue to subject the susceptible species to high infection risk, even as it declines to very low numbers.

Lastly, evolutionary theory also suggests that when hosts are exposed to novel pathogens, they are likely to experience particularly severe impacts (Altizer et al. 2001; McCallum 2012). When a host and pathogen have interacted for an extended period of evolutionary time, selection will have acted on the host population to increase resistance to infection. Less obviously, there is often selection for those strains of the parasite that cause less disease in their hosts, as strains that kill hosts more quickly may have less opportunity to spread to new hosts (see Chapter 3, Figure 3.1) (Anderson and May 1982, but also O'Hanlon et al. 2018).

There are many examples that support these theoretical predictions. First, consider the prediction that host-specific pathogens are more likely to cause the extinction of their hosts if transmission is relatively insensitive to declines in population density. The Tasmanian devil (*Sarcophilus harrisii*) is the largest

surviving marsupial carnivore, and is threatened with extinction by an infectious cancer, Tasmanian devil facial tumor disease (McCallum et al. 2009). Although the tumor, essentially a lethally parasitic mammalian cell line, only infects Tasmanian devils, it threatens to cause their extinction because transmission does not depend much on high host density. The tumor is spread by biting, which frequently occurs during sexual contact. The tumor therefore behaves similarly to a sexually transmitted disease—high rates of biting within the species occur even at low population densities.

Second, theory predicts that extinction of highly susceptible species is more likely if there are alternate host species that are able to carry high levels of infection with limited impact on their own survival or fecundity. Two key case studies support this generalization. The amphibian chytrid fungus *Batrachochytrium dendrobatidis* is a generalist frog pathogen and has caused large-scale declines and extinctions worldwide, particularly in the Americas and in Australia (Berger et al. 1998; Lips et al. 2006; Briggs et al. 2010; O'Hanlon et al. 2018; Scheele et al. 2019) (Figure 1.1). The fungus affects a wide variety of frog species, some of which are highly susceptible and have since gone extinct, while other species are capable of carrying high infection burdens without major deleterious effects, therefore continuing to disseminate the pathogen into the populations of the disappearing species (Voyles et al. 2018). As another example, the Hawaiian avifauna has been devastated by two mosquito-borne pathogens, avian malaria and bird pox (Warner 1968; van Riper et al. 1986). As with the amphibian chytrid fungus, both of these pathogens have a broad host range and while some bird species are highly susceptible, others are able to carry and spread the infections while experiencing only limited deleterious effects.

The third and final generalization is that pathogens that are novel are particularly deleterious. This is supported by all three of these key case studies. The Tasmanian devil facial tumor disease was first detected only in 1996, the amphibian chytrid fungus first appeared in the Americas and Australia in the past few decades, and avian malaria and bird pox caused major declines in Hawaii shortly after the introduction of mosquitoes capable of transmitting the diseases in the nineteenth century. These key case studies are particularly well understood and will be returned to at numerous points in this book.

While the previous examples dealt directly with the role of host–parasite interactions in host extinction, it is also possible that a pathogen causes the extinction of a species indirectly, for example, by destroying its habitat. Relatively few examples have been recorded of this process, but the North American limpet (*Lottia alveus*) appears to belong in this category. The species went extinct after its key habitat, the coastal Atlantic eelgrass meadows, formed by the eel grass *Zostera marina*, disappeared following a wasting disease epidemic caused by the slime mold *Labyrinthula* (Carlton et al. 1991).

1.4.3 Detecting infectious disease threats to populations or communities

There are four distinct circumstances under which we may need to recognize potential or actual infectious disease impacts on populations or ecological communities. They correspond broadly to four stages of establishment of a new parasite/pathogen in a host population.

1. *A pathogen is not already present in a particular geographical area, but there may be a risk of its introduction.* There are reasons to believe that if the pathogen did arrive, there would be serious consequences for one or more native species. The threat is usually recognized because the pathogen is present in the same or related species in other regions, where it has had a serious impact. For example, the fungus *Batrachochytrium salamandrivorans* is known to be causing substantial population declines in several salamander species in Europe (Martel et al. 2013), but at the time of writing is not known to be present in North America (Waddle et al. 2020). Given the impact of the pathogen in Europe, and given that North America has a very high diversity of salamanders, there is clearly a very large threat posed by its introduction (Martel et al. 2014) and there is an urgent need to develop proactive strategies to manage this threat (Grant et al. 2017). Approaches to

preventing diseases from entering regions or populations in which they do not currently occur are discussed in Chapter 12.

2. *A pathogen has been introduced into a region or population from which it was previously absent and is in the process of epidemic spread.* The key indication that such a pathogen is having serious impacts on one or more host populations is a pattern of spatially spreading population decline, sometimes called an "extinction wave." For example, Laurance et al. (1996) hypothesized that infectious disease was responsible for the decline and extinction of a number of Australian rainforest species on the basis of a pattern of sudden declines spreading northward up the coast of eastern Australia. At the time that paper was published, an infectious agent had not yet been identified. Subsequently, it was recognized that the aforementioned amphibian chytrid fungus *B. dendrobatidis* was the causal agent (Berger et al. 1998) and that similar spatially spreading population declines in the Americas could also be attributed to the same fungus (Lips et al. 2006). Spatially spreading population declines in the Tasmanian devil *S. harrisii* were associated with facial tumors (Hawkins et al. 2006), before it was clear that the tumor was in fact infectious (Pearse and Swift 2006). Identifying the causative agent is helpful for management, but if there is a reasonable suspicion that an infectious agent is responsible, a number of strategies can be put in place to limit further spread before the agent is unequivocally identified (McCallum and Jones 2006). Provided the extent of spatial spread is limited, it may be feasible to attempt to eliminate a pathogen in this epidemic phase. Strategies to do this are discussed in Chapter 13.

3. *A pathogen is well established in the population or community and is substantially reducing the population of one or more species, affecting their viability.* This is a difficult situation both to detect and manage (Stallknecht 2007; Preece et al. 2017). If infection is endemic, then spatial spread is likely to have stopped. An endemic infectious agent present at high prevalence is in fact unlikely to have a major impact on its host population, because, if pathogenicity were high, infected individuals would be rapidly removed by death,

resulting in reduced prevalence (McCallum and Dobson 1995). For a widespread agent, even high prevalence among dead or morbid animals is not a good indication of impact on the host population. If prevalence is high in the population as a whole, it will also be high in dying animals, even if infection has no impact on mortality. A substantial difference between prevalence in morbid animals and the general population may be an indication of disease impact (McCallum and Dobson 1995), although some third factor (e.g., nutritional or other stress) may be responsible both for mortality and a high prevalence of a putative pathogen. For macroparasites, attempts have been made to infer disease-induced mortality on the basis of truncation of the frequency distribution of parasites among hosts (Crofton 1971). Indeed, if a macroparasite has a negative effect on the host, one would expect that the individual hosts most affected would be those carrying the largest helminth burdens, meaning that these individuals would die and not appear in a population sampled by a disease investigator. What s/he would observe instead is a distribution curve of parasite loads in the host population from which the most heavily infected animals would be missing. However, numerous other factors may also cause truncation of the upper tail of a parasite frequency distribution (Anderson and Gordon 1982). To detect or evaluate the impact of pathogens or parasites on individuals in a population in the field, some form of experimental manipulation of infection is often necessary, and to detect impacts at a population level, experimental manipulation should ideally also be at the population level. Many of the considerations that go into first detecting a pathogen in a population, and then quantifying the impact on its hosts, are discussed in Chapters 4 and 6. Strategies for managing endemic infections on host populations are considered in Chapter 14.

4. *A pathogen has already caused the loss of one or more host species from the area of interest.* This is the most challenging situation to detect. If a species has been driven to extinction, then no extant animals are available for investigation. Identifying infectious disease as a primary agent of extinction

is often speculative (MacPhee and Marx 1997; Lyons et al. 2004; Abbott 2006) and may rely on detecting evidence of infection from museum specimens (Wyatt et al. 2008) or "ancient" DNA (Reid et al. 1999), and only few such examples are known to date. In addition to the previously discussed case of the Christmas Island rats, there is some evidence that *Plasmodium relictum* malaria, as well as avian pox (*Poxvirus avium*) were involved in the extinction of Hawaiian bird species as early as the nineteenth century (Warner 1968; MacPhee and Greenwood 2013) (see also discussion in Chapter 3). Attributing global extinctions to parasites and pathogens may seem to be largely of academic interest. However, there are also practical applications of this knowledge. For example, if reintroduction is considered after a species population has disappeared, it is important to determine whether infectious disease may have been responsible for local extinction and whether the pathogen is still present in the environment. If the factors leading to the disappearance of a species from an area are not addressed, any reintroduction is likely to fail (Caughley 1994).

1.5 Endangerment and conservation of parasites

Parasites are part of our biosphere and we, as biologists, must accord them the same respect we exhibit for their hosts. If we truly appreciate biological diversity, we must advocate that all species are precious, even parasites.
Equal Rights for Parasites!
Windsor (1990, 1995)

1.5.1 Are parasites threatened?

Most of this book will be concerned with managing the threats posed by parasites and pathogens to free-living species. Nevertheless, it is also important to recognize from the outset that parasites are an important component of earth's biodiversity, and that the general arguments made for preserving biodiversity also apply to parasites.

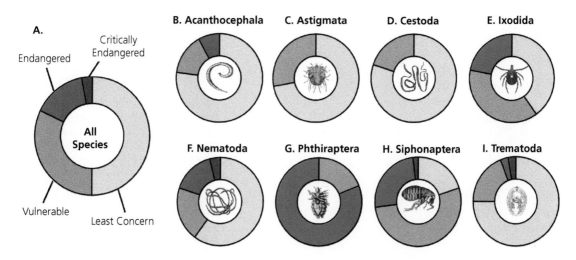

Figure 1.2 (A) Levels of endangerment (based on standard IUCN "Red List" categories) for various groups of parasites according to projected losses of their habitat owing to global climate change. Breakdowns are given by extent of habitat loss between now and 2070: 0 to 25%—Least Concern; 25 to 50%—Vulnerable; 50 to 80%—Endangered; 80 to 100%—Critically Endangered. (B–I) Individual results for eight major clades of parasitic organisms. (Figure modified from Carlson et al. 2017.)

The bulk of the world's biodiversity consists of species that obtain their resources through some sort of parasitism (May 1988; Windsor 1998). Although no exact numbers are available, it is safe to assume that parasites constitute the most diverse group of metazoa (Whiteman and Parker 2005). At the same time, this diversity is very poorly characterized, with the majority of species remaining undescribed (Dobson et al. 2008). Despite this taxonomic ambiguity, it is becoming clear that the global diversity of parasitic and pathogenic organisms is declining and some species are increasingly threatened with extinction (Friderici 1997; Dunn et al. 2009; Carlson et al. 2017) (Figure 1.2). While this has led to calls for their protection (Dougherty et al. 2016), it has also precipitated a spirited discussion as to the overall need for parasite conservation.

A common response to concerns about parasite conservation is that such organisms are intrinsically undesirable, and that their loss from an ecosystem is either a beneficial, or at least, an inconsequential issue. While such opinions appear to be quite widespread, even among biologists, there are several reasons why the continued survival of both parasites and pathogens needs to be an issue of interest for environmental managers. In the next

section of this chapter, we will discuss reasons why the preservation of parasite and pathogen diversity is such an important conservation topic.

1.5.2 Why conserve parasites—Ethical reasons

The inherent right of an organism to exist has been well recognized (Naess 2008) and extends also to those that, in the eyes of the general public, are considered either repulsive, or unimportant, or both, including pathogens or parasites. From a moral perspective it is therefore wrong to cause the extinction of a species, irrespective of whether it is parasitic or not. The discussion surrounding the prevention of terminal extinction of particular pathogens arose during the last stages of the global eradication campaigns of two important diseases: smallpox and rinderpest (a disease of grazing mammals) (Normile 2008, 2010; Henderson and Klepac 2013). Currently two other human pathogens, the Guinea worm (*Dracunculus medinensis*) and the poliovirus, are close to final eradication, while five other human diseases (measles, mumps, rubella, lymphatic filariasis, and cysticercosis) are considered to be realistic eradication targets, meaning that this discussion is likely to resurface again in the intermediate future.

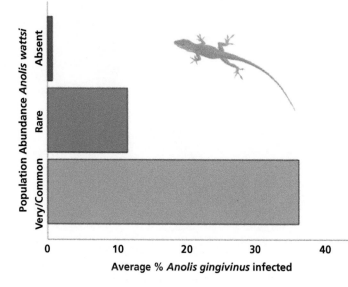

Figure 1.3 Parasites can determine the outcome of competitive interactions between species, as seen in this example of two competing lizard taxa. On the Caribbean island of St. Maarten two species of *Anolis* lizard are found. *Anolis wattsi*, the competitive inferior species, can be found only in those locations where its competitively dominant species *Anolis gingivinus* is infected with the lizard malaria *Plasmodium azurophilum*. The parasite, which does not appear to affect very much the competitively inferior *A. wattsi*, takes enough of a toll on *A. gingivinus* populations to prevent it from outcompeting *A. wattsi*. (Figure by author, based on data from Schall 1992.)

1.5.3 Why conserve parasites—Applied reasons

1.5.3.1 Ecological significance of parasites

Ecologists have argued that, beyond ethical considerations, there are also practical reasons to protect parasite taxa. A growing body of evidence suggests that parasites—despite their frequently obscured presence—play a critical role in the function of healthy ecosystems and that their disappearance can have unexpected ecological consequences (LoGiudice 2003; Hudson et al. 2006). Parasites are, for example, known to affect the competitive interactions between different host species (Hudson and Greenman 1998; Lafferty et al. 2005). On the islands of the Lesser Antilles, coexistence between two species of *Anolis* lizards occurs only where the competitively dominant species is limited by lizard malaria infections (Schall 1992) (Figure 1.3). Conversely, the distributions of forest caribou (*Rangifer tarandus caribou*) and moose (*Alces alces*) are limited at the southern edge of their range by lethal infections caused by the brainworm *Parelaphostrongylus tenuis*; this species commonly parasitizes white-tailed deer (*Odocoileus virginianus*) without much ill effect (Lankester 2010). Counterintuitively, overall high host and parasite diversity appears to also be protective against the damaging emergence of catastrophic single pathogen epidemics: experimental wetland sites with high amphibian host and parasite diversity had lower incidence of the virulent trematode *Ribeiroia ondatrae* (Johnson et al. 2013). Such dampening of epidemics may stem from (i) between-parasite competition/predation occurring inside infected hosts (Sousa 1993), (ii) enhanced immune responses in heavily parasitized host populations (Froeschke and Sommer 2012), or (iii) the fact that in species-rich habitats, individual host species tend on average to be relatively rare, therefore reducing transmission rates of host-specific pathogens.

The presence or absence of even a single parasitic species can also shape distributions and population dynamics of whole species communities. The trematodes that infect the foot muscle of the New Zealand cockle *Austrovenus stutchburyi* also affect the way cockles disturb the sediment, which in turn influences the structure and functioning of the rest of the invertebrate species community colonizing the particular sediment (Mouritsen and Poulin 2004; Hudson et al. 2006). As this example suggests, parasites and pathogens can have profound impacts on the distribution and population dynamics of natural food webs and are therefore important ecological agents (Kwak et al. 2020).

The critical role that parasitic organisms play in the function of healthy ecosystems becomes evident in those situations where they have disappeared following human activities. A major reason

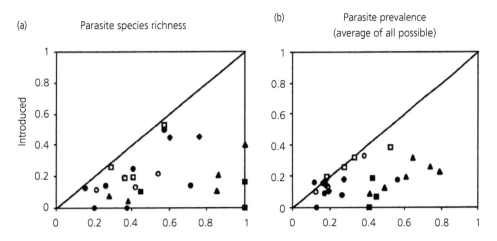

Figure 1.4 Invasive species harbor fewer parasites in their new as opposed to their original range. Hence, because invasive species are relatively free from the burden of disease and parasitism relative to the native species, they are able to outcompete them and expand their geographic range. (a) Standardized parasite species richness in the native (x axis) and introduced (y axis) range. (b) Average prevalence (% of hosts infected) in the native (x axis) and introduced (y axis) range. Each point represents an invasive species. A point located on the diagonal line would indicate no difference in parasitism between the native versus the new, introduced range. Different symbols represent different types of hosts. (For additional details see source, Torchin et al. 2003.)

responsible for the success of invasive species in areas where they have been newly introduced is that they have been released from the burden of ever-present parasitism that impacts all resident taxa (Torchin et al. 2003) (Figure 1.4). Because the initial introduction of an exotic species into a new region often involves very small numbers of colonizers, parasites that were native to the original range, and which often have patchy distributions there, may fail to make the transfer into the new region as well. As a result, parasite communities living on invasive host taxa tend to be more impoverished in the new range relative to the old. These invaders, operating now under a reduced parasite burden, have a critical competitive advantage against the native species, which in turn facilitates their spread into the new ecosystem.

1.5.3.2 Beneficial effects of parasites on host physiology

Paralleling the important role of parasites for whole ecosystems, there is increasing evidence suggesting that parasites are also critical for the proper function of their individual hosts' physiology (de Ruiter et al. 2017). For instance, some intestinal helminths bioaccumulate heavy metals to concentrations ca.

2,000% above environmental levels (Kwak et al. 2020; Sures 2003, 2004) and may concentrate up to 50% of the cadmium and zinc found in coastal salt marshes (Dobson et al. 2008). Vertebrate hosts appear to be so closely adapted to the presence of various parasites that disappearance of the parasites can precipitate various physiological and metabolic pathologies. In humans, where this issue has been best studied, there is an inverse correlation between the worldwide distribution of helminth infections and the presence of autoimmune disorders (Lorimer 2017; Kahl et al. 2018). Mounting evidence indicates that the presence of parasitic helminths in the gastrointestinal tract modulates not just immune function, but also host metabolism, including the occurrence and impacts of type 1 and type 2 diabetes (Bhattacharjee et al. 2017; Surendar et al. 2017). Furthermore, recent studies indicate that infection with helminths can have a positive effect on the resolution of neuroinflammation in neuropsychiatric disorders like multiple sclerosis (Dixit et al. 2017; Abdoli and Ardakani 2019). But perhaps the strongest evidence points to the protective effects that helminth infections can have against inflammatory bowel diseases such as celiac disease (Lerner et al. 2017) and Crohn's disease (Maruszewska-Cheruiyot et al. 2018). In the absence of any professional advice, significant numbers of patients with these diseases have resorted to self-treatment by infecting themselves with helminths, with the majority apparently reporting positive outcomes (Liu et al. 2017). While this is a very active field of research, more information is needed, and it is not known whether similar effects occur in non-human animals, though they are to be anticipated.

1.5.3.3 Parasites as indicator species and sources of information

Extinction of parasites is expected to lead to loss of important medical opportunities to advance human health. Parasites are powerhouses of biochemical production and have been used regularly to identify and isolate biomedically important agents. As an example of such bioprospecting, tick saliva includes numerous anti-inflammatory, anticoagulant, and immunosuppressive agents, and is the focus of intense pharmaceutical research (Ribeiro et al. 1985; Durden and Keirans 1996; Steen et al. 2006; Hovius et al. 2008). Similarly, leech saliva contains a plethora of biochemically active compounds and has a long history of being used for therapeutic purposes (Michalsen et al. 2003; Hildebrandt and Lemke 2011).

Parasites can also be very useful by providing information that helps us better understand and manage the natural world. Because parasitic organisms live in close proximity, and are frequently very specialized to their host's physiology, they can provide key insights into the evolutionary relationships and demographic history of their host species. Since they evolve faster than their hosts, parasites are more likely to reveal in their genome patterns of past divergence that may not be readily detectable in the genome of their relatively invariant hosts. For example, the genome of the avian louse (*Degeeriella regalis*, Phthiraptera) contains much more phylogenetic information than that of its host, the Galapagos hawk (*Buteo galapagoensis*) (Whiteman and Parker 2005). What is often particularly instructive is to analyze and compare the phylogenetic relationships of both host and parasite species (Brooks and McLennan 1993). Beyond this, parasites can also provide important information about the past ecology of their hosts; for example, dating

the rise of the human body lice (*Pediculus humanus corporis* and *P.h. humanus*) has helped determine when humans first started wearing clothing, which these ectoparasites require for their survival (Kittler et al. 2003). Lastly, because parasites are often very adept at finding their hosts they can be used to identify the presence of otherwise cryptic hosts: blood meals found in leeches are an excellent source of information on difficult-to-detect wildlife taxa (Schnell et al. 2015).

Mounting evidence shows that parasitic organisms can respond closely and rapidly to ecological change and can therefore serve as sensitive indicators of environmental change. Environmental stressors can affect parasite and pathogen populations directly or indirectly via their hosts; ultimately both positive and negative effects on parasite communities are possible (Lafferty and Holt 2003; Wood et al. 2020). When considering the various effects of environmental stressors on host–parasite interactions it is useful to distinguish between: (i) effects that undermine the host's physiology and hence its ability to defend itself against infection, thereby resulting in increased parasite/pathogen prevalence (Ross et al. 2000; Van Loveren et al. 2000); and (ii) effects that lead to declines in host population size or in transmission rates, and where one would expect parasite populations to decline as well (Lafferty and Kuris 1999).

Many parasites, particularly helminths, have complex life cycles, with each parasite life history stage utilizing hosts that may be very different taxonomically or ecologically. The latter is especially true for macroparasites, which require a range of different hosts and utilize a spectrum of distinct habitats. This means that parasite species will likely be more sensitive indicators of environmental changes than any single free-living species. Helminths with complex life cycles are thought to be among the first species to disappear in stressed ecosystems and could serve as the proverbial "canaries in a coalmine." An example of this comes from digenean helminths living in the salt marshes of southern California (Lafferty 1997). There, birds act as the final hosts and disseminators of digenean parasites into the resident salt marsh snail populations. In areas where human activities have displaced native birds, transmission has collapsed and the snails remain uninfected by digeneans.

An example of the aforementioned complexity of relationships becomes evident when trying to answer the seemingly simple question "How are parasites affected by pollution?" A review of the literature on the effects of pollutants on aquatic parasites paints a nuanced picture in which different categories of parasitic organisms respond individualistically to various types of chemicals (Lafferty 1997). The most consistent pattern observed has been that waterway eutrophication leads to broad increases in the prevalence of various micro- and macroparasites, presumably because of the elevated availability of nutrients. In contrast, the effects of other pollutants varied much more depending on the individual group of parasites investigated.

1.5.4 How do parasites become endangered?

Elucidating how parasites become endangered requires us first to understand the relationship of dependence between parasites and their host populations. In general, because parasites are so intimately tied to the continued existence of intact host populations, any host disappearance is likely to lead to secondary parasite declines (Rósza 1992), although other counterintuitive dynamics are at least hypothetically possible. Both empirical and theoretical evidence support this notion of dependence. For example, a striking, recently observed, field example of this reliance comes from the successful eradication of rats, an invasive species on Palmyra Atoll (Lafferty et al. 2018). Extermination of the island's rats led to the unexpected disappearance of the island's Asian tiger mosquito (*Aedes albopictus*) population, another exotic species that apparently subsisted on the resident rodent population.

Host–parasite dependence patterns are however rarely this straightforward because there is not necessarily a one-to-one relationship between host and parasite species. Hence, it is possible that a parasite may infect and be able to survive in multiple host species; conversely, a single wildlife species may serve as the only host for multiple parasite taxa. Our understanding of the functional dependence of parasitic organisms on their hosts at the community level can therefore be advanced by asking the questions: What is the relationship between host species diversity and parasite diversity? How will parasite diversity change if host species diversity declines? On a broad, ecosystem-wide scale, one would expect that aggregate parasite diversity would correlate positively with host species diversity and that the number of parasite species will decline following any extinctions of host populations. Since the degree of host specialization varies, the exact functional relationship between parasite and host diversities in an area remains under debate, although food web modeling points toward a quasi-linear relationship between host and parasite species richness (Figure 1.5; Lafferty 2012). Furthermore, there is evidence that small and declining host populations, such as those of endangered primate species, also harbor lower parasite diversities (Altizer et al. 2007).

A parasite's disappearance may even precede the extinction of its host if the host population declines to a density where the transmission rate fails to counterbalance parasite deaths (Lafferty and Kuris 1999). As a result, it is possible that numerous parasite species may have already gone extinct even though their declining hosts are still surviving (Powell 2011; Rósza and Vas 2015; Bulgarella and Palma 2017). This susceptibility is likely more pronounced in parasites with complex life cycles, such as tapeworms, that require passage through several distinct species of host to complete the life cycle (see Chapter 2). Consequently, the loss of any one of these hosts (or their sufficient decline)—if no alternative host exists within the particular ecological community—will also lead to the loss of the parasite.

The notion that parasites do well when their hosts do well, can, in at least in some circumstances, be extended beyond the population to the individual level. Therefore, parasites may benefit from their individual host's good condition: studies have shown that avian fleas living on well-nourished great tits (*Parus major*) had enhanced reproductive

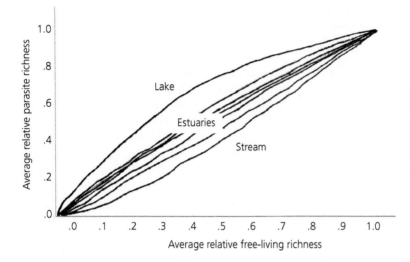

Figure 1.5 The number of parasite species in a community depends on the number of host species found in that area. Here, associations between free-living host species richness and parasite richness were obtained from the modeled disassembly of various aquatic food webs. The one lake in this sample is the concave down curve, and the one stream is the concave up curve. The various estuaries show quasi-linear relationships. (Source: Lafferty 2012.)

success (Tschirren et al. 2007). Similarly, mountain spiny lizards (*Sceloporus jarrovi*) that were in better condition were able to support larger loads of ectoparasitic mites (Foufopoulos 1998).

Parasite populations may become endangered by any factor that either undermines their host populations or disrupts the process of parasite reproduction, that is, their transmission from host to host, for example, habitat loss, invasive species, or climate change. Prosecution and extirpation of the hosts by humans can result in the concomitant extinction of their parasites. Several species of avian lice have disappeared following overkill of their hosts (e.g. the louse *Psittacobrosus bechsteini* that used to live on the now extinct Cuban macaw (*Ara tricolor*); see Mey 2005). Others have become endangered together with their overhunted hosts (e.g., the giant rhinoceros stomach bot fly *Gyrostigma rhinocerontis* living on black rhinoceros (*Diceros bicornis*) Colwell et al. 2012) (Figure 1.6).

Beyond direct overkill, destruction of natural habitats often results in the reduction of free-ranging host populations and in the isolation of the surviving subpopulations. While these landscape-level processes can have unexpected effects on the dynamics of transmission (e.g., García-Ramírez et al. 2005), in practice they mostly lead to the deterioration of natural parasite communities (for examples, see de Bellocq et al. 2002; Lindstrom et al. 2004; Bush et al. 2013). On the other hand, there is some evidence that restoration of degraded habitats can also help recovery of parasite communities: salt marshes that were degraded experienced recovery of their native trematode communities when the habitat was restored (Huspeni and Lafferty 2004).

Lastly, global climate change is thought to be another major driver of parasite extinction, although its effects on different parasite groups diverge, with some species being winners and others losers (Altizer et al. 2013; Carlson et al. 2013). From a conservation perspective, however, it is projected that up to 10% of all parasites are likely to go extinct by the year 2070 by climate-driven habitat loss alone, while up to 30% of all parasitic worm species are likely to be lost under the combined effects of direct and indirect drivers (Carlson et al. 2017; see also Figure 1.2). Methods that can assist in a more accurate evaluation of the conservation status of parasitic organisms, as well as their effective protection, are presently under development (Carlson et al. 2020; Kwak et al., 2020).

1.5.5 What are the effects of conservation management activities on parasite communities?

Species conservation activities tend to have a narrow focus on the recovery of a target species and typically do not extend to the preservation of any parasites and pathogens inhabiting

(a)   (b)

Figure 1.6 An example of co-endangerment of a host and parasite. (a) The black rhinoceros (*Diceros bicornis*) has declined across most of its range in Africa. (b) *Girostigma rhinocerontis*, the African rhino stomach bot fly is one of the largest flies in the world and a co-endangered species that depends on black rhinos. (Sources: (a), Karl Stromayer/United States Fish and Wildlife Service; (b), Vida van der Walt.)

this focal organism. Consequently, any effects of conservation management activities on the resident parasite community, whether positive or negative, are incidental. So how do species conservation activities affect the diversity of associated parasites? Everything else being equal, the recovery of a host species' free-ranging population is most likely to have a positive effect on the attendant parasite/pathogen communities because larger host populations tend to support bigger and more diverse parasite populations (Dobson and Carper 1996; Roca et al. 2009). Similarly, (re)introductions of a host species into new areas are also likely to expand the range of the host's parasite community. Translocations and reintroductions of wildlife have the potential to also reintroduce or translocate parasites and pathogens and can have detrimental effects on both the reintroduced host species and also other species within the community where the introduction has taken place (Kock et al. 2010).

Captive breeding efforts typically have negative impacts on the parasites associated with a species for two reasons. On the one hand, commensal, specialized parasites frequently cannot tolerate conditions of captivity and progressively disappear from their captive hosts (Windsor 1990). Indeed, a recent review found that parasites with complex life cycles and environmental or vector-borne transmission were less common in captive hosts (Milotic et al. 2020). On the other hand, if they can survive in captivity, then—because of the high-density populations typical of captive environments—they often will build up high parasite burdens leading to host pathogenicity. Conservation managers often view them as a problem that will impede recovery of the host and remove them through drug administration (Pachaly et al. 2016). This is despite the fact that parasites are a critical part of the natural environment that the focal species has evolved to survive and may even be required for its long-term well-being. This means that conservation managers face a difficult trade-off in suppressing potentially pathogenic parasite burdens while maintaining the natural parasite community within captive host populations.

1.5.6 Which parasite species are most likely to become endangered?

Identifying which life history traits predispose any species to extinction is a well-established and successful strategy that conservation biologists utilize to prioritize scarce management funds and guide appropriate preservation actions (Foufopoulos and Ives 1999; Henle et al. 2004; Primack 2012). While there is mounting recognition that parasitic organisms are now threatened by human activities, little is known about which types of parasites are most likely to be affected by anthropogenic change (Bush and Kennedy 1994; Lafferty 1997). At the simplest level, parasites do become endangered if their host species decline

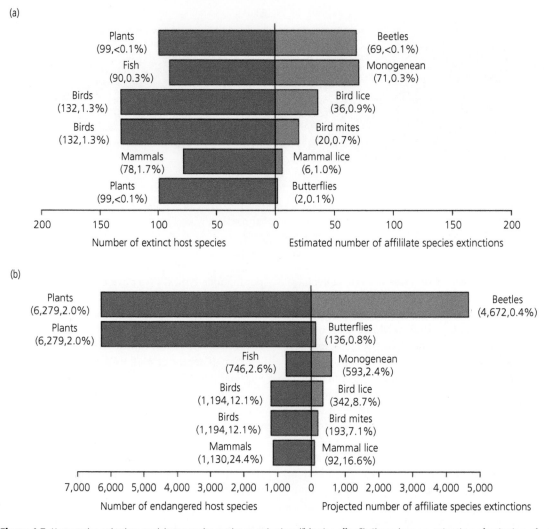

Figure 1.7 Host species extinctions precipitate parasite species co-extinctions ("domino effect"). Shown here are estimations of extinctions of affiliate species given extinctions of endangered hosts. (a) Past extinctions: estimated numbers of historically extinct affiliate species based on the number of host species that are known to have gone extinct. (b) Future extinctions: projected numbers of affiliate species extinctions, if all currently endangered hosts were to go extinct. The first number in the parentheses represents the number of species, and the second value reflects the percentage of species that have gone extinct or are endangered based on predictions by a nomographic model. (Figures redrawn and simplified from Koh et al. 2004.)

or if their natural habitat is destroyed (so-called co-extinctions; Koh et al. 2004; Doña and Johnson 2020; Moir and Brennan, 2020; Figure 1.7). Beyond such trivial patterns, other life history characteristics may also determine a parasitic organism's susceptibility to anthropogenic change. Epidemiological data from the field suggest that—at least for simple-lifecycle microparasites—narrow host range, brief infection period, lack of persistent stages, lack of environmental reservoirs, as well as complete, lifelong host immunity, all make a pathogen more susceptible to disruption. In contrast, microparasites that establish chronic infections, infect multiple host species, have environmental reservoirs, and are transmitted even at low host densities (e.g., frequency-transmitted pathogens) are probably less likely to decline or become extirpated.

Parallel research in macroparasites suggests that species with a narrower host spectrum, smaller geographic ranges, and specialized habitat requirements are more likely to disappear. Similarly, parasites with complex life cycles, that is, those that require passage through multiple intermediate hosts to complete their development, are more likely to face disruption relative to those that have only simple life cycles (Roca et al. 2009; Pickles et al. 2013). Furthermore, large-bodied macroparasites, which tend to infect also large-bodied hosts (Bush and Clayton 2006), as well as parasites that infect hosts at higher trophic levels (Estes et al. 2011), two host categories that tend to be at higher conservation risk, are expected to also be more imperiled (Colwell et al. 2012). For example, among the species of European unionid freshwater mussels, which are dependent on certain fish species as hosts for their glochidia larvae, endangerment status appears to be determined by the number of fish species that can serve as hosts (Lopez and Altaba 2006; Lopes-Lima et al. 2017).

1.6 Conclusion

Ultimately, the most direct way to safeguard parasite diversity is to maintain large and ecologically functional populations of hosts. In addition, because transmission is often a process that is shaped by the presence of alternative and intermediate hosts, the full complement of functional communities of host species needs to be preserved. Lastly, since successful transmission often depends on prevailing environmental conditions, it is important that these host species communities continue to exist under environmental conditions that facilitate natural ecosystem processes.

References

Abbott, I. (2006). Mammalian faunal collapse in Western Australia, 1875–1925: The hypothesized role of epizootic disease and a conceptual model of its origin, introduction, transmission, and spread. *Australian Zoologist*, 33:530–61.

Abdoli, A. and Ardakani, H.M. (2019). Potential application of helminth therapy for resolution of neuroinflammation in neuropsychiatric disorders. *Metabolic Brain Disease*, 35:95–110.

Alroy, J. (2008). Dynamics of origination and extinction in the marine fossil record. *Proceedings of the National Academy of Sciences*, **105** (Suppl. 1):11536–42.

Altizer, S., Foufopoulos, J., and Gager, A. (2001). Disease and conservation. In: Levin, S. (ed.), *Encyclopedia of Biodiversity*, Vol. 2, pp. 109–29. Academic Press, San Diego, CA.

Altizer, S., Nunn, C.L., and Lindenfors, P. (2007). Do threatened hosts have fewer parasites? A comparative study in primates. *Journal of Animal Ecology*, **76**: 304–14.

Altizer S., Ostfeld, R.S., Johnson, P.T., Kutz, S., and Harvell, C.D. (2013). Climate change and infectious diseases: From evidence to a predictive framework. *Science* **341**:514–19.

Anderson, R.M. and Gordon, D.M. (1982). Processes influencing the distribution of parasite numbers within host populations, with special emphasis on parasite-induced host mortalities. *Parasitology*, **85**:373–98.

Anderson, R.M. and May, R.M. (1982). Coevolution of hosts and parasites. *Parasitology*, **85**:411–26.

Barnosky, A.D., Matzke, N., Tomiya, S., Wogan, G.O., Swartz, B., Quental, T.B., et al. (2011). Has the earth's sixth mass extinction already arrived? *Nature*, **471**:51–57.

Berger, L., Speare, R., Daszak, P., Green, D.E., Cunningham, A.A., Goggin, C.L., et al. (1998). Chytridiomycosis causes amphibian mortality associated with population declines in the rain forests of Australia and Central America. *Proceedings of the National Academy of Sciences*, **95**:9031–36.

Bhattacharjee, S., Kalbfuss, N., and Prazeres Da Costa, C. (2017). Parasites, microbiota and metabolic disease. *Parasite Immunology*, 39:e12390.

Briggs, C.J., Knapp, R.A., and Vredenburg, V.T. (2010). Enzootic and epizootic dynamics of the chytrid fungal pathogen of amphibians. *Proceedings of the National Academy of Sciences*, **107**:9695–700.

Brook, B.W., Sodhi, N.S., and Bradshaw, C.J. (2008). Synergies among extinction drivers under global change. *Trends in Ecology & Evolution*, **23**:453–60.

Brooks, D. and McLennan, D.A. (1993). *Parascript: Parasites and the Language of Evolution*. Smithsonian Institution Press, Washington, DC.

Bulgarella, M. and Palma, R.L. (2017). Coextinction dilemma in the Galápagos Islands: Can Darwin's finches and their native ectoparasites survive the control of the introduced fly *Philornis downsi*? *Insect Conservation and Diversity*, **10**:193–99.

Bush, A.O. and Kennedy, C.R. (1994). Host fragmentation and helminth parasites: Hedging your bets against extinction. *International Journal for Parasitology*, **24**:1333–43.

Bush, S.E. and Clayton, D.H. (2006). The role of body size in host specificity: Reciprocal transfer experiments with feather lice. *Evolution*, **60**:2158–67.

Bush, S.E., Reed, M., and Maher, S. (2013). Impact of forest size on parasite biodiversity: Implications for conservation of hosts and parasites. *Biodiversity and Conservation*, **22**:1391–404.

Cardinale, B., Primack, R., and Murdock, J. (2019). *Conservation Biology*. Oxford University Press, Oxford.

Carlson, C.J., Burgio, K.R., Dougherty, E.R., Phillips, A.J., Bueno, V.M., Clements, C.F., et al. (2017). Parasite biodiversity faces extinction and redistribution in a changing climate. *Science Advances*, **3**:e1602422.

Carlson, C.J., Cizauskas, C.A., Burgio, K.R., Clements, C.F., and Harris, N.C. (2013). The more parasites, the better? *Science*, **342**:1041.

Carlson, C.J., Hopkins, S., Bell, K.C., Doña, J., Godfrey, S.S., Kwak, M.L., et al. (2020). A global parasite conservation plan. *Biological Conservation*, **248**:108596.

Carlton, J.T., Vermeij, G.J., Lindberg, D.R., Carlton, D.A., and Dubley, E.C. (1991). The first historical extinction of a marine invertebrate in an ocean basin: The demise of the eelgrass limpet Lottia alveus. *The Biological Bulletin*, **180**:72–80.

Caughley, G. (1994). Directions in conservation biology. *Journal of Animal Ecology*, **63**:215–44.

Ceballos, G., Ehrlich, P.R., Barnosky, A.D., García, A., Pringle, R.M., and Palmer, T.M. (2015). Accelerated modern human-induced species losses: Entering the sixth mass extinction. *Science Advances*, **1**: e1400253.

Ceballos, G., Ehrlich, P.R., and Dirzo, R. (2017). Biological annihilation via the ongoing sixth mass extinction signaled by vertebrate population losses and declines. *Proceedings of the National Academy of Sciences*, **114**:e6089–96.

Coates, D.J., Byrne, M., and Moritz, C. (2018). Genetic diversity and conservation units: Dealing with the species-population continuum in the age of genomics. *Frontiers in Ecology and Evolution*, **6**:13.

Colwell, R.K., Dunn, R.R., and Harris, N.C. (2012). Coextinction and persistence of dependent species in a changing world. *Annual Review of Ecology, Evolution, and Systematics*, **43**:183–203.

Coyne, J. and Orr, H.A. (2004). *Speciation*. Sinauer Associates, Sunderland, MA.

Cracraft, J. (1983). Species concepts and speciation analysis. In: Johnston R.F. (ed.), *Current Ornithology*, Vol. 1, pp. 159–87. Springer, Boston, MA.

Crofton, H.D. (1971). A quantitative approach to parasitism. *Parasitology*, **62**:179–94.

De Castro, F. and Bolker, B. (2005). Mechanisms of disease-induced extinction. *Ecology Letters*, **8**:117–26.

De Queiroz, K. (1998). The general lineage concept of species: Species criteria and the process of speciation. In: Howard, D.J. and Berlocher, S.H. (eds.), *Endless Forms: Species and Speciation*, pp. 57–75. Oxford University Press, New York.

De Queiroz, K. (2005). Different species problems and their resolution. *BioEssays*, **27**:1263–69.

De Queiroz, K. (2007). Species concepts and species delimitation. *Systematic Biology*, **56**:879–86.

De Ruiter, K., Tahapary, D.L., Sartono, E., Soewondo, P., Supali, T., Smit, J.W., et al. (2017). Helminths, hygiene hypothesis and type 2 diabetes. *Parasite Immunology*, **39**:e12404.

De Vos, J.M., Joppa, L.N., Gittleman, J.L., Stephens, P.R., and Pimm, S.L. (2015). Estimating the normal background rate of species extinction. *Conservation Biology*, **29**:452–62.

Diamond, J. (1984). "Normal" extinctions of isolated populations. In: Nitecki, M.H. (ed.), *Extinctions*, pp. 191–246. Chicago University Press, Chicago, IL.

Díaz, S., Pascual, U., Stenseke, M., Martín-López, B., Watson, R.T., Molnár, Z., et al. (2018). Assessing nature's contributions to people. *Science* **359**:270–72.

Diaz, S., Settele, J., Brondizio, E.S., Ngo, H.T., Agard, J., Arneth, A., et al. (2019). Pervasive human-driven decline of life on earth points to the need for transformative change. *Science* **366**:6471.

Dirzo, R., Young, H.S., Galetti, M., Ceballos, G., Isaac, N.J., and Collen, B. (2014). Defaunation in the Anthropocene. *Science*, **345**:401–06.

Dixit, A., Tanaka, A., Greer, J.M., and Donnelly, S. (2017). Novel therapeutics for multiple sclerosis designed by parasitic worms. *International Journal of Molecular Sciences*, **18**:2141.

Dobson, A.P. and Carper, E.R. (1996). Infectious diseases and human population history. *Bioscience*, **46**:115–26.

Dobson, A.P., Lafferty, K.D., Kuris, A.M., Hechinger, R.F., and Jetz, W. (2008). Homage to Linnaeus: How many parasites? How many hosts? *Proceedings of the National Academy of Sciences*, **105**:11482–89.

Doña, J. and Johnson, K.P. (2020). Assessing symbiont extinction risk using cophylogenetic data. *Biological Conservation*, **248**:108705.

Dougherty, E.R., Carlson, C.J., Bueno, V.M., Burgio, K.R., Cizauskas, C.A., Clements, C.F., et al. (2016). Paradigms for parasite conservation. *Conservation Biology*, **30**:724–33.

Dunlop, J. (2015). The ecology and host-parasite dynamics of a fauna translocation in Australia. Unpublished PhD thesis, Murdoch University, Australia.

Dunn, R.R., Harris, N.C., Colwell, R.K., Koh, L.P., and Sodhi, N.S. (2009). The sixth mass coextinction: Are most endangered species parasites and mutualists? *Proceedings of the Royal Society B: Biological Sciences,* **276**:3037–45.

Durden, L.A. and Keirans, J.E. (1996). Host-parasite coextinction and the plight of tick conservations. *American Entomologist,* **42**:87–91.

Durham, H.E. (1908). Notes on Nagana and on some Haematozoa observed during my travels. *Parasitology,* **1**:227–35.

Dybing, N.A., Jacobson, C., Irwin, P., Algar, D., and Adams, P.J. (2016). Ghosts of Christmas past? Absence of trypanosomes in feral cats and black rats from Christmas Island and Western Australia. *Parasitology Open,* **2**:e4.

Elewa, A.M. and Joseph, R. (2009). The history, origins, and causes of mass extinctions. *Journal of Cosmology,* **2**:201–20.

Estes, J.A., Terborgh, J., Brashares, J.S., Power, M.E., Berger, J., William, J.B., et al. (2011). Trophic downgrading of planet earth. *Science,* **333**:301–06.

Foufopoulos, J. (1998). Host-parasite interactions in the mountain spiny lizard Sceloporus jarrovi. Unpublished PhD thesis, University of Wisconsin, Madison, WI.

Foufopoulos, J. and Ives, A.R. (1999). Reptile extinctions on land-bridge islands: Life history attributes and vulnerability to extinction. *American Naturalist,* **153**: 1–25.

Fraser, D. J. and Bernatchez, L. (2001). Adaptive evolutionary conservation: Towards a unified concept for defining conservation units. *Molecular Ecology,* **10**:2741–52.

Friderici, P. (1997). Passenger pigeon chewing louse (*Columbicola extincta*). *Wild Earth,* **7**:37–38.

Froeschke, G. and Sommer, S. (2012). Insights into the complex associations between MHC class II DRB polymorphism and multiple gastrointestinal parasite infestations in the striped mouse. *PLoS ONE,* **7**:e31820.

García-Ramírez, A., Delgado-García, J.D., Foronda-Rodríguez, P., and Abreu-Acosta, N. (2005). Haematozoans, mites and body condition in the oceanic island lizard *Gallotia atlantica* (Peters and Doria, 1882) (Reptilia: Lacertidae). *Journal of Natural History,* **39**:1299–305.

Goüy De Bellocq, J.G., Morand, S., and Feliu, C. (2002). Patterns of parasite species richness of Western Palaearctic micro-mammals: Island effects. *Ecography,* **25**:173–83.

Grant, E.H., Muths, E., Katz, R.A., Canessa, S., Adams, M.J., Ballard, J.R., et al. (2017). Using decision analysis to support proactive management of emerging infectious wildlife diseases. *Frontiers in Ecology and the Environment,* **15**:214–21.

Green, P. (2014). Mammal extinction by introduced infectious disease on Christmas Island (Indian Ocean): The historical context. *Australian Zoologist,* **37**:1–14.

Hawkins, C.E., Baars, C., Hesterman, H., Hocking, G.J., Jones, M.E., Lazenby, B., et al. (2006). Emerging disease and population decline of an island endemic, the Tasmanian devil Sarcophilus harrisii. *Biological Conservation,* **131**:307–24.

Henderson, D.A. and Klepac, P. (2013). Lessons from the eradication of smallpox: An interview with D.A. Henderson. *Philosophical Transactions of the Royal Society B: Biological Sciences,* **368**:20130113.

Henle, K., Davies, K.F., Kleyer, M., Margules, C., and Settele, J. (2004). Predictors of species sensitivity to fragmentation. *Biodiversity and Conservation,* **13**: 207–51.

Hildebrandt, J.P. and Lemke, S. (2011). Small bite, large impact—Saliva and salivary molecules in the medicinal leech, Hirudo medicinalis. *Naturwissenschaften,* **98**: 995–1008.

Hovius, J.W., Levi, M., and Fikrig, E. (2008). Salivating for knowledge: Potential pharmacological agents in tick saliva. *PLoS Medicine,* **5**:e43.

Hudson, P. and Greenman, J. (1998). Competition mediated by parasites: Biological and theoretical progress. *Trends in Ecology & Evolution,* **13**:387–90.

Hudson, P.J., Dobson, A.P., and Lafferty, K.D. (2006). Is a healthy ecosystem one that is rich in parasites? *Trends in Ecology & Evolution,* **21**:381–85.

Huspeni, T.C. and Lafferty, K.D. (2004). Using larval trematodes that parasitize snails to evaluate a salt-marsh restoration project. *Ecological Applications,* **14**:795–804.

IPBES. (2019). *Summary for Policymakers of the Global Assessment Report on Biodiversity and Ecosystem Services of the Intergovernmental Science-Policy Platform on Biodiversity and Ecosystem Services.* Díaz, S., Settele, J., Brondízio, E.S., Ngo, H.T., Guèze, M., Agard, J., et al. (eds.). IPBES Secretariat, Bonn. https://doi.org/10.5281/zenodo.3553579 (accessed February 18, 2020).

Johnson, P.T., Preston, D.L., Hoverman, J.T., and Richgels, K.L. (2013). Biodiversity decreases disease through predictable changes in host community competence. *Nature,* **494**:230–33.

Kahl, J., Brattig, N., and Liebau, E. (2018). The untapped pharmacopeic potential of helminths. *Trends in Parasitology,* **34**:828–42.

Keesing, F., Belden, L.K., Daszak, P., Dobson, A., Harvell, C.D., Holt, R.D., et al. (2010). Impacts of biodiversity on the emergence and transmission of infectious diseases. *Nature,* **468**:647–52.

Keesing, F., Holt, R.D., and Ostfeld, R.S., (2006). Effects of species diversity on disease risk. *Ecology Letters,* **9**:485–98.

Kittler, R., Kayser, M., and Stoneking, M. (2003). Molecular evolution of *Pediculus humanus* and the origin of clothing. *Current Biology*, 13:1414–17.

Kock, R.A., Woodford, M.H., and Rossiter, P.B. (2010). Disease risks associated with the translocation of wildlife. *Revue Scientifique et Technique-International Office of Epizootics*, 29:329–50.

Koh, L.P., Dunn, R.R., Sodhi, N.S., Colwell, R.K., Proctor, H.C., and Smith, V.S. (2004). Species coextinctions and the biodiversity crisis. *Science*, 305:1632–34.

Kwak, M.L., Heath, A.C., and Cardoso, P. (2020). Methods for the assessment and conservation of threatened animal parasites. *Biological Conservation*, 248: 108696.

Lafferty, K.D. (1997). Environmental parasitology: What can parasites tell us about human impacts on the environment? *Parasitology Today*, 13:251–55.

Lafferty, K.D. (2012). Biodiversity loss decreases parasite diversity: Theory and patterns. *Philosophical Transactions of the Royal Society B: Biological Sciences*, 367: 2814–27.

Lafferty, K.D. and Holt, R.D. (2003). How should environmental stress affect the population dynamics of disease? *Ecology Letters*, 6:654–64.

Lafferty, K.D. and Kuris, A.M. (1999). How environmental stress affects the impacts of parasites. *Limnology and Oceanography*, 44:925–31.

Lafferty, K.D., McLaughlin, J.P., Gruner, D.S., Bogar, T.A., Bui, A., Childress, J.N., et al. (2018). Local extinction of the Asian tiger mosquito (*Aedes albopictus*) following rat eradication on Palmyra Atoll. *Biology Letters*, 14:20170743.

Lafferty, K.D., Smith, K.F., Torchin, M.E., Dobson, A.P., and Kuris, A.M. (2005). The role of infectious diseases in natural communities. In: Sax, D.F., Stachowicz, J.J., and Gaines, S.D. (eds.), *Species invasions: Insights into Ecology, Evolution, and Biogeography*, pp. 111–34. Sinauer Associates, Sunderland, MA.

Lankester, M.W. (2010). Understanding the impact of meningeal worm, *Parelaphostrongylus tenuis*, on moose populations. *Alces*, 46:53–70.

Laurance, W., McDonald, K., and Speare, R. (1996). Catastrophic declines of Australian rain forest frogs: Support for the epidemic disease hypothesis. *Conservation Biology*, 10:406–13.

Lawton, J.H. and May, R.M. (2005). *Extinction Rates*. Oxford University Press, Oxford.

Lerner, A., Arleevskaya, M., Schmiedl, A., and Matthias, T. (2017). Microbes and viruses are bugging the gut in celiac disease. Are they friends or foes? *Frontiers in Microbiology*, 8:1392.

Lindström, K.M., Foufopoulos, J., Pärn, H., and Wikelski, M. (2004). Immunological investments reflect parasite abundance in island populations of Darwin's finches. *Proceedings of the Royal Society B: Biological Sciences*, 271:1513–19.

Lips, K.R., Brem, F., Brenes, R., Reeve, J.D., Alford, R.A., Voyles, J., et al. (2006). Emerging infectious disease and the loss of biodiversity in a Neotropical amphibian community. *Proceedings of the National Academy of Sciences*, 103:3165–70.

Liu, J., Morey, R.A., Wilson, J.K., and Parker, W. (2017). Practices and outcomes of self-treatment with helminths based on physicians' observations. *Journal of Helminthology*, 91:267–77.

LoGiudice, K. (2003). Trophically transmitted parasites and the conservation of small populations: Raccoon roundworm and the imperiled Allegheny woodrat. *Conservation Biology*, 17:258–66.

Lopes-Lima, M., Sousa, R., Geist, J., Aldridge, D.C., Araujo, R., Bergengren, J., et al. (2017). Conservation status of freshwater mussels in Europe: State of the art and future challenges. *Biological Reviews*, 92:572–607.

Lopez, M.A. and Altaba, C.R. (2006). Fish host determination for *Margaritifera Auricularia* (Bivalvia: Unionoida): Results and implications. *Bollettino Malacologico*, 41:89–98.

Lorimer, J. (2017). Parasites, ghosts and mutualists: A relational geography of microbes for global health. *Transactions of the Institute of British Geographers*, 42: 544–58.

Lyons, K.S., Smith, F.A., Wagner, P.J., White, E.P., and Brown, J.H. (2004). Was a "hyperdisease" responsible for the late Pleistocene megafaunal extinction? *Ecology Letters*, 7:859–68.

MacPhee, R.D. and Greenwood, A.D. (2013). Infectious disease, endangerment, and extinction. *International Journal of Evolutionary Biology*, 2013: 571939.

MacPhee, R.D. and Marx, P.A. (1997). Humans, hyperdisease, and first-contact extinctions. In: Goodman, S.M. and Patterson, B.D. (eds.), *Natural Change and Human Impact in Madagascar*, pp. 169–217. Smithsonian Institution Press, Washington, DC.

Mallet, J. (2001). The speciation revolution (Commentary). *Journal of Evolutionary Biology*, 14:887–88.

Martel, A., Blooi, M., Adriaensen, C., Van Rooij, P., Beukema, W., Fisher, M.C., et al. (2014). Recent introduction of a chytrid fungus endangers Western Palearctic salamanders. *Science*, 346:630–31.

Martel, A., Spitzen-van Der Sluijs, A., Blooi, M., Bert, W., Ducatelle, R., Fisher, M.C., et al. (2013). *Batrachochytrium salamandrivorans* sp. nov. causes lethal chytridiomycosis in amphibians. *Proceedings of the National Academy of Sciences*, 113:15325–29.

Maruszewska-Cheruiyot, M., Donskow-Łysoniewska, K., and Doligalska, M. (2018). Helminth therapy: Advances in the use of parasitic worms against inflammatory bowel diseases and its challenges. *Helminthologia*, **55**: 1–11.

May, R.M. (1988). How many species are there on earth? *Science*, 241:1441–49.

Mayr, E. (1963). *Animal Species and Evolution*. Harvard University Press,, London.

McCallum, H. (2012). Disease and the dynamics of extinction. *Philosophical Transactions of the Royal Society B: Biological Sciences*, **367**:2828–39.

McCallum, H. and Dobson, A. (1995). Detecting disease and parasite threats to endangered species and ecosystems. *Trends in Ecology and Evolution*, **10**: 190–94.

McCallum, H. and Jones, M. (2006). To lose both would look like carelessness: Tasmanian devil facial tumour disease. *PLoS Biology*, **4**:1671–74.

McCallum, H., Jones, M., Hawkins, C., Hamede, R., Lachish, S., Sinn, D.L., et al. (2009). Transmission dynamics of Tasmanian devil facial tumor disease may lead to disease-induced extinction. *Ecology*, **90**:3379–92.

Mey, E. (2005). *Psittacobrosus bechsteini*: A new extinct chewing louse (Insecta, Phthiraptera, Amblycera) off the Cuban macaw *Ara tricolor* (Psittaciiformes), with an annotated review of fossil and recently extinct animal lice. *Anzeiger des Vereins Thüringer Ornithologen*, **5**:201–17.

Michalsen, A., Klotz, S., Lüdtke, R., Moebus, S., Spahn, G., and Dobos, G.J. (2003). Effectiveness of leech therapy in osteoarthritis of the knee: A randomized, controlled trial. *Annals of Internal Medicine*, **139**:724–30.

Milotic, M., Lymbery, A., Thompson, A., Doherty, J.F., and Godfrey, S. (2020). Parasites are endangered by the conservation of their hosts: Meta-analyses of the effect of host captivity on the odds of parasite infection. *Biological Conservation*, **248**:108702.

Moir, M.L. and Brennan, K.E. (2020). Incorporating coextinction in threat assessments and policy will rapidly improve the accuracy of threatened species lists. *Biological Conservation*, **249**:108715.

Mordecai, E.A., Cohen, J.M., Evans, M.V., Gudapati, P., Johnson, L.R., Lippi, C.A., et al. (2017). Detecting the impact of temperature on transmission of Zika, dengue, and Chikungunya using mechanistic models. *PLoS Neglected Tropical Diseases*, 11:e0005568.

Mordecai, E.A., Paaijmans, K.P., Johnson, L.R., Balzer, C., Ben-Horin, T., Moor, E., et al. (2013). Optimal temperature for malaria transmission is dramatically lower than previously predicted. *Ecology Letters*, **16**:22–30.

Mouritsen, K.N. and Poulin, R. (2004). Parasites boosts biodiversity and changes animal community structure by trait-mediated indirect effects. *Oikos*, **108**:344–50.

Naess, A. (2008). *The Ecology of Wisdom: Writings by Arne Naess*. Drengston, A. and Devall, B. (eds.). Counterpoint Press, Berkeley, CA.

Normile, D. (2008). Driven to extinction. *Science*, **319**:1606–09.

Normile, D. (2010). Rinderpest, deadly for cattle, joins smallpox as a vanquished disease. *Science*, 330:435.

O'Hanlon, S.J., Rieux, A., Farrer, R.A., Rosa, G.M., Waldman, B., Bataille, A., et al. (2018). Recent Asian origin of chytrid fungi causing global amphibian declines. *Science*, 360:621–27.

Pachaly, J.R., Monteiro-Filho, L.P., Gonçalves, D.D., and Voltarelli-Pachaly, E.M. (2016). *Gyrostigma rhinocerontis* (Diptera: Oestridae, Gasterophilinae) in white rhinoceroses (*Ceratotherium simum*) imported from South Africa: Occurrence in Itatiba, São Paulo, Brazil. *Pesquisa Veterinária Brasileira*, **36**:749–52.

Paterson, S., Wilson, K., and Pemberton, J.M. (1998). Major histocompatibility complex variation associated with juvenile survival and parasite resistance in a large unmanaged ungulate population (*Ovis aries* L.). *Proceedings of the National Academy of Sciences*, **95**:3714–19.

Pearse, A.-M. and Swift, K. (2006). Allograft theory: Transmission of devil facial-tumour disease. *Nature*, **439**:549.

Pedersen, A. B., Jones, K.E., Nunn, C.L., and Altizer, S. (2007). Infectious diseases and extinction risk in wild mammals. *Conservation Biology*, **21**:1269–79.

Pickering, J. and Norris, C.A. (1996). New evidence concerning the extinction of the endemic murid *Rattus macleari* from Christmas Island, Indian Ocean. *Australian Mammalogy*, **19**:19–25.

Pickles, R.S., Thornton, D., Feldman, R., Marques, A., and Murray, D.L. (2013). Predicting shifts in parasite distribution with climate change: A multitrophic level approach. *Global Change Biology*, **19**:2645–54.

Powell, F.A. (2011). Can early loss of affiliates explain the coextinction paradox? An example from *Acacia*-inhabiting psyllids (Hemiptera: Psylloidea). *Biodiversity and Conservation*, **20**:1533–44.

Preece, N.D., Abell, S.E., Grogan, L., Wayne, A., Skerratt, L.F., Van Oosterzee, P., et al. (2017). A guide for ecologists: Detecting the role of disease in faunal declines and managing population recovery. *Biological Conservation*, **214**:136–46.

Primack, R.B. (2012). *A Primer of Conservation Biology*. 5th ed. Sinauer Associates, Sunderland, MA.

Raup, D.M. and Sepkoski, J.J. (1982). Mass extinctions in the marine fossil record. *Science*, **215**:1501–03.

Reid, A.H., Fanning, T.G., Hultin, J.V., and Taubenberger, J.K. (1999). Origin and evolution of the 1918 "Spanish" influenza virus hemagglutinin gene. *Proceedings of the National Academy of Sciences*, **96**:1651–56.

Ribeiro, J.M., Makoul, G.T., Levine, J., Robinson, D.R., and Spielman, A. (1985). Antihemostatic, antiinflammatory, and immunosuppressive properties of the saliva of a tick, Ixodes dammini. *Journal of Experimental Medicine*, **161**:332–44.

Roca, V., Foufopoulos, J., Valakos, E., and Pafilis, P. (2009). Parasitic infracommunities of the Aegean wall lizard *Podarcis erhardii* (Lacertidae, Sauria): Isolation and impoverishment in small island populations. *Amphibia-Reptilia*, **30**:493–503.

Rohde, R.A. and Muller, R.A. (2005). Cycles in fossil diversity. *Nature*, **434**:208–10.

Ross, P.S., Vos, J.G., Birnbaum, L.S., and Osterhaus, A.D. (2000). PCBs are a health risk for humans and wildlife. *Science*, **289**:1878–79.

Rósza, L. (1992). Points in question: Endangered parasite species. *International Journal of Parasitology*, **22**:265–66.

Rósza, L. and Vas, Z. (2015). Co-extinct and critically co-endangered species of parasitic lice, and conservation-induced extinction: Should lice be reintroduced to their hosts? *Oryx*, **49**:107–10.

Schall, J.J. (1992). Parasite-mediated competition in *Anolis* lizards. *Oecologia*, **92**:58–64.

Scheele, B.C., Pasmans, F., Skerratt, L.F., Berger, L., Martel, A., Beukema, W., et al. (2019). Amphibian fungal panzootic causes catastrophic and ongoing loss of biodiversity. *Science*, **363**:1459–63.

Schmidt, K. A. and Ostfeld, R.S. (2001). Biodiversity and the dilution effect in disease ecology. *Ecology*, **82**: 609–19.

Schnell, I.B., Sollmann, R., Calvignac-Spencer, S., Siddall, M.E., Douglas, W.Y., Wilting, A., et al. (2015). iDNA from terrestrial haematophagous leeches as a wildlife surveying and monitoring tool—Prospects, pitfalls and avenues to be developed. *Frontiers in Zoology*, **12**:24.

Smith, K. F., Sax, D.F., and Lafferty, K.D. (2006). Evidence for the role of infectious disease in species extinction and endangerment. *Conservation Biology*, **20**:1349–57.

Soulé, M. E. (1985). What is conservation biology? Bioscience, **35**:727–734.

Soulé, M.E. (1986). Conservation Biology: The Science of Scarcity and Diversity. Sinauer Associates, Sunderland, MA.

Sousa, W.P. (1993). Interspecific antagonism and species coexistence in a diverse guild of larval trematode parasites. *Ecological Monographs*, **63**:103–28.

Spielman, D., Brook, B.W., Briscoe, D.A., and Frankham, R. (2004). Does inbreeding and loss of genetic diversity decrease disease resistance? *Conservation Genetics*, **5**:439–48.

Stallknecht, D. (2007). Impediments to wildlife disease surveillance, research, and diagnostics. In: Childs, J.E., Mackenzie, J.S., and Richt, J.A (eds.), *Wildlife and Emerging Zoonotic Diseases: The Biology, Circumstances and Consequences of Cross-Species Transmission*, pp. 445–61. Springer, Berlin, Heidelberg.

Steen, N.A., Barker, S.C., and Alewood, P.F. (2006). Proteins in the saliva of the Ixodida (ticks): Pharmacological features and biological significance. *Toxicon*, **47**: 1–20.

Surendar, J., Indulekha, K., Hoerauf, A., and Hübner, M.P. (2017). Immunomodulation by helminths: Similar impact on type 1 and type 2 diabetes? *Parasite Immunology*, **39**:e12401.

Sures, B. (2003). Accumulation of heavy metals by intestinal helminths in fish: An overview and perspective. *Parasitology*, **126**:S53–60.

Sures, B. (2004). Environmental parasitology: Relevancy of parasites in monitoring environmental pollution. *Trends in Parasitology*, **20**:170–77.

Torchin, M.E., Lafferty, K.D., Dobson, A.P., McKenzie, V.J., and Kuris, A.M. (2003). Introduced species and their missing parasites. *Nature*, **421**:628–30.

Tschirren, B., Bischoff, L.L., Saladin, V., and Richner, H. (2007). Host condition and host immunity affect parasite fitness in a bird–ectoparasite system. *Functional Ecology*, **21**:372–78.

United Nations (1992). Convention on Biological Diversity. United Nations, New York. https://www.cbd.int/doc/legal/cbd-en.pdf (accessed February 14, 2020).

Van Loveren, H., Ross, P.S., Osterhaus, A.D.M.E., and Vos, J.G. (2000). Contaminant-induced immunosuppression and mass mortalities among harbor seals. *Toxicology Letters*, **112**:319–24.

Van Riper, C., Van Riper, S.G., Lee Goff, M., and Laird, M. (1986). The epizootiology and ecological significance of malaria in Hawaiian land birds. *Ecological Monographs*, **56**:327–44.

Veron, J.E. (2008). Mass extinctions and ocean acidification: Biological constraints on geological dilemmas. *Coral Reefs*, **27**:459–72.

Voyles, J., Woodhams, D.C., Saenz, V., Byrne, A.Q., Perez, R., Rios-Sotelo, G., et al. (2018). Shifts in disease dynamics in a tropical amphibian assemblage are not due to pathogen attenuation. *Science*, **359**:1517–19.

Waddle, J.H., Grear, D.A., Mosher, B.A., Grant, E.H.C., Adams, M.J., Backlin, A.R., et al. (2020). *Batrachochytrium salamandrivorans* (Bsal) not detected in an intensive survey of wild North American amphibians. *Scientific Reports*, **10**:1–7.

Wake, D.B. and Vredenburg, V.T. (2008). Are we in the midst of the sixth mass extinction? A view from the world of amphibians. *Proceedings of the National Academy of Sciences*, **105**:11466–73.

Warner, R.E. (1968). The role of introduced diseases in the extinction of the endemic Hawaiian avifauna. *Condor*, **70**:101–20.

Whiteman, N.K. and Parker, P.G. (2005). Using parasites to infer host population history: a new rationale for parasite conservation. *Animal Conservation*, **8**:175–81.

Wikelski, M., Foufopoulos, J., Vargas, H., and Snell, H. (2004). Galápagos birds and diseases: Invasive pathogens as threats for island species. *Ecology and Society*, **9**:5.

Wilson, E.O. (ed.). (1988) *Biodiversity*. National Academy of Sciences/Smithsonian, Washington, DC.

Windsor, D.A. (1990). Heavenly hosts. *Nature*, **348**:104.

Windsor, D.A. (1995). Equal rights for parasites. *Conservation Biology*, **9**:1–2.

Windsor, D.A. (1998). Most of the species on earth are parasites. *International Journal of Parasitology*, **28**:1939–41.

Wood, C.L., Summerside, M., and Johnson, P.T. (2020). How host diversity and abundance affect parasite infections: Results from a whole-ecosystem manipulation of bird activity. *Biological Conservation*, **248**:108683.

Wyatt, K.B., Campos, P.F., Gilbert, M.T., Kolokotronis, S.-O., Hynes, W.H., DeSalle, R., et al. (2008). Historical mammal extinction on Christmas Island (Indian Ocean) correlates with introduced infectious disease. *PLoS ONE*, **3**:e3602.

CHAPTER 2

Disease Epidemiology in Natural Systems

To be ill is to be human.
Anonymous

2.1 Introduction

Infection and disease are facts of life. Nonetheless, it is important to distinguish conceptually from the outset between the infection with a parasitic organism and the resulting disease. Infection is "the presence of a parasite within a host individual or host population" (Scott 1988), whereas disease is defined as "an impairment that interferes with, or modifies the performance of normal functions" (Wobeser 2007). Not all diseases are a result of infection (e.g., very few cancers are infectious in origin). Conversely, disease may be, but is not necessarily, an outcome of infection, and animals are often infected—perhaps in the vast majority of cases—without being overtly diseased. It is important to keep in mind, however, that parasitic infection without visible or readily detectable symptoms of disease can still have a detrimental effect on the fitness of an animal by increasing its susceptibility to predation, decreasing foraging ability, decreasing reproductive output, or reducing ability to find a mate (see, e.g., Schall et al. 1982, 1996). "Disease" is sometimes used as shorthand for "an infectious agent that causes disease," and we may do this ourselves at some points in the book where it doesn't cause confusion.

The parasitic lifestyle is part of the continuum of exploiter–victim interactions. The usual definition of a parasite is an organism that lives in close association with an individual organism of another species (the host) and which has a detrimental effect on the host. This definition covers a range of different strategies by which the parasite might gain resources from its host. Lafferty and Kuris (2002) proposed a taxonomy of the exploiter–victim spectrum using four dichotomies: (i) whether more than one victim is attacked per exploiter life history stage; (ii) whether the fitness of the victim is reduced to zero; (iii) whether the exploiter life cycle necessarily requires the death of the victim; and (iv) whether information on the number of exploiters per victim is necessary to understand the impact of exploitation (see Figure 2.1). According to this classification, organisms usually regarded as parasites can utilize at least seven different strategies. Parasites are ubiquitous in natural ecosystems, with estimates that up to 40% of all species have a parasitic lifestyle (Dobson et al. 2008). There are often more parasite–host links in food webs than predator–prey links (Lafferty et al. 2006) and in some ecosystems parasites may even represent a substantial proportion of the total biomass (Kuris et al. 2008). Despite this,

Infectious Disease Ecology and Conservation. Johannes Foufopoulos, Gary A. Wobeser and Hamish McCallum, Oxford University Press.
© Johannes Foufopoulos, Gary A. Wobeser and Hamish McCallum (2022). DOI: 10.1093/oso/9780199583508.003.0002

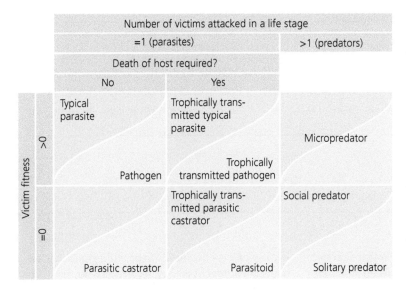

Figure 2.1 Seven types of parasitism and three types of predation separated by four life-history dichotomies. Intensity-dependent relationships are above the diagonal line; intensity-independent ones are below. This results in 10 trophic strategies of natural enemies. Note that the three predation strategies combine to four types of predator: micropredators, facultative micropredators, solitary predators, and facultative social predators. Victim fitness = 0 indicates that the interaction either kills or blocks victim reproduction, whereas victim fitness >1 indicates that the victim typically survives the interaction and can reproduce. (Source: Lafferty and Kuris 2002.)

parasitic interactions have been widely neglected by ecologists until relatively recently.

Throughout this book, we will use both the terms "parasite" and "pathogen." Pathogens are simply a subset of parasites: they are microorganisms and have substantial detrimental effects on their host. A more useful differentiation however, and one that has become increasingly established in recent years, is the distinction between micro- and macroparasites, a grouping that can be made based on the life history characteristics of an organism (see Table 2.1, which summarizes these differences). As the names suggest, microparasites are unicellular microorganisms (viruses, bacteria, fungi, or protozoa) whereas macroparasites are multicellular organisms (e.g., worms or arthropods). In contrast to macroparasites, which tend to infect a typical host in relatively small numbers, microparasites are present in very large numbers (well in excess of 10^6 individuals) inside an average host; this stark difference in parasite numbers present is used as a distinguishing criterion in much of the parasitological literature. This distinction in numbers present also has repercussions on how population biologists model an epizootic: for macroparasites only, it is important to quantify how many parasites an individual host harbors in order to understand the population dynamics of the parasitic infection. Microparasites are usually capable of rapid

multiplication within their host and, particularly in vertebrates, the host is typically capable of mounting an immune response adequate to entirely remove an infection. In contrast, macroparasites typically do not multiply within an individual host and any immune response from the host is usually incomplete; as a result, infections tend to be chronic. These distinctions are not clear-cut; exceptions include some microorganisms that cannot multiply without leaving their individual host, and some multicellular parasites that are capable of multiplying within an individual. Nevertheless, these generalizations have important implications for how parasite infections in a host population should be quantified and how they can be modeled.

For microparasites, classifying hosts as "susceptible," "infected," or "recovered" is extremely useful in understanding their epidemiology. Because microparasites can multiply rapidly within a host, the level of infection within an individual host is less determined by the level of the infective dose and more by the ability of the host's immune response to control the microparasite population. Prevalence, which is the proportion of individuals in the host population that is infected by a microparasite, rather than parasite load, is therefore the key quantity that needs to be estimated.

Table 2.1 Basic differences between micro- and macroparasites.

Trait	Microparasites	Macroparasites
Size	Very small (generally unicellular)	Larger (multicellular, generally visible)
Generation time	Very short (typically hours and days)	Much longer (generally days to months)
Life cycle	Most often simple (one host)	Frequently complex (multiple stages in several host spp.)
Reproduction	Within the host (typically asexual); rapid	Often outside primary host; slower
Parasite load (number of individual parasites in host)	Extremely high	Much lower
Duration of infection	Generally short (but can be chronic)	Generally long (sometimes life-long)
Host immunity	Strong, often life-long	Weak and transient

As mentioned previously, macroparasites cannot multiply within an individual host. Their numbers within a host are instead determined by the number of infective stages to which the host was exposed. The resulting quantity of parasites in a host determines the impact that they will have on the host (collectively termed the parasite burden). In many natural populations of hosts, almost all individuals carry at least some parasites. This means that for macroparasites, prevalence is often not very informative as a measure of the parasites' impact on the host population. Instead, it is necessary to estimate the number of parasites each host harbors. In most host–macroparasite interactions, the distribution of parasites among hosts is strongly nonrandom as most of the parasite population is harbored by a small proportion of the host population (Figure 2.2). Consequently, it is in this relatively small proportion of heavily infected hosts that the parasite has a substantial effect on survival and fecundity. In turn, these heavily infected hosts will generate most of the infective stages and therefore any control measures directed at such hosts will have a much greater effect at the population level than indiscriminately applied control measures. For macroparasites, it is therefore necessary to estimate not only the mean parasite burden, but also the variance in parasite burden between individual hosts.

2.2 Transmission and parasite life cycles

All parasites require a means to spread from one host individual to another. Transmission is therefore the key process that needs to be understood in parasite and pathogen ecology. It can be a complex affair involving more than one host species and multiple parasite life history stages. In general however, and especially for directly transmitted parasites, the ease of transmission is positively dependent on the density of the host population. The reason is that in more dense populations, the probability of two hosts encountering each other and successfully transmitting a parasite between them is higher. Because the density of a host population both determines the rate at which a pathogen is being transmitted and is also shaped by a disease (e.g., through pathogen-induced mortality), epidemiologists are greatly interested in the exact relationship between host density and pathogen transmission rate. Shedding light on the transmission–density relationship is critical for understanding not only the spread of a pathogen, but also the management of an epizootic. Transmission is typically categorized into *density-dependent transmission*, in which the rate at which uninfected hosts become infected depends on the density of nearby infected hosts, and *frequency-dependent transmission*, in which this rate depends on the proportion of the host population in the vicinity of the susceptible host that is infected (McCallum et al. 2001).

2.2.1 Transmission in single host–single parasite infections

In the simplest case, disease transmission involves only one host species and one parasite species.

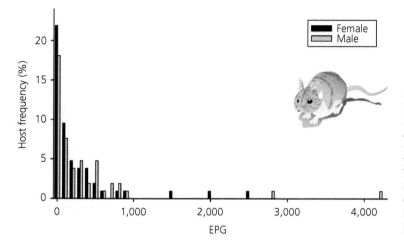

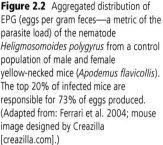

Figure 2.2 Aggregated distribution of EPG (eggs per gram feces—a metric of the parasite load) of the nematode *Heligmosomoides polygyrus* from a control population of male and female yellow-necked mice (*Apodemus flavicollis*). The top 20% of infected mice are responsible for 73% of eggs produced. (Adapted from: Ferrari et al. 2004; mouse image designed by Creazilla [creazilla.com].)

Even in this situation, there are four different common modes of transmission that need to be considered: horizontal, venereal (sexual), vertical, and environmental. The way in which host density influences the rate of transmission varies between these modes.

Horizontal transmission is likely the most familiar, occurring when parasites are transferred directly between individuals, possibly via a free-living infectious stage. Aerosol transmission of a virus is an example of horizontal transmission, as is the transmission of tapeworm eggs via a fecal–oral route. Because direct contacts increase in frequency in dense host aggregations, horizontally transmitted pathogens can be transmitted at higher rates in dense populations. Horizontal transmission is therefore typically considered to be density dependent.

Sexually transmitted diseases are familiar from human medicine, but are also widespread in other organisms (Lockhart et al. 1996). Close contact between males and females during reproduction is ubiquitous in terrestrial vertebrates, providing the ideal circumstances for the transmission of parasites or pathogens. It is useful to distinguish sexual transmission from other types of horizontal transmission for two reasons. First, sexually transmitted pathogens tend to be very host-specific, simply because copulation usually occurs between members of the same species. Second, the number of sexual contacts an individual has is determined more by the mating system of the species concerned than

by population density. Therefore, the transmission rate of sexually transmitted diseases is usually only weakly related to population density and is likely to be frequency dependent. Sexually transmitted pathogens tend to generate long-lasting, relatively low-virulence infections, since the presence of overt disease symptoms is likely to undermine the infected host's mating success and, by extension, the pathogen's chances of being transmitted (Lockhart et al. 1996).

Vertical transmission occurs when parasites are transferred directly from parent to offspring. This may take place through females feeding or grooming their young, or, specifically in mammals, transplacentally, or during birth. Retroviruses, such as the koala retrovirus (KoRV) group (Denner and Young 2013), can become incorporated into the host genome and therefore can be transmitted vertically through gametes in a way similar to any other hereditary gene. If the primary means of transmission of a pathogen is vertical, then the transmission rate will not be affected by population density (Anderson and May 1979).

Finally, since some parasites may persist as free-living organisms, that can be transferred via environmental transmission. For example, nematodes of the genus *Halicephalobus* usually persist saprophytically in the soil, but can cause parasitic infections in mammals, especially equids, by producing what can be fatal encephalitis (Ondrejka et al. 2010). Because facultative parasites do not require hosts in order to persist, transmission may not depend at all on host density (Kuris et al. 2014). Some pathogens, for example the anthrax bacterium, form extremely long-lived soil spores and therefore may be able to persist in the environment for long periods in the absence of hosts (Hugh-Jones and Blackburn 2009).

2.2.2 Transmission in multiple host species–single parasite infection

Many parasites are not host-specific and can infect a number of different host species. This has profound implications for understanding transmission and managing infections. It is necessary from the outset to differentiate between two types of transmission in this case. The first type of parasite must utilize sequentially several different host species in order to successfully complete their life cycle. The second type can complete its life cycle in a single host but is able to infect several (or many) different host taxa. An example of the second category is rabies (Rupprecht et al. 2002), which can—and does—infect a broad range or different mammalian species (even though different strains may specialize on specific host groups, e.g., bats).

Parasites that must utilize several different host species in order to complete their life cycle are described as having complex life cycles. Distinct life history stages of this type of parasite occur in different hosts. The host in which the sexual reproduction of the parasite takes place is known as the definitive host, while all other hosts are known as intermediate hosts. Some examples of complex life cycles are shown in Figure 2.3. Within the wide variety of parasites with complex life cycles, two types are of particular concern: vector-borne pathogens and trophically transmitted parasites.

Vectors are small, mobile hosts that transmit infection between larger, less mobile hosts (Wilson et al. 2017). In terrestrial environments, the most common vectors are arthropods such as mosquitoes and ticks. Although some infections may be transmitted by bloodsucking arthropods essentially acting as "dirty needles," more often the arthropods act as an intermediate, or even definitive host for the obligatory developmental stages of the pathogen (see Figure 2.3c). An important characteristic of most vector-transmitted infections is that the rate of transmission is not strongly dependent on the density of the nonvector host. This is because most vectors are capable of feeding on a number of different host species and their numbers are more frequently determined by environmental conditions. Vector-borne infections are so important in epidemiology that they are often considered as a separate category from other pathogens with complex life histories.

Trophically transmitted parasites are those in which an intermediate host must be consumed by the next host in the life cycle for transmission to occur (see Figure 2.3). This type of transmission is important for conservation biologists for a variety of reasons. First, it is a very common mode of transmission, particularly for macroparasites (Lafferty 1999). Because the very first step in managing a parasite threat in any population is to break the chain of transmission, it is essential to understand the nature of the parasite life cycle, including the predator–prey relationships among all species involved. Second, because for this type of parasite, host predation is typically necessary to complete the life cycle, there is strong selective pressure on the parasites infecting the intermediate host to render that host more susceptible to predation by the definitive host. This means that such parasites are likely to have substantial effects on the intermediate hosts' behavior in a way that will increase the rate of predator-induced mortality (Moore 2002). For example, rats infected by the protozoan *Toxoplasma gondii* have been shown to actively seek out parts of the environment scented with urine of cats, the parasite's definitive host (Berdoy et al. 2000). Trophically transmitted parasites are therefore likely to have major effects on their intermediate host populations when predators are present. Furthermore, because any parasite effect on the host fitness is mediated through increased predation mortality, it is important to conduct any study assessing the parasite's impacts on the host, in the presence of the normal suite of predators. Table 2.2 summarizes the relationships between mode of transmission, host density, and virulence.

2.2.3 The basic reproductive number R_0

The most important single concept in epidemiology is the basic reproductive number (R_0) of a parasite

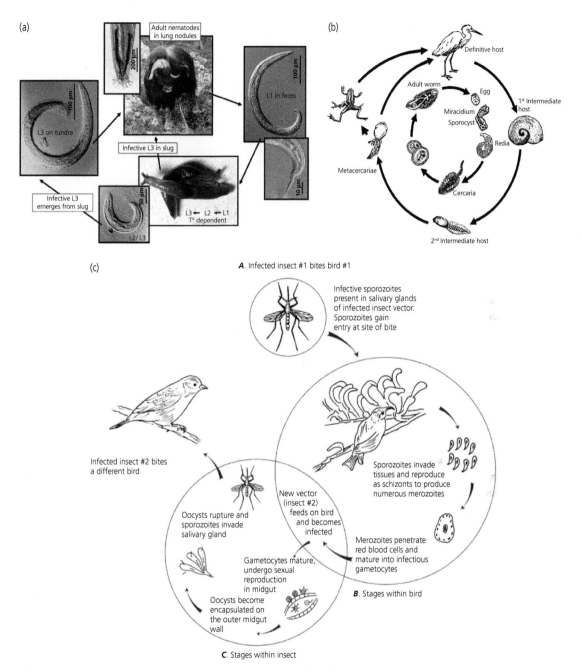

Figure 2.3 (a) Complex life cycle of an emerging wildlife parasitic helminth. The protostrongylid nematode *Umingmakstrongylus pallikuukensis* uses muskoxen (*Ovibos moschatus*) as its definitive hosts and utilizes snails as intermediate hosts. Adult nematodes live in the lungs of muskoxen and lay eggs, which hatch to first-stage larvae (L1). L1 move up the airways, are swallowed, and after traveling through the muskoxen GI tract are excreted with the feces. Then they must find and invade the foot of their snail intermediate hosts for further development into the infective third-stage larvae (L3). Muskoxen become then infected again either by ingesting during grazing a parasitized gastropod containing L3, or by ingesting L3 that have emerged from the gastropods and are free in the environment. (Source: Hoberg et al. 2008.) (b) The complex life cycle of *Ribeiroia ondatrae*. Clockwise from the top (*outer circle*), is the definitive host (avian or semi-aquatic mammal), followed by the first intermediate host (aquatic snail) and finally the second intermediate host (ranid tadpoles), where *R. ondatrae* preferentially encysts in the developing limb bud system. The inner circle depicts the various life stages of the parasite as it is transferred from host to host. (Source: adapted from Szuroczki and Richardson 2009, with image courtesy of B. Ballengée, *ibid*.) (c) Life cycle of avian malaria parasites (using *Plasmodium relictum* in Hawaiian birds as an example). The complex life cycle of hemosporidian parasites begins with (*A*), an infected insect biting a susceptible bird. Separate infectious and developmental stages occur in (*B*), the bird host, and (*C*), the insect vectors. (Source: Atkinson 1999.)

Table 2.2 Different types of parasite transmission.

Form of transmission	Density or frequency dependent?	Virulence	Examples	References
Horizontal (direct)	Often density dependent	Usually low	*Mycoplasma gallisepticum* in house finches (*Carpodacus mexicanus*)	Adelman et al. 2013, 2015
Vector-borne	Usually frequency dependent	Typically high relative to horizontal	Avian malaria (*Plasmodium relictum*) in Hawaiian drepanids	Warner 1968; Van Riper et al. 1986; Samuel et al. 2011)
Sexual	Frequency dependent	Mortality low, but often sterilizing	Chlamydial infection (*Chlamydia pecorum*) in koalas (*Phascolarctos cinereus*)	Lockhart et al. 1996; Nyari et al. 2017
Vertical	Usually frequency dependent	Typically low	Hookworm (*Uncinaria sanguinis*) in Australian sea lion (*Neophoca cinerea*)	Marcus et al. 2014
Cadavers	Often density dependent	High	Botulism in various waterfowl species	Rocke and Friend 1999
Persistent infective stages	Weakly density dependent	High	Chronic wasting disease and anthrax in various ungulate species	Hugh-Jones and Blackburn 2009; Almberg et al. 2011; Turner et al. 2016
Waterborne	Weakly density dependent	Often high	*Giardia* and *Cryptosporidium* in free-ranging mammals, avian influenza in waterfowl	Appelbee et al. 2005; Roche et al. 2009; Oates et al. 2012
Trophic transmission	Complex, as it depends on predator–prey dynamics	High to intermediate depending on host	*Ribeiroia ondatrae* in amphibians	Szuroczki and Richardson 2009
Environmental	Independent of host density	Often high	Mucormycosis in platypus	Gust and Griffiths 2009; Kuris et al. 2014

(Heersterbeek 2002). If, on average, each infected individual in a population successfully transmits the infection to more than one other individual before it dies or recovers (i.e. $R_0 > 1$), then infection will spread into the susceptible population and an epizootic will occur. Alternatively, if each infected host on average transmits the infection to fewer than one other host before it dies or recovers (i.e. $R_0 < 1$), then the epizootic will eventually fizzle out. Understanding R_0 and the factors that contribute to it is therefore critical to managing parasites and pathogens at a population level. If R_0 can be driven to below one for a sustained period of time, then disease will eventually be eliminated from a population; conversely, changes that lead to an increase of R_0 from below one to above one will lead to disease emergence.

For microparasites, the basic reproductive number is the average number of secondary infections per primary infection when a pathogen is first introduced into a fully susceptible population. It is useful to think of it as being made up of two components: the rate at which primary infections generate secondary infections per unit of time, multiplied by the average length of time an individual remains infectious. It is crucial to recognize that, for most microparasites, there are two ways in which a host carrying a pathogen can cease to become infectious: it may recover or it may die. Hence, the death of a vertebrate host usually results in the death of all the parasites it is carrying, although there are some important exceptions (e.g., anthrax; Hugh-Jones and Blackburn 2009).

For macroparasites, the same general principle applies. However, R_0 must be defined from the point of view of an individual parasite, rather than an infected individual (Anderson and May 1982a). If each parasite individual replaces itself by more than one individual parasite in the next generation when the parasite is first introduced into a host population, then the parasite population will increase. On the other hand, if each individual parasite replaces itself by less than one new parasite in the next generation, then the infection will die out. One obvious complication is that one needs to be consistent about the life history stage at which the parasite population per generation is censused. It does not matter whether it is the egg, larva, or adult stage that is censused, as long as the same stage is accounted for each generation. For macroparasites, then, R_0 is defined as the number of second-generation parasites per first-generation parasite when the parasite is first introduced into a fully susceptible population.

An important point, whether dealing with macroparasites or microparasites, is that R_0 is defined based on what happens when a parasite is first introduced into an immunologically naive population. Later in an epidemic, the number of recovered and immune hosts may increase so that the average number of secondary infections per primary infection decreases to below one. Since the epidemic has passed its peak, this stage does not affect the definition of R_0. The effective reproductive number (R_e) is a generalization of R_0 and is the average number of secondary infections generated per infected host at any time during an epidemic (Anderson et al. 2004).

For complex life histories, defining R_0 is slightly more difficult, as it becomes necessary to determine, at the same point of the life cycle, the number of second-generation infections per primary infection. There are also complications when there are multiple host species. To define R_0 in either of these cases, it is necessary to build models based on the relevant parasite life history. Details on constructing models and estimating R_0 from epidemiological data are discussed in later chapters.

2.2.4 Natural selection and host–parasite interactions

Understanding how evolutionary forces act upon host–parasite interactions, particularly through coevolution, provides important guidance on the circumstances in which parasites and pathogens are likely to be important in conservation biology. All things being equal (with a few important exceptions, discussed later), a parasite strain with a higher R_0 will be able to outcompete a parasite strain with a lower R_0 and may eventually replace it (Anderson and May 1982b). Natural selection therefore usually acts on parasites to maximize R_0. On the other hand, selection usually acts on hosts to minimize the impact that parasite-caused disease has on fitness. This can occur through increased resistance to infection—the ability to withstand infection when exposed to transmission stages—and/or increased tolerance to infection—the ability to harbor a given level of infection with reduced impact on fitness (Carval and Ferriere 2010).

The relationship between a pathogen and a host can be viewed as a struggle or arms race, in which natural selection acts on the host to resist infection and on the pathogen to overcome host defenses (Ebert and Hamilton 1996). This is certainly the case with regard to resistance, but the situation with tolerance is more nuanced. Selection on hosts will clearly favor those hosts that can tolerate infection, but increased tolerance may also be in the "interest" of the pathogen. All other things being equal, the longer an infected host remains alive, the more time the parasite has to generate secondary infections in other hosts, effectively increasing R_0. This reasoning leads to the maxim that a "well-adapted parasite" does not harm its host, which is a basic concept found in numerous introductory texts (e.g., Putman and Wratten 1984). However, all things are rarely equal in the natural environment. There are usually trade-offs between virulence, transmissibility, and host recovery rate, which are affected by many factors including the mode of transmission (May and Anderson 1983; Ewald 1994; Alizon and Baalen 2008). This means that pathogen strains of intermediate virulence are usually selected for, rather than the most avirulent strains (May and Anderson 1983).

Nevertheless, many of the most severe impacts of infectious diseases on wildlife occur when a pathogen, which has been refining its aggressive skills for centuries in an arms race with its normal host in one area, is introduced to a population of naive hosts. These hosts are especially susceptible because they have had no prior contact with the pathogen, and thus have not been selected for enhanced resistance. The situation is analogous to an invasion by combat-hardened veterans (pathogens) into a country defended by new recruits and unarmed citizens (hosts)—the resultant disease occurrence has been called a "virgin soil epidemic" (e.g., Crosby 1976) (see also Section 3.2).

Epizootics in naive hosts may occur through three sorts of introductions. First, a vector may be introduced, as was the case with the introduction of the mosquito *Culex pipiens* into Hawaii, causing epizootics of avian malaria and birdpox (Samuel et al. 2018). Second, a pathogen itself may also be introduced, as in the case of the chytrid fungus *Batrachochytrium dendrobatidis*, which has been carried on muddy boots into remote wetlands, and has devastated local frog populations (Berger et al. 1998; Skerratt et al. 2007). Alternatively a pathogen may be brought in along with an introduced host as a "biological weapon" enabling it to displace a competitor species that is less tolerant of the pathogen (Vilcinskas 2015). Third, a highly competent reservoir host may be introduced, which then allows a native, pre-existing pathogen to expand its host range and geographic distribution into susceptible local host species. Such was the case of the Omsk hemorrhagic fever virus, which spread through Siberian wildlife populations following the introduction of muskrats (*Ondatra zibethicus*), a highly susceptible amplification host (see Chapter 12, Section 12.3; Figure 12.5). This also happened after the introduction of salmon farms on the Pacific coast of Canada. Very large numbers of farmed salmon, crammed into sea pens, became infected with native sea lice (*Lepeophtheirus salmonis*), amplifying their populations and disseminating them to native juvenile pink salmon (*Oncorhynchus gorbuscha*) swimming by during migration (Krkosek et al. 2007; Peacock et al. 2013). Even more complex processes, where multiple host species are involved, are also possible (Hoyer et al. 2017). Table 2.3 summarizes the factors that may cause parasite infections to become an important threatening process.

2.3 Factors determining outcome of infection

The outcome of a parasitic infection is affected by many factors, including the dynamic relationship between the immune system of the host and the physiology of the parasite, all under the modulating influence of prevailing environmental conditions. This relationship is captured in the "disease triangle" (Figure 2.4), and its outcome can vary tremendously from one host to another and between different environmental contexts.

2.3.1 Host effects

A number of host traits can shape the outcome of an infection. First and foremost, taxonomic identity and genetic makeup can determine the ability of a host to control the presence of a parasite. For example, the impact of invasive *Plasmodium relictum* malaria on the avian endemics of Hawaii has been quite variable, with some species such as the omao (*Myadestes obscurus*) and the apapane (*Himatione sanguinea*) being at least partially resistant to experimental infections that have quickly killed other endemic honeycreepers such as the 'I'iwi (*Vestiaria coccinea*) (Atkinson et al. 2001a, 2001b). Similarly, the recent global chytridiomycosis (*B. dendrobatidis*) epizootic appears to have widely varying impacts on the hosts, with some taxa surviving infection well, such as bullfrogs (*Rana catesbeiana*) (Schloegel et al. 2010), while others, such as the members of the genera *Atelopus* (Lips et al. 2008) and *Rheobatrachus* (Hero et al. 2006), have been quickly pushed to the brink of extinction. Even within species, there may be substantial differences in resistance or tolerance of infection. They depend on factors such as genetic makeup, including heterozygosity and genetic diversity, age, reproductive status, and stress. In particular, several studies in free-ranging animals have shown that loss of genetic diversity is associated with increases in parasitism both for macroparasites (e.g., lungworms in bighorn sheep, Luikart et al. 2008; ectoparasites on island hawks, Whiteman et al. 2006) and microparasites (e.g. haematozoans in white-crowned sparrows, MacDougall-Shackleton et al. 2005).

Table 2.3 List of factors or circumstances that can result in parasite-induced host extinction (modified from de Castro and Bolker 2005).

Factor	Reason	Examples
1. When transmission is frequency dependent	No threshold host density	Tasmanian devil facial tumor disease (also incl. factor 4) (McCallum et al. 2009)
2. When there are reservoir species	High force of infection from reservoirs, even as threatened host becomes rare	Chytridiomycosis in frogs (also factor 4) (Kilpatrick, Briggs, and Daszak 2010)
3. When the endangered species is the intermediate host of a trophically transmitted parasite	Selection for infected intermediate host to be highly susceptible to predation	Toxoplasmosis (plus factors 2 and 4) (Innes 1997)
4. When the host–parasite interaction is evolutionarily young	Hosts have not evolved resistance or tolerance	Malaria and birdpox in Hawaiian birds (also factor 2) (van Riper et al. 1986). Many other examples exist
5. When there are multiple stressors	Parasite may act synergistically with stressors	*Pasteurella* infection in saiga antelopes (Kock et al. 2018)

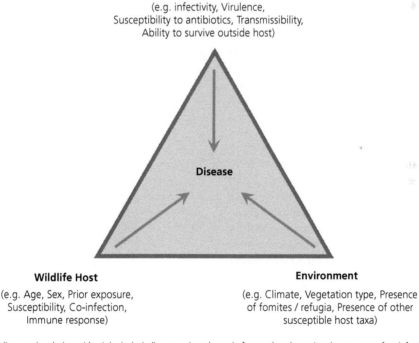

Pathogen / Parasite

(e.g. infectivity, Virulence, Susceptibility to antibiotics, Transmissibility, Ability to survive outside host)

Disease

Wildlife Host

(e.g. Age, Sex, Prior exposure, Susceptibility, Co-infection, Immune response)

Environment

(e.g. Climate, Vegetation type, Presence of fomites / refugia, Presence of other susceptible host taxa)

Figure 2.4 The disease triangle (or epidemiological triad) summarizes the main factors that determine the outcome of an infection. (Source: J. Foufopoulos.)

2.3.2 Pathogen effects

Different strains of the same parasite may vary substantially in pathogenicity. For example, research has identified a number of different strains of the fungus *B. dendrobatidis* that vary in virulence; some are hypervirulent enough to cause widespread frog extinctions on at least four continents (Farrer et al. 2011).

2.3.3 Environmental effects

The environment can modulate the impact of a pathogen on individual hosts in a variety of ways. One way may be through effects on the pathogen—for example, the growth rate and survival of *B. dendrobatidis* on its frog hosts is strongly dependent on temperature, with infection cleared at temperatures in excess of $37°C$ (Woodhams et al. 2003). In an analogous example, deadly *Pasteurella* outbreaks in saiga antelope herds appear to be triggered by extreme environmental conditions (Kock et al. 2018) (see also Chapter 7, Section 7.2; Figure 7.1a, b). Beyond this, both resistance and tolerance of a host to infection are shaped by nutritional stress and food availability, which in turn are determined by prevailing environmental conditions.

2.3.3.1 Multi-species parasitism—co-infection and interactions between parasites

Free-ranging vertebrates host a community of parasitic organisms, the composition of which is determined not only by the taxonomic affiliation of the host, but also by particular life history characteristics of the individual and the prevailing environmental conditions. The likelihood that a host will be infected by a parasite, as well as the effect that this specific parasitic infection may have on the host are influenced by the community of parasites already co-occurring as well as any previous exposure to other parasites. For example, rabbit hemorrhagic disease virus (RHDV) was introduced into Australia in an attempt to control pest rabbit populations (see Chapter 15). Its success as a biological control agent was variable and one factor limiting its impact in some areas was the presence of a related non-pathogenic calicivirus, rabbit calicivirus Australia 1 (RCV-A1), which, for a limited period after exposure, provides temporary cross-immunity to RHDV (Strive et al. 2013; Cooke et al. 2018). Immune responses to one parasite can also increase susceptibility to another. For example, the immune system's response to helminth infection can increase susceptibility to pathogens such as *Mycobacterium bovis*, the causative agent of bovine tuberculosis (BtB). Ezenwa and Jolles (2015) found that anthelminthic treatment decreased the mortality of wild BtB-positive African buffalo ninefold compared with negative controls, although the anthelminthic treatment had no impact on whether buffalo became infected with BtB.

In some cases, research has even demonstrated direct predation events between two species of parasite inside the host (Sousa 1993). In a study of the helminth community within wild rabbits in Scotland, evidence was found of both positive and negative interactions between parasites, mediated by a variety of factors including direct competition for resources or space and effects on the host physiology and immune system (Lello et al. 2004). Both theoretical models and empirical data suggest such interactions can have profound effects on the population dynamics of their hosts. Therefore it is critical to take the potential effects of co-infections into account when investigating or attempting to manage the impact of parasites on has populations (Lello et al. 2008).

References

Adelman, J.S., Carter, A.W., Hopkins, W.A., and Hawley, D.M. (2013). Deposition of pathogenic *Mycoplasma gallisepticum* onto bird feeders: Host pathology is more important than temperature-driven increases in food intake. *Biology Letters*, **9**: 20130594.

Adelman, J.S., Moyers, S.C., Farine, D.R., and Hawley, D.M. (2015). Feeder use predicts both acquisition and transmission of a contagious pathogen in a North American songbird. *Proceedings of the Royal Society B: Biological Sciences*, **282**: 20151429.

Alizon, S. and Van Baalen, M. (2008). Multiple infections, immune dynamics, and the evolution of virulence. *The American Naturalist*, **172**: E150–68.

Almberg, E.S., Cross, P.C., Johnson, C.J., Heisey, D.M., and Richards, B.J. (2011). Modeling routes of chronic wasting disease transmission: Environmental prion persistence promotes deer population decline and extinction. *PloS ONE*, **6**: p.e19896.

Anderson, R.M., Fraser, C., Ghani, A.C., Donnelly, C.A., Riley, S., Ferguson, N.M., et al. (2004). Epidemiology, transmission dynamics and control of SARS: The 2002–2003 epidemic. *Philosophical Transactions of the Royal Society of London B: Biological Sciences*, **359**: 1091–105.

Anderson, R.M. and May, R.M. (1979). Population biology of infectious diseases. Part I. *Nature*, **280**: 361–67.

Anderson, R.M. and May, R.M. (1982a). Population dynamics of human helminth infections: Control by chemotherapy. *Nature*, **297**: 557–63.

Anderson, R.M. and May, R.M. (1982b). Coevolution of hosts and parasites. *Parasitology*, **85**: 411–26.

Appelbee, A.J., Thompson, R.A., and Olson, M.E. (2005). *Giardia* and *Cryptosporidium* in mammalian wildlife—current status and future needs. *Trends in Parasitology*, **21**: 370–76.

Atkinson, C. (1999). Hemosporidiosis. In: Friend, M. and Franson, J.C. (eds.), *Field Manual of Wildlife Diseases. General Field Procedures and Diseases of Birds*. (No. ITR-1999-001), pp. 193–201. Geological Survey, Biological Resources Div., Madison, WI.

Atkinson, C.T., Dusek, R.J., and Lease, J.K. (2001a). Serological responses and immunity to superinfection with avian malaria in experimentally-infected Hawaii Amakihi. *Journal of Wildlife Diseases*, **37**: 20–27.

Atkinson, C.T., Lease, J.K., Drake, B.M., and Shema, N.P. (2001b). Pathogenicity, serological responses, and diagnosis of experimental and natural malarial infections in native Hawaiian thrushes. *The Condor*, **103**: 209–18.

Berdoy, M., Webster, J.P., and Macdonald, D.W. (2000). Fatal attraction in rats infected with *Toxoplasma gondii*. *Proceedings of the Royal Society of London B: Biological Sciences*, **267** (1452):1591–94.

Berger, L., Speare, R., Daszak, P., Green, D.E., Cunningham, A.A., Goggin, C.L., et al. (1998). Chytridiomycosis causes amphibian mortality associated with population declines in the rain forests of Australia and Central America. *Proceedings of the National Academy of Sciences*, **95**: 9031–36.

Carval, D. and Ferriere, R. (2010). A unified model for the coevolution of resistance, tolerance, and virulence. *Evolution*, **64**: 2988–3009.

Cooke, B.D., Duncan, R.P., McDonald, I., Liu, J., Capucci, L., Mutze, G.J., et al. (2018). Prior exposure to non-pathogenic calicivirus RCV-A1 reduces both infection rate and mortality from rabbit haemorrhagic disease in a population of wild rabbits in Australia. *Transboundary and Emerging Diseases*, **65**: e470–77.

Crosby, A.W. (1976). Virgin soil epidemics as a factor in aboriginal depopulation in America. *The William and Mary Quarterly*, **33**: 289–99.

De Castro, F. and Bolker, B. (2005). Mechanisms of disease-induced extinction. *Ecology Letters*, **8**: 117–26.

Denner, J. and Young, P.R. (2013). Koala retroviruses: Characterization and impact on the life of koalas. *Retrovirology*, **10**: 108.

Dobson, A., Lafferty, K.D., Kuris, A.M., Hechinger, R.F., and Jetz, W. (2008). Homage to Linnaeus: How many parasites? How many hosts? *Proceedings of the National Academy of Sciences*, **105**: 11482–89.

Ebert, D. and Hamilton, W.D. (1996). Sex against virulence: The coevolution of parasitic diseases. *Trends in Ecology and Evolution*, **11**: 79–82.

Ewald, P.W. (1994). *Evolution of Infectious Disease*. Oxford University Press, New York.

Ezenwa, V.O. and Jolles, A.E. (2015). Opposite effects of anthelmintic treatment on microbial infection at individual versus population scales. *Science*, **347**: 175–77.

Farrer, R.A., Weinert, L.A., Bielby, J., Trenton, W.J., Balloux, F., Clare, F., et al. (2011). Multiple emergences of genetically diverse amphibian-infecting chytrids include a globalized hypervirulent recombinant lineage. *Proceedings of the National Academy of Sciences*, **108** (18): 732–36.

Ferrari, N., Cattadori, I.M., Nespereira, J., Rizzoli, A., and Hudson, P.J. (2004). The role of host sex in parasite dynamics: Field experiments on the yellow-necked mouse *Apodemus flavicollis*. *Ecology Letters*, **7**: 88–94.

Gust, N. and Griffiths, J. (2009). Platypus mucormycosis and its conservation implications. *Australasian Mycologist*, **28**: 1–8.

Heesterbeek, J.A.P. (2002). A brief history of R_0 and a recipe for its calculation. *Acta Biotheoretica*, **50**: 189–204.

Hero, J., Morrison, C., Gillespie, G., Roberts, D., McDonald, K.R., Mahony, M., et al. (2006). Overview of the conservation status of Australian frogs. *Pacific Conservation Biology*, **12**: 313–20.

Hoberg, E.P., Polley, L., Jenkins, E.J., Kutz, S.J., Veitch, A.M., and Elkin, B.T. (2008). Integrated approaches and empirical models for investigation of parasitic diseases in northern wildlife. *Emerging Infectious Diseases*, **14**: 10–17.

Hoyer, I.J., Blosser, E.M., Acevedo, C., Thompson, A.C., Reeves, L.E., and Burkett-Cadena, N.D. (2017). Mammal decline, linked to invasive Burmese python, shifts host use of vector mosquito towards reservoir hosts of a zoonotic disease. *Biology Letters*, **13**: 20170353.

Hugh-Jones, M. and Blackburn, J. (2009). The ecology of *Bacillus anthracis*. *Molecular Aspects of Medicine*, **30**: 356–67.

Innes, E.A. (1997). Toxoplasmosis: Comparative species susceptibility and host immune response. *Comparative Immunology, Microbiology, and Infectious Diseases*, **20**: 131–38.

Kilpatrick, A.M., Briggs, C.J., and Daszak, P. (2010). The ecology and impact of chytridiomycosis: An emerging disease of amphibians. *Trends in Ecology and Evolution*, **25**: 109–18.

Kock, R.A., Orynbayev, M., Robinson, S., Zuther, S., Singh, N.J., Beauvais, W., et al. (2018). Saigas on the brink: Multidisciplinary analysis of the factors influencing mass mortality events. *Science Advances*, **4**: eaao2314.

Krkošek, M., Ford, J.S., Morton, A., Lele, S., Myers, R.A., and Lewis, M.A. (2007). Declining wild salmon populations in relation to parasites from farm salmon. *Science*, **318**: 1772–75.

Kuris, A.M., Hechinger, R.F., Shaw, J.C., Whitney, K.L., Aguirre-Macdeo, L., Boch, C.A., et al. (2008). Ecosystem energetic implications of parasite and free-living biomass in three estuaries. *Nature*, **454**: 515–18.

Kuris, A.M., Lafferty, K.D., and Sokolow, S.H. (2014). Sapronosis: A distinctive type of infectious agent. *Trends in Parasitology*, **30**: 386–93.

Lafferty, K.D. (1999). The evolution of trophic transmission. *Parasitology Today*, **15**: 111–15.

Lafferty, K.D. and Kuris, A.M. (2002). Trophic strategies, animal diversity and body size. *Trends in Ecology and Evolution*, **17**: 507–13.

Lafferty, K.D., Dobson, A.P., and Kuris, A.M. (2006). Parasites dominate food web links. *Proceedings of the National Academy of Sciences*, **103**: 11211–16.

Lello, J., Boag, B., Fenton, A., Stevenson, I.R., and Hudson., P.J. (2004). Competition and mutualism among the gut helminths of a mammalian host. *Nature*, **428**: 840–44.

Lello, J., Norman, R., Boag, B., Hudson, P.J., and Fenton., A. (2008). Pathogen interactions, population cycles, and phase shifts. *The American Naturalist*, **171**: 176–82.

Lips, K.R., Diffendorfer, J., Mendelson III, J.R., and Sears, M.W. (2008). Riding the wave: Reconciling the roles of disease and climate change in amphibian declines. *PLoS Biology*, **6**: e72.

Lockhart, A.B., Thrall, P.H., and Antonovics, J. (1996). Sexually transmitted diseases in animals: Ecological and evolutionary implications. *Biological Reviews*, **71**: 415–71.

Luikart, G., Pilgrim, K., Visty, J., Ezenwa, V.O., and Schartz., M.K. (2008). Candidate gene microsatellite variation is associated with parasitism in wild bighorn sheep. *Biology Letters*, **4**: 228–31.

MacDougall-Shackleton, E.A., Derryberry, E.P., Foufopoulos, J., Dobson, A.P., and Hahn, T.P. (2005). Parasite-mediated heterozygote advantage in an outbred songbird population. *Biology Letters*, **1**: 105–07.

Marcus, A.D., Higgins, D.P., and Gray, R. (2014). Epidemiology of hookworm (*Uncinaria sanguinis*) infection in free-ranging Australian sea lion (*Neophoca cinerea*) pups. *Parasitology Research*, **113**: 3341–53.

May, R.M. and Anderson, R.M. (1983). Epidemiology and genetics in the coevolution of parasites and hosts. *Proceedings of the Royal Society of London B: Biological Sciences*, **219**: 281–313.

McCallum, H., Barlow, N., and Hone, J. (2001). How should transmission be modelled? *Trends in Ecology & Evolution*, **16**: 295–300.

McCallum, H., Jones, M., Hawkins, C., Hamede, R., Lachish, S., Sinn, D.L., et al. (2009). Transmission dynamics of Tasmanian devil facial tumor disease may lead to disease-induced extinction. *Ecology*, **90**: 3379–92.

Moore, J. (2002). *Parasites and the Behavior of Animals*. Oxford University Press, Oxford.

Nyari, S., Waugh, C.A., Dong, J., Quigley, B.L., Hanger, J., Loader, J., et al. (2017). Epidemiology of chlamydial infection and disease in a free-ranging koala (*Phascolarctos cinereus*) population. *PLoS ONE*, **12**: p.e0190114.

Oates, S.C., Miller, M.A., Hardin, D., Conrad, P.A., Melli, A., Jessup, D.A., et al. (2012). Prevalence, environmental loading, and molecular characterization of *Cryptosporidium* and *Giardia* isolates from domestic and wild animals along the Central California Coast. *Applied and Environmental Microbiology*, **78** (24):8762–72.

Ondrejka, S.L., Procop, G.W., Lai, K.K., and Prayson. R.A. (2010). Fatal parasitic meningoencephalomyelitis caused by *Halicephalobus deletrix*: A case report and review of the literature. *Archives of Pathology and Laboratory Medicine*, **134**: 625–29.

Peacock, S.J., Krkošek, M., Proboszcz, S., Orr, C., and Lewis, M.A. (2013). Cessation of a salmon decline with control of parasites. *Ecological Applications*, **23**: 606–20.

Putman, R.J. and Wratten, S.D. (1984). *Principles of Ecology*. University of California Press, Berkeley, CA.

Roche, B., Lebarbenchon, C., Gauthier-Clerc, M., Chang. C., Thomas, F., Renaud, F., et al. (2009). Water-borne transmission drives avian influenza dynamics in wild birds: The case of the 2005–2006 epidemics in the Camargue area. *Infection Genetics and Evolution*, **9**: 800–05.

Rocke, T. and Friend, M. (1999). Botulism. In: Friend, M. and Franson, J.C. (eds.), *Field Manual of Wildlife Diseases. General Field Procedures and Diseases of Birds* (No. ITR-1999-001), pp. 271–81. Geological Survey, Biological Resources Div., Madison, WI.

Rupprecht, C.E., Hanlon, C.A., and Hemachudha, T. (2002). Rabies re-examined. *The Lancet Infectious Diseases*, **2**: 327–43.

Samuel, M.D., Hobbelen, P.H.F., DeCastro, F., Ahumada, J.A., LaPointe, D.A, Atkinson, C.T., et al. (2011). The dynamics, transmission, and population impacts of avian malaria in native Hawaiian birds: A modeling approach. *Ecological Applications*, **21**: 2960–73.

Samuel, M.D., Woodworth, B.L., Atkinson, C.T., Hart, P.J., and LaPointe, D.A. (2018). The epidemiology of avian pox and interaction with avian malaria in Hawaiian forest birds. *Ecological Monographs*, **88**: 621–37.

Schall, J.J. (1982). Lizards infected with malaria: Physiological and behavioral consequences. *Science*, **217**: 1057–59.

Schall, J.J. (1996). Malarial parasites of lizards: Diversity and ecology. *Advances in Parasitology*, **37**: 255–333.

Schloegel, L.M., Ferreira, C.M., James, T.Y., Hipolito, M., Longcore, J.E., Hyatt, A.D., et al. (2010). The North

American bullfrog as a reservoir for the spread of *Batrachochytrium dendrobatidis* in Brazil. *Animal Conservation*, **13**: 53–61.

Scott, M.E. (1988). The impact of infection and disease on animal populations: Implications for conservation biology. *Conservation Biology*, **2**: 40–56.

Skerratt, L.F., Berger, L., Speare, R., Cashins, S., McDonald, K.R., Phillott, A.D., et al. (2007). Spread of chytridiomycosis has caused the rapid global decline and extinction of frogs. *EcoHealth*, **4**: 125–34.

Sousa, W.P. (1993). Interspecific antagonism and species coexistence in a diverse guild of larval trematode parasites. *Ecological Monographs*, **63**: 103–28.

Strive, T., Elsworth, P., Liu, J., Wright. J.D., Kovaliski, J., and Capucci, L. (2013). The non-pathogenic Australian rabbit calicivirus RCV-A1 provides temporal and partial cross protection to lethal Rabbit Haemorrhagic Disease Virus infection which is not dependent on antibody titres. *Veterinary Research*, **44**: 51.

Szuroczki, D. and Richardson, J.M. (2009). The role of trematode parasites in larval anuran communities: An aquatic ecologist's guide to the major players. *Oecologia*, **161**: 371–85.

Turner, W.C., Kausrud, K.L., Beyer, W., Easterday, W.R., Barandongo, Z.R., Blaschke, E., et al. (2016). Lethal exposure: An integrated

approach to pathogen transmission via environmental reservoirs. *Scientific Reports*, **6**: 27311.

Van Riper III, C., Van Riper, S.G., Goff, M.L., and Laird, M. (1986). The epizootiology and ecological significance of malaria in Hawaiian land birds. *Ecological Monographs*, **56**: 327–44.

Vilcinskas, A. (2015). Pathogens as biological weapons of invasive species. *PLoS Pathogens*, **11**: e1004714.

Warner, R.E. (1968). The role of introduced diseases in the extinction of the endemic Hawaiian avifauna. *The Condor*, **70**: 101–20.

Whiteman, N.K., Matson, K.D., Bollmer, J.L., and Parker, P.G. (2006). Disease ecology in the Galapagos hawk (*Buteo galapagoensis*): Host genetic diversity, parasite load and natural antibodies. *Proceedings of the Royal Society of London B: Biological Sciences*, **273**: 797–804.

Wilson, A.J., Morgan, E.R., Booth, M., Norman, R., Perkins, S.E., Hauffe, H.C., et al. (2017). What is a vector? *Philosophical Transactions of the Royal Society B: Biological Sciences*, **372**: 1719.

Wobeser, G.A. (2007). *Disease in Wild Animals: Investigation and Management*, 2nd edn. Springer-Verlag, Berlin.

Woodhams, D.C., Alford, R.A., and Marantelli, G. (2003). Emerging disease of amphibians cured by elevated body temperature. *Diseases of Aquatic Organisms*, **55**: 65–67.

CHAPTER 3

Anthropogenic Effects and Wildlife Diseases

Almost all the wise world is little else, in nature, but parasites or sub-parasites.

Ben Jonson (1606) *Volpone*, Act 3, Scene 1

3.1 Introduction

As seen in Chapter 2, the occurrence and impact of parasitic organisms in natural ecosystems depend greatly on prevailing environmental conditions, as well as host and parasite characteristics. Because human activities have increasingly profound impacts on natural ecosystems, and because epidemics are fundamentally ecological processes, it is reasonable to expect that human activities will also affect the dynamics of pathogens and parasites in natural populations. Indeed, mounting evidence from across the planet suggests that humans, through a broad range of processes, are undermining wildlife and ecosystem health. In this chapter, we will discuss some of the most important human activities that affect disease occurrence in wildlife populations.

3.2 Human introduction of pathogens or hosts

Pathogenic organisms, like every other living species, have a natural geographic range that is defined by a combination of abiotic factors (such as climate, weather, soil type, and topography) and biotic circumstances, including host and vector availability. There are many examples, both in humans and wildlife, where an epidemic has occurred as the result of movement of either pathogens or hosts into areas beyond their normal range. Such movements can create at least three types of disease situation: (i) introduction of a pathogen into an area where it did not occur previously (sometimes called "pathogen pollution," Daszak et al. 2000); (ii) introduction of susceptible hosts into an area where a pathogen is indigenous in other species; and (iii) introduction of a new host species that alters the transmission dynamics of a pre-existing disease.

Why can exotic pathogens be so dangerous? As described previously, the relationship between a host and its parasites is best viewed as a dynamic equilibrium between the reproductive potential of the parasite and the ability of the host's immune system to control parasite establishment. The prior history of contact is crucial in determining the outcome of this interaction; it has been noted that when a host species has coexisted with a particular pathogen for a long time the outcome of infection is often less damaging (Bolker et al. 2009). This pattern is neither universal nor predetermined, as several old, yet quite lethal pathogens like anthrax and rabies demonstrate; such species have evolved special life history "tricks" to translate high virulence into superior transmission. While virulence to a host is ultimately shaped by several interacting factors, it is often the case that newly

Infectious Disease Ecology and Conservation. Johannes Foufopoulos, Gary A. Wobeser and Hamish McCallum, Oxford University Press.
© Johannes Foufopoulos, Gary A. Wobeser and Hamish McCallum (2022). DOI: 10.1093/oso/9780199583508.003.0003

established pathogens have more severe impacts on the host population and are therefore a greater conservation concern for managers. Several factors may be responsible for this pattern: in some cases, the invading pathogen simply appears to be ill-adapted to the physiology of a newly colonized host, resulting in death of the host before the pathogen can become established. This seems to be the case for the hyper-lethal infections observed in primates infected with Ebola virus (Walsh et al. 2003). Thus, while a pathogen may cause little or no overt disease in one species (its "normal" host), it may be extremely injurious to other ("abnormal") host species. For example, the normal definitive host for the nematode *Parelaphostrongylus tenuis* is the white-tailed deer (*Odocoileus virginianus*), in which infection rarely results in clinical disease. In contrast, infection in many other ungulates ("abnormal hosts") is debilitating or fatal. Often the actual basis for such disparities in resistance between "normal" and "abnormal" hosts is unknown, though in the case of *P. tenuis* part of the difference is related to the success of larvae in reaching the sensitive central nervous system (Lankester 2001). In white-tailed deer, only about 5% of larvae reach the spinal cord, whereas in moose (*Alces alces*) (an abnormal host) about 20% of larvae reach the spinal cord. Because the death of the host usually results in death of the parasites it is carrying, rapidly killing the host—either directly, or by debilitating it to the point where it succumbs to exposure or predation—is not likely to be advantageous for most pathogens. Typically, several pathogen strains exist simultaneously and compete within a given host population. In most cases, those strains with the highest reproductive rate (R_0) will spread most rapidly. Since highly virulent strains may kill their host before it has the chance to transmit their infection, it is likely that strains of lower virulence will eventually prevail.

In the host population, prolonged exposure to a pathogen will eventually select for those individuals that exhibit at least partial resistance to infection. Because vertebrate hosts have much longer generation times than any pathogen, this process is expected to occur over more extended periods of time.

The epidemiological changes that occur as the result of host–parasite coevolution have been best studied in the case of the myxoma virus in rabbits (*Oryctolagus cuniculus*). This South American pathogen was introduced in the early 1950s into Australia to control rabbit populations that were causing extensive ecological and economic damage (Fenner and Fantini 1999). The strains selected for introduction were extremely virulent and led to a massive epizootic and a rapid collapse of the rabbit populations. However, within a few years the impact of the disease weakened to the point where rabbit populations in many areas of Australia began to recover (Fenner and Fantini 1999). Careful laboratory studies revealed that this reduction in field mortality was the result of two interacting factors. First, soon after the release, there was rapid evolution in the pathogen toward reduced virulence, as measured by infecting a standard stock of fully susceptible laboratory rabbits with virus samples collected over the years (Figure 3.1a). At the same time, field rabbit populations, descending from the few survivors of the first epizootic, soon evolved partial resistance to the virus. Accordingly, field rabbits brought into the lab over the years showed increasing resistance to standardized myxomatosis strains (Figure 3.1b, Chapter 15, Section 15.2).

From a practical perspective, the evolution of resistance is welcome news to conservationists concerned with species decline owing to invasive diseases. However, this process is slow, may confer only partial protection, and depends on a number of underlying factors including initial virulence of the parasite, amount of genetic diversity within the host population, host generation time, and spatial heterogeneity. Host populations that are genetically diverse, have short generation times, and can survive infection in certain locations are likely to evolve resistance sooner.

3.2.1 Introduction of a novel disease agent into a naive population

Introduction of a novel pathogen into a naive indigenous population is the predominant cause of a so-called "virgin soil" epidemic. Such introductions are typically due to human actions (Table 3.1), and some have been deliberate, such as the previously mentioned release of myxoma virus (Fenner and Fantini 1999) and rabbit hemorrhagic virus

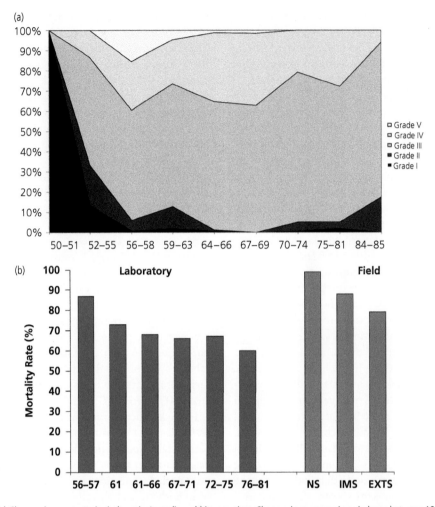

Figure 3.1 (a) Changes in myxomatosis virulence in Australian rabbits over time. Changes in myxoma virus virulence between 1950 (initial introduction of virus into the wild) and 1985 are expressed as percentages of viral field samples causing infections of varying severity in groups of experimentally infected, fully susceptible laboratory rabbits. Severity ranges from Grade I (extreme virulence, case fatality >99.5%, black) to Grade V (very low virulence, case fatality <50%, light gray). Within <10 years of introduction into the wild, myxoma virus virulence dropped rapidly from causing mostly Grade I infections to mostly Grade III and IV (moderate to low severity) infections. (Data from Fenner and Fantini 1999, table 8.1.) (b) Increases in resistance to myxoma infection in wild Australian rabbits between 1956 and 1981. Resistance is expressed as the percentage of animals surviving infection with the Standard Laboratory Strain (Grade I) of the myxoma virus (*Left*). Similar increases in resistance were seen in laboratory rabbits originating from selective breeding experiments. (NS, no selection; IMS, intermediate selection; EXTS, extensive selection) (*Right*). Individuals used in the tests were previously unexposed to the virus. (Data from Fenner and Fantini 1999, table 14.6.)

(Cooke and Fenner 2002) into Australia. More often, introductions have been unintentional, either as a result of unknowingly translocating the pathogen (e.g., in contaminated cargo), because of blatant disregard for the potential effects of a recognized pathogen, or because of the failure of measures designed to prevent pathogen translocation.

Animal pathogens are most frequently translocated through the commercial transport of domestic and exotic animals that happen to be infected. It is nearly impossible to separate a living animal from its parasites, so it is very likely that potential pathogens will be moved whenever animals are being translocated. Increased use of air

Table 3.1 Selected examples of pathogens translocated into new areas through the movement of infected animals by humans.

Pathogen	Animal translocated	Origin	Destination	Effects	Reference
Rabies virus	Raccoon	Florida/Georgia	Virginia	Epizootic in local racoon populations	Baer 1985
Aleutian mink disease virus	American mink	North America	Europe	Infection of native wildlife	Fournier-Chambrillon et al. 2004
Squirrel parapoxvirus	Gray squirrel	North America	United Kingdom	Extermination of local red squirrel populations	Tompkins et al. 2003
Histomoniasis (*Histomonas meleagridis*)	Domestic turkeys	North American mainland	Martha's Vineyard Isl.	Extinction of heath hen (*Tympanuchus c. cupido*)	Day 1981
Trypanosoma sp.	Black rats (*Rattus rattus*)	Unclear?	Christmas Island (Australia)	Extinction of the endemic rat species *Rattus macleari* and *R. nativitatus*	Wyatt et al. 2008
Elaphostrongylus cervi	Red deer	Europe	New Zealand	Infection of native deer populations	Watson and Gill 1985
Elaphostrongylus cervi	Elk	New Zealand	Australia	Infection of native deer populations	Presidente 1986
Elaphostrongylus rangiferi	Reindeer	Norway	Newfoundland, Canada	Infection of native deer populations	Lankester 2001

transport for moving both wildlife and livestock means that movement now occurs more rapidly than the incubation period of many pathogens, making detection of infection and prevention of introduction very difficult. The global trade in wildlife occurs at an astounding scale, with over 1.5 billion wild animals having been imported into the United States since the year 2000. Many of these animals are imported for the pet trade, with no mandatory testing for pathogen infection (Smith et al. 2009b). Several methods for assessing the risk of disease translocation have been described, each with limitations (Leighton 2002). Although the screening and treatment of animals prior to movement may reduce the risk of translocation, these methods are ineffective against pathogens that are not known to occur in a species. Properly validated tests do not exist for many agents, and tests in general are rarely 100% effective in detecting pathogens. Furthermore, the effectiveness of a treatment is often unknown, particularly in wild animals. Because the full range of pathogens that might be present in wild animals is seldom known, unrecognized pathogens are unlikely to be detected or treated successfully.

Pathogens can also be introduced with animals that move into new areas by other means. For example, a virulent strain of canine distemper virus likely entered Santa Catalina Island with raccoons (*Procyon lotor*) that stowed away on boats coming from mainland California. The virus quickly became established and is thought to have caused a catastrophic decline in the population of the indigenous Santa Catalina Island fox (*Urocyon littoralis catalinae*) (Timm et al. 2009). Disease agents may also be introduced by the natural expansion of the range of host animals (either vertebrates or invertebrates) as a result of environmental change such as habitat loss or climate change. For instance, the aforementioned *P. tenuis* is thought to have spread northward with white-tailed deer into the range of moose as a result of habitat change (Anderson 1992). Pathogens can also be moved by species that have no active role in their life history. For example, ticks infected with *Borrelia burgdorferi*, the causative agent of Lyme disease, are dispersed by migrating songbirds (Morshed et al. 2005) and by simple events such as inclement weather. Furthermore, West Nile virus is thought to have been introduced into Israel with white storks

(*Ciconia ciconia*) blown off course during migration (Malkinson et al. 2002). Last but not least, pathogens are introduced into wildlife populations through infected humans and their pets as has been the case with SARS-CoV-2 (see Box 3.1).

From a conservation practitioner perspective, introduced pathogens likely pose the biggest threat to the survival of an endangered population. As a result, it is of paramount importance to prevent the accidental introduction of such organisms through scrupulous enforcement of quarantine processes and screening and treatment protocols.

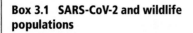

Box 3.1 SARS-CoV-2 and wildlife populations

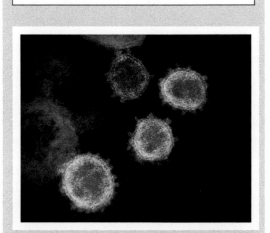

Figure 1 This transmission electron microscope image shows SARS-CoV-2—also known as 2019-nCoV, the virus that causes COVID-19—isolated from a patient in the US. Virus particles are shown emerging from the surface of cells cultured in the lab. Protruding from each virion are the uniquely shaped viral spike proteins that allow binding to the host cell. Image captured and colorized at the National Institute of Allergy and Infectious Diseases' (NIAID) Rocky Mountain Laboratories (RML) in Hamilton, Montana, USA. (Source: NIAID-RML, public domain image.)

Most of the public discourse about the impacts of SARS-CoV-2, the causative agent of the COVID-19 pandemic, has centered on the effects of infection on humans. What has received less attention is the potentially serious impact of the disease on wildlife populations. Recent studies have shown that the pathogen is able to infect a diverse group of mammals that appear to be susceptible because they

have appropriately shaped angiotensin-converting enzyme 2 (ACE-2) receptors, allowing the viral spike proteins to dock on the host cells. This mammalian group includes mustelids (the weasel, ferret, marten, and otter family), felids (the cat family), great apes, and of course Microchiroptera hosts. Indeed, at this point there have been numerous examples of captive or domesticated animals that have contracted COVID-19 infections from humans. However, many of these species are dead-end hosts, since not all of these species are able to successfully transmit the virus to other animals. The best evidence for successful animal-to-animal transmission has come from mustelids. Indeed, severe COVID epizootics have broken out in mink farms, resulting in massive culling efforts in Denmark associated with significant economic costs. More worrisome, there is increasing evidence that COVID infection has spread from captive to free-ranging mink. If this is true, it could represent the beginning of significant and hard-to-control outbreaks in wildlife populations. In anticipation of such outbreaks at least one endangered species of mustelid, the black-footed ferret (*Mustela nigripes*) is being vaccinated against SARS-CoV-2.

Sources: Cahan (2020), Mallapaty (2020), and Aleccia (2021).

3.2.2 Introduction of naive hosts to an area where an indigenous pathogen exists

When naive animals are moved or naturally expand into a new area, they run a risk of being severely affected by any disease agent existing in that region. This situation may be less severe than the previous one, as it only affects the translocated animals. However, it may cause an introduction to fail, or may prevent the successful recovery of endangered species. Examples of this situation can be seen in the multiple failed attempts to introduce reindeer and caribou (*Rangifer* spp.), as well as black-tailed deer (*Odocoileus hemionus columbianus*) into areas where *P. tenuis* is enzootic in white-tailed deer (*O. virginianus*) (Lankester 2001).

3.2.3 Introduction of a host that alters the ecology of an indigenous disease

A new animal species entering an area may change the ecology of an existing disease. This appears to have been the case with the tapeworm *Echinococcus multilocularis*, a parasite of concern because it can cause fatal infections in humans. *Echinococcus multilocularis* is endemic to Siberia, but may have become established on the arctic island of Spitsbergen in the Svalbard archipelago through the introduction of the sibling vole (*Microtus levis*). The parasite has an obligate two-host life cycle: adult worms live in the intestine of certain carnivores, while the larval stage occurs in rodents (Figure 3.3a). Carnivores are infected by eating parasitized rodents, and rodents in turn become infected by consuming vegetation contaminated with the parasite eggs shed in the carnivore's feces. Arctic foxes (*Vulpes lagopus*) are a suitable carnivore host for the parasite and are common in Svalbard. During winter they roam on pack ice, sometimes reaching Siberia where they may become infected with *E. multilocularis*. It is likely that such infected foxes returned to Svalbard. However, without native rodents, the parasite could not complete its life cycle and thus never became established on the islands. This changed with the introduction of the sibling vole. These rodents may have been introduced in livestock forage imported from mainland Russia, and eventually became established in a restricted area of Spitsbergen. With the introduction of a suitable rodent host, *E. multilocularis* could now complete its life cycle. Indeed, the first cases of the parasite in voles were reported in 1999, and it was subsequently found in arctic foxes, sled dogs, and humans (Henttonen et al. 2001). Since the voles have a restricted spatial distribution, infected foxes have only been found in areas where the vole is established (Fuglei et al. 2008).

Bovine tuberculosis (infection by *Mycobacterium bovis*) was likely introduced to New Zealand with imported domestic livestock. Similar introductions of this pathogen have occurred in many parts of the world, but tuberculosis has taken a very different course in New Zealand than in most other areas because of the involvement of introduced wild species, the most important of which is the brushtail possum (*Trichosurus vulpecula*). This species was intentionally introduced from Australia as a potential wild fur bearer. It thrived in the new environment, reaching densities as high as 25 individuals/hectare (O'Neil and Pharo 1995). Because possums are particularly susceptible to tuberculosis, they quickly became a new

maintenance host for the disease and have since played an important role in the epidemiology of this infection.

3.2.3.1 Avian malaria: A model of translocation problems

Probably no other single disease illustrates the multitude of problems that may result from the movement of pathogens or hosts better than avian malaria (infection by blood parasites of the genus *Plasmodium*). In birds, malaria is a mosquito-transmitted infection that occurs in all zoogeographic regions of the world, except in locations where the specific mosquito species required for transmission do not exist, such as in Antarctica and on isolated islands. More than 40 species of *Plasmodium* that infect birds have been identified, although the taxonomy of the group remains "in a continual state of flux" (Atkinson 2008; Atkinson et al. 2001). In areas of the world where the disease is endemic, there is usually a high prevalence of *Plasmodium* infections among native birds. Infection generally causes little mortality in adult native birds, but can be fatal to fledglings, affecting reproductive success (Gilman et al. 2007), and increasing vulnerability to predation (Valkiunas 2005; Møller and Nielsen 2007). Because malaria is a vector-transmitted disease, both suitable avian and mosquito hosts must be present for *Plasmodium* spp. to become established in an area. The most dramatic effects of malaria have been documented in the Hawaiian Islands, where both the pathogen (*Plasmodium relictum*) and a suitable mosquito vector (*Culex quinquefasciatus*) were introduced. Although *Plasmodium* spp. were believed to be present in birds migrating through Hawaii, mosquitoes didn't exist there before the Europeans arrived, and so the parasite was not transmitted to native birds (Warner 1968). Accidental introduction of *C. quinquefasciatus* in 1826 satisfied the conditions for establishment of malaria and led to severe epizootics in native honeycreepers. Since then, the disease has contributed to the extinction of about 50% of the endemic honeycreeper species (van Riper et al. 1986; Samuel et al. 2015). Today, honeycreepers are predominantly found in the higher elevations of the islands where conditions are too cool to allow effective transmission of the pathogen. As a result, there has been increasing concern over the potential for avian malaria to have similar impacts on the native birds of the Galapagos Islands (Miller et al. 2001; Wikelski et al. 2004; but see Levin et al. 2013) and in New Zealand (Tompkins and Gleeson 2006).

Outbreaks of avian malaria among immunologically naive birds that were introduced into areas where *Plasmodium* infection is endemic in indigenous bird communities have been known for many years and are best exemplified by the occurrence of disease in captive penguins. Penguins "as a group, are generally naive, both evolutionarily and physiologically, to blood haemosporidian parasites" (Bennett et al., 1993) because, like Hawaiian honeycreepers, they evolved in regions free of *Plasmodium*-transmitting mosquitoes. Penguins held in open-air pens and bitten by local mosquitoes develop rapidly fatal *Plasmodium* spp. disease, while exposure to the same pathogen causes only asymptomatic infection in resident wild birds. Avian malaria has been described as the "the most important cause of death of penguins displayed in open air exhibits around the world" (Stoskopf and Beier 1979). Similar situations have occurred when grey partridge (*Perdix perdix*) were introduced from Hungary to France (Garnham 1966), and when New Zealand keas (*Nestor notabilis*)—an endemic New Zealand parrot species—died from massive *Plasmodium* infections within 3 weeks of movement to a zoo in Malaysia (Bennett et al. 1993).

3.3 Interactions between habitat degradation/loss and infectious disease

Habitat degradation in general, and habitat fragmentation/loss in particular, are probably the most important drivers of species extinction today (Groombridge 1992; Henle et al. 2004; Bregman et al. 2014). Habitat-degrading human activities range from logging, grazing, urbanization, and mining, to soil erosion and alteration of normal hydrological and fire cycles. Such activities can also affect the occurrence and prevalence of pathogenic organisms to the point where the overall health and functionality of an ecosystem become impacted (Aguirre et al. 2002; Hudson et al. 2002). Habitat degradation is

relevant to the role that diseases play in conservation biology in several ways. On one side, exotic pathogens can act as powerful forces that shape vegetation structure and availability and can therefore act as agents of habitat degradation or destruction for wildlife. On the other side, anthropogenic habitat fragmentation can alter the occurrence and dynamics of naturally occurring parasitic organisms, through direct changes in the density of hosts or the size of vector populations. In addition, human activities can also indirectly influence the effect of naturally occurring parasites by impacting the susceptibility of hosts to infection. We will explore some of these processes in the next few sections.

3.3.1 Parasites/pathogens modifying habitat suitability

Certain exotic parasites and pathogens can have significant conservation effects not by directly infecting a protected species per se, but rather by impacting the structure and function of the species' natural habitat (Aguirre et al. 2002; Hudson et al. 2002). Some of the most dramatic examples of ecosystem alteration come from plant pathogens that shape vegetation community composition. Since most consumer species are directly or indirectly dependent on vegetation cover, this can have strong, positive or negative, effects on animal populations of conservation importance.

Many bird species can only excavate nesting cavities in trees infected with native wood-softening fungi. For example, in the monospecific aspen stands (*Populus tremuloides*) of the US Rocky Mountains, a whole guild of cavity-nesting bird species depends on the ability of red-naped sapsuckers (*Sphyrapicus nuchalis*) to excavate nesting cavities in aspen trunks that were previously softened by aspen heart rot fungus (*Phellinus tremulae*) infections (Losin et al. 2006). While this particular case is well described, the importance of wood-infecting fungi for tree cavity-using vertebrates is now being documented globally (Remm and Lohmus 2011; Cockle et al. 2012). In these studies, the presence of tree-infecting fungi enhances reproduction of resident wildlife; however, there are other situations where pathogen presence can have negative repercussions for biodiversity. The introduction of chestnut blight

(*Cryphonectria parasitica*) into North America via the importation of infected ornamental plants in 1904 resulted in massive die-offs and the eventual ecological extinction of the American chestnut (*Castanea dentata*), which at that time was perhaps the most common tree in eastern North American forests (Anagnostakis 1987). Eight species of Lepidoptera that depended on this species went extinct shortly after (Meffe and Carroll 1997). Several regions in southern Australia are currently facing severe problems stemming from the spread of *Phytophthora cinnamomi*, an exotic fungus that infects and kills a very broad range of native plants. Infection with this pathogen, which is spread through the use of contaminated logging and soil-moving equipment, has resulted in the long-term devastation of natural plant communities. These dramatic and apparently permanent shifts in vegetation cover have been accompanied by the rapid impoverishment of the native bird and mammal communities, which rely on the original vegetation cover for survival (Laidlaw and Wilson 2006).

3.3.2 Effects of habitat degradation/fragmentation on parasites and pathogens

Effects of habitat degradation can effect disease dynamics through changes in the community of species occurring in an area, or through changes directly in the host or parasite physiology.

3.3.2.1 Effects on species communities

Studies on the effects of habitat fragmentation on infectious diseases are hindered by the fact that fragmentation rarely occurs alone, but is usually accompanied by other transmission-relevant processes such as invasion by exotic species or intensification of hunting. Even when considered alone, the effects of fragmentation are complex and can lead to either increases or decreases in the incidence of parasitism and disease. It is therefore difficult to make specific predictions about the outcome of infections in fragmented populations.

Many of the strongest effects of habitat fragmentation occur on vegetation structure and habitat

quality; this in turn can affect pathogen transmission rates (Figure 3.2). For example, studies in sub-Saharan Africa have shown that forests degraded through partial logging can contain more infectious parasite stages and thus drive higher rates of parasitism in surviving primate populations (Gillespie et al. 2005; Gillespie and Chapman 2006).

In practice, habitat fragmentation will most likely lead to changes in disease prevalence in those areas closest to the edges of a habitat fragment. It is along such edges that native wildlife will most likely encounter invasive reservoir host species residing in the surrounding matrix of human-dominated habitats (Gates and Evans 1998; Power and Mitchell 2004; Faust et al. 2018). Empirical data suggest that such widespread hosts (often feral species) frequently have partial resistance to infection while retaining their long-term infectiousness. As a result, they are often the drivers behind epizootics in endangered wildlife. The presence of partially resistant reservoir hosts "pumping" a pathogen into a susceptible wildlife population is one of the few scenarios in which parasites can cause the extinction of their hosts without first becoming extinct themselves (de Castro and Bolger 2005).

From a community ecology perspective, habitat fragmentation typically results in a progressive reduction in the number of native species found in a habitat fragment (Newmark 1987). Such species loss, termed "community relaxation," produces simplified communities of hosts and parasites as well (Meffe and Carroll 1997; Roca et al. 2009; Foufopoulos et al. 2016). At the same time, collapses in host communities are often accompanied by marked shifts in individual species population density, which in turn can affect the probability of disease transmission. Studies of primates found in forest fragments suggest that the initial crowding of surviving populations into small patches of forest can promote disease owing to a direct increase in density-dependent transmission rates (e.g., Lebarbenchon et al. 2007; Mbora and McPeek 2009). Because animals flee into the remaining habitat and pathogen transmission from one host to another is easier in denser populations, infection rates may rise in fragmented populations. In western North America, the ongoing destruction

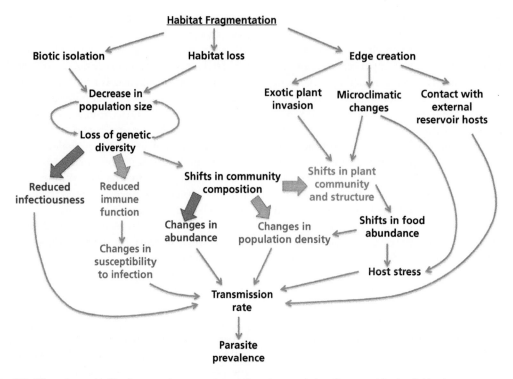

Figure 3.2 Effects of natural habitat fragmentation on parasite prevalence in a population of hosts surviving in a habitat fragment. Fragmentation leads through biotic isolation, habitat loss, and edge creation to population extinctions and a rapid loss of species (both hosts and parasites) remaining in a patch of habitat. Changes in host population density, host stress and immune function, availability of alternative host species, as well as shifts in environmental conditions, all lead to changes in transmission rates. Such shifts in transmission rates—which can either increase or decrease—can ultimately lead to concomitant changes in parasite prevalence.

of wetlands has led to the periodic aggregation of large numbers of migrating waterfowl into a limited number of small, degraded wetlands. The crowded conditions appear to have promoted outbreaks of various exotic pathogens including avian cholera (Friend et al. 1999).

Ultimately, the effects of fragmentation on wildlife health will depend on the characteristics of both the host and pathogen. Studies investigating the effects of forest fragmentation on tropical forest mammals have found divergent patterns that depend on factors like host diet and pathogen host range (Gillespie et al. 2005; Trejo-Macias et al. 2007; Wells et al. 2007; Puettker et al. 2008). Reviews of the effects of anthropogenic land use change on infectious diseases found that while most studies reported an increase in pathogen transmission in disturbed landscapes, there was also recognition

of the complex interactions between different disease-generating processes (Brearley et al. 2013; Gottdenker et al. 2014).

3.3.2.2 Effects on host physiology

While habitat fragmentation or degradation can cause wildlife health problems by directly altering the distribution of hosts and pathogens, it can also exacerbate the impacts of native pathogens simply by leaving their hosts more susceptible to infection. By tilting the balance of long-established host–parasite interactions in favor of the disease, human activities can create conservation problems where they did not exist before.

What processes might render wildlife more susceptible to infection? One of the primary outcomes of habitat fragmentation is decreases in wildlife population size, which, over the medium and long term, will result in the erosion of genetic diversity in surviving individuals (Hurston et al. 2009). Low genetic diversity in turn appears to be associated with attenuated immune functions, which can leave host populations susceptible to infection (Luikart et al. 2008). Although published information on these issues is still incomplete, it appears that small, genetically depauperate populations are more likely to suffer when exposed to either native or exotic parasites (Leberg and Vrijenhoek 1994; Whiteman et al. 2006; Hale and Briskie 2007; McCallum 2008).

Chronic stress stemming from a variety of anthropogenic sources can also increase host susceptibility to infection (Norris and Evans 2000). Mounting evidence suggests that wildlife populations living in fragmented or degraded environments often suffer from chronically elevated levels of stress hormones when compared with others found further away from human activities (Wasser et al. 1997; Suorza et al. 2003). As mentioned earlier, fragmentation can also cause wildlife crowding in remaining habitat patches. Preliminary evidence from captive populations (Oppliger et al. 1998; Hawley et al. 2005) and free-ranging primates (Martinez-Mota et al. 2007) indeed suggests that crowding, in addition to facilitating transmission, can also elevate both stress levels and susceptibility to disease (Lebarbenchon et al. 2007).

Immune function is related to nutritional status, and seasonal food shortages, which can be caused by habitat fragmentation or environmental degradation, can also exacerbate the impacts of normal parasite infections (Appleby et al. 1999; Pedersen and Greives 2008; Kock et al. 2018). Lastly, the presence of humans, even in the form of nonintrusive tourism, can also elevate circulating stress levels (Muellner et al. 2004; Ellenberg et al. 2007; Thiele et al. 2008; Palacios et al. 2018; but see van Meter et al. 2009) and affect the ability of wildlife to control parasite loads (Amo et al. 2006).

3.3.3 Conclusion

Does fragmentation always result in increased presence of parasites or pathogens? As the earlier discussion indicates, this is not always the case. As a matter of fact, mounting evidence suggests

that healthy ecosystems harbor the highest parasite diversity (Hudson et al. 2006). Nevertheless, this does not mean that increases in parasite diversity necessarily lead to increases in overt disease, as many cases are attributed to relatively rare macroparasites causing asymptomatic infections. In one case, the restoration of degraded salt marsh habitats led to a gradual increase in parasitic helminth prevalence and diversity to levels seen in nondegraded habitats (Huspeni and Lafferty 2004). Similarly, habitat size, whether measured as area of habitat or host population size, is often positively correlated with parasite diversity. For example, smaller island host populations tend to also have impoverished parasite communities and reduced rates of infection (Lindström et al. 2004; Roca et al. 2009). In an analogous inter-specific analysis comparing parasitism across primate taxa, Altizer and colleagues (2007) found that endangered species, which by definition have smaller and generally more fragmented populations, also have lower levels of parasitism.

Ultimately, anthropogenic habitat degradation likely leads to reductions in certain parasites (Bradley and Altizer 2006) such as specialized helminths with narrow host ranges and complex life cycles (see Chapter 1, Section 1.5). At the same time, such degradation probably also promotes generalist microparasites with high reproductive rates and broad host ranges. Such "weedy" species, typically viruses, bacteria, and fungi—which can invade from the surrounding disturbed matrix— have a long history of causing rapid and catastrophic host population die-offs and precipitating conservation crises (Cleaveland et al. 2001; Dobson and Foufopoulos 2001). Clearly, additional studies are needed to elucidate these relationships.

3.4 Parasitism and predation

Distribution of pathogens in nature is strongly affected by the extent and type of prevailing predation. Because parasitism so often interacts with predation in nature, it is important to consider both processes in tandem when trying to understand and manage disease threats to natural populations. Furthermore, any effects of humans on predator

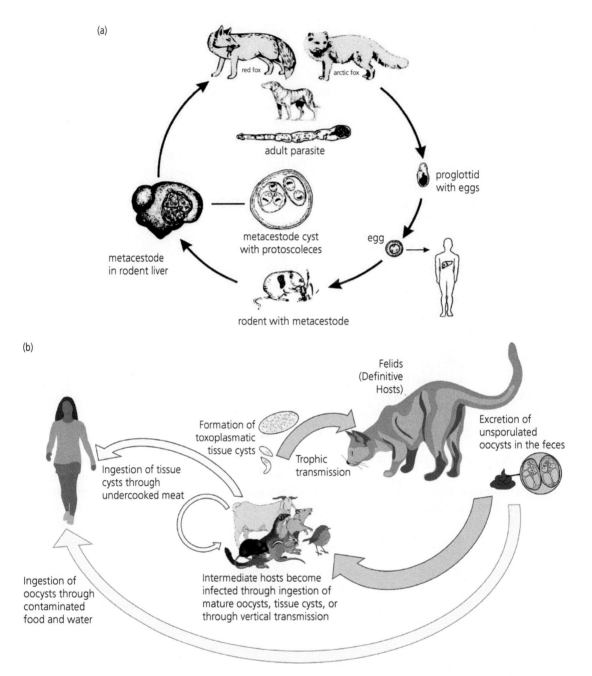

Figure 3.3 (a) Life cycle of *Echinococcus multilocularis*. This parasite, like *E. granulosus*, requires canids (such as various fox species) as its definitive hosts but also utilizes a variety of other species as intermediate hosts. Definitive hosts become exposed by consuming a parasitized intermediate host. Intermediate hosts, including humans, become infected when ingesting parasite eggs. Much of the conservation importance of this parasite stems from the fact that the parasite forms large hydatid cysts inside intermediate or accidental hosts—these can be debilitating or lethal to the animal. (Source: Torgerson et al. 2010.) (b) Life cycle of *Toxoplasma gondii*. This protozoan parasite requires members of the cat family (felids) as its definitive hosts. They become exposed by consuming parasitized prey (intermediate hosts). These in turn become infected by ingesting oocysts shed in the fecal material of the definitive host. Toxoplasmosis in intermediate hosts is associated with pregnancy failures and behavioral abnormalities that often predispose the host to predation by felids.

communities can be expected to have downstream effects on wildlife disease.

Predators are, by virtue of their biology, particularly susceptible to human activities. Top predators require large home ranges and exist at low population densities, so they are particularly vulnerable to habitat fragmentation (Woodroffe and Ginsberg 1998). They are regularly viewed as threats to livestock, and are consequently still persecuted in many areas of the world. However, human actions can also, directly or indirectly, lead to increases in levels of predation. Human-introduced cats and dogs have become feral on many islands and the entire continent of Australia. Even in areas where felids and canids are indigenous, population densities may increase greatly in association with human settlement. Some fur-bearing predators, such as the American mink (*Neovison vison*) in the United Kingdom, were accidentally released and have become feral. Predators have also been introduced as game species (e.g., the European fox (*Vulpes vulpes*) in Australia) or as ill-conceived biological control agents (mongooses on Hawaii and other islands). Perhaps counterintuitively, human extermination of top predators has led to substantial increases in the numbers of smaller predators and resulting levels of wildlife predation (also known as "mesopredator release"; Crooks and Soule 1999). The reason is that top predators compete, and even hunt, smaller predator species and ultimately control smaller predator populations.

3.4.1 Trophic transmission

So how does predation affect disease transmission in natural ecosystems? As explained previously (Chapter 2), many parasite species require several host species for their development. In most cases, for the life cycle of the parasite to be completed, the definitive host, most often a top predator, needs to consume the intermediate hosts (Figure 3.3a and b). This fact makes predators disproportionately important both for the existence of trophically transmitted parasites and for the population of an intermediate host that may be both consumed and parasitized. If the definitive host of a parasite is an introduced predator species or its numbers have substantially increased owing to human activities,

trophically transmitted parasites may become a substantial conservation problem for endangered species. In essence, elevated numbers of the definitive host "pump" large numbers of parasites into the population of the susceptible wildlife species. Two important parasites in this category are the hydatid tapeworm *Echinococcus granulosus*, which uses carnivores and especially canids as its definitive hosts (Jenkins and MacPherson 2003), and the protozoan *Toxoplasma gondii*, which has felids as its definitive hosts (Dubey 2008). Both of these pathogens are also zoonotic, causing large number of wildlife and human cases worldwide. As both domesticated dog and cat numbers are surging across the planet and both have become feral in many parts of the world, these two parasites are rapidly gaining importance for conservation managers.

Echinococcus is responsible for substantial mortality and morbidity in several endangered Australian mammals, including the bridled nailtail wallaby *Onychogalea fraenata* (Turni and Smales 2001) and the brushtailed rock wallaby *Petrogale penicillata* (Barnes et al. 2008) (Figure 3.4a and b). In Africa, cysts containing intermediate stages of the parasite have been reported in at least 19 species of wild herbivores, with adult parasites being reported from a range of predators, including canids, hyenas, and lions (Jenkins and MacPherson 2003). Suitable definitive hosts have therefore existed in Africa for millennia, whereas in Australia the origin of the parasite is less clear. It is debatable whether the parasite was introduced to Australia with the dingo (*Canis lupus dingo*) ca. 4,000 years ago or more recently with European domestic dogs (Jenkins and Morris 2003). Nevertheless, on both continents it is highly likely that recent human disturbance of natural habitats, particularly fragmentation, and rising numbers of feral domestic dogs have led to increased impacts on wildlife (Jenkins and MacPherson 2003; Davidson et al. 2012). Appropriate management strategies in wildlife are similar to those proposed to reduce transmission to domestic animals and humans (Torgerson and Heath 2003). In particular, managing transmission to dogs through the disposal of livestock carcasses (especially sheep) and the vaccination or treatment of dogs and other wild canids is likely to reduce prevalence in wildlife.

(a)

(b)

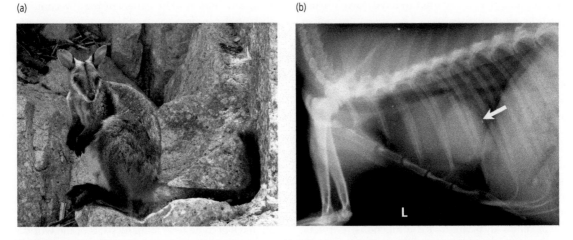

Figure 3.4 (a) Brushtailed rock wallaby (*Petrogale penicillata*) in typical boulder habitat. Populations of the species have been declining owing to disease, predation by introduced foxes, competition by exotic grazers, and habitat fragmentation, which leads to increasing isolation of the surviving populations. (Photo: Anne Goldizen, U. Queensland.) (b) Radiograph of a large hydadid cyst (arrow) in a brushtailed rock wallaby (*P. penicillata*) taken with a portable X-ray imaging machine in the field. Infection with *Echinococcus granulosus* has contributed to the decline of this species in many areas of Queensland (Australia). (Photo: Anne Goldizen, U. Queensland.)

Toxoplasma gondii is a protozoan parasite with a worldwide distribution. While it affects a very wide range of mammalian hosts, including humans, with infection being acquired either by ingestion of oocysts or by carnivory, only members of the cat family shed oocysts in their feces (Dubey 2008). Although infection in humans is usually asymptomatic, toxoplasmosis does result in human mortality; the disease is also a major cause of abortion in sheep (Dubey 2008). It can also cause disease in a range of wildlife species, including members of the kangaroo family (Reddacliff et al. 1993), sea otters (Conrad et al. 2005), and bottlenose dolphins (Inskeep et al. 1990).

In the case of sea otter toxoplasmosis, the source of infection is oocysts washed into the marine environment. This has widely been believed to be a disease attributable to human impacts, with oocysts from domestic cat feces entering the marine environment via stormwater or the sewage system (Conrad et al. 2005). However, more recent work suggests that wild felids in California may be important definitive hosts (Lafferty 2015). In contrast, toxoplasmosis in terrestrial Australian mammals is likely derived from oocysts shed by feral cats. Australian mammals may be particularly susceptible to toxoplasmosis because, owing to the original absence of native felids on this continent, they had no previous exposure to the parasite (Parameswaran et al. 2009), although there is little firm evidence to support the hypothesis that toxoplasmosis has been responsible for declines of wild marsupials (Hillman et al. 2016). Control of toxoplasmosis in marine mammals relies on management of stormwater and sewage and appropriate disposal of cat litter, whereas control of the pathogen in terrestrial systems requires effective management of feral cat populations.

3.4.2 Effects of predators on herd health

In a system in which a species of conservation interest is subjected to both predation and infection with a parasite, these two processes can interact in complex ways. By reducing host density and by preferentially preying on diseased members of a population, predators may play an important role in suppressing disease and maintaining "healthy herds" (Packer et al. 2003; Wild et al. 2011). Packer et al. concluded that the presence of predators leads to lower levels of parasite infection, regardless of the extent to which predators select infected hosts or whether the parasite in question is a microparasite or a macroparasite. However, it is important to note that the models used by Packer et al. assumed that parasite transmission will increase with rising animal density (i.e., is density dependent). In contrast, if transmission is frequency dependent, then predation will only reduce parasitism if predators preferentially select infected hosts. This effect is likely to be particularly pronounced in macroparasites, because it is almost invariably the case that the majority of the parasite population is aggregated within a relatively small proportion of the host population (Woolhouse et al. 1997), and it is those heavily infected hosts that should be expected to be particularly subject to predation.

3.4.3 Combined effects of predation and parasitism on host population size and stability

Generally, the presence of predators reduces the presence of infection and parasitism in its prey. However, the actual effects on host/prey population size can be more complex and depend on the selectivity of the predator and the virulence of the parasite. Although it seems intuitive that the presence of predators would have a negative influence on prey species populations, there are exceptions where, in the presence of parasites, a decrease in predator density may paradoxically lead to a decrease in prey abundance. The reason for this is that if a parasite is sufficiently virulent and predators prey preferentially on infected hosts, absence of predation will result in an increase in parasite-induced mortality that can exceed the mortality caused by the predators alone.

Beyond the effects on overall population size, the combined effect of parasitism and predation has important implications for the population stability of a host/prey species. Currently available evidence indicates that in many cases parasitism can destabilize the population dynamics of prey species. One of the best-studied cases of this is in populations of red grouse (*Lagopus lagopus scoticus*), an important game species in the United Kingdom, which have long been known to go through dramatic boom-and-bust cycles. Red grouse are regularly preyed upon by foxes and are also heavily infected with the cecal nematode *Trichostrongylus tenuis*. Empirical evidence indicates that foxes prey

disproportionally on more parasitized grouse, most likely because these individuals emit a stronger smell, making detection possible. Early research had shown that grouse population fluctuations were particularly pronounced in estates where fox populations were controlled by gamekeepers (Hudson et al. 1992). The initial conclusion of the study was that predators dampen grouse population fluctuations caused by parasites. This was further supported by subsequent large-scale field investigations: the experimental removal of *Trichostongylus* from wild birds through antihelminthic treatment prevented the collapse observed in control grouse populations (Hudson et al. 1998). Similar results were seen in Canadian snowshoe hare (*Lepus americanus*) populations, where the presence of parasites appears to exacerbate the regular predator–prey cycles (Ives and Murray 1997). Because of these links between predation and disease, traditional practices of predator reduction to increase game populations can backfire, causing smaller, less stable, and less healthy game populations.

In summary, there are important interactions that occur between parasites infecting the population of a prey species and the predators preying on it. As a result, human actions that alter predator communities or populations can have profound effects on the impact of parasites or infectious diseases of their host populations. In the short term, increases in predation pressure can magnify the effects of disease especially when the parasite renders its host more susceptible to predation. In the long term, however, the presence of an intact natural predator guild likely results in healthier prey populations. How this relationship translates into actual population sizes will depend on the particular circumstances of each system. Conservation managers need to be aware that reducing normal predator pressure on populations may cause the emergence of disease and parasite problems where they had not previously occurred.

3.5 Climate change and wildlife disease

Much attention has been given to the likely effects of anthropogenic climate change on human diseases (Kovats et al. 2001; Epstein 2002; Sutherst 2004; Haines et al. 2006; Tamerius et al. 2007; Randolph 2008; Reiter 2008; Paaijmans et al. 2009). More recently, there has been increasing attention toward the effects of climate change on wildlife disease (Black et al. 2008; de la Roque et al. 2008; Hoberg et al. 2008; Harvell et al. 2009; Lafferty 2009; Rohr et al. 2011; Cable et al. 2017). The effect of climate change on wildlife infections is complex and contingent on the life history of the host, as well as the parasite or pathogen in question. Changes in precipitation and moisture are likely at least as important for the survival of infective stages and vectors as increases in temperature (Lafferty 2009). It is therefore an oversimplification to suggest that climate change will cause an increase in the distribution or impact of all infectious diseases of wildlife. While there is a higher diversity of human pathogens in the tropics compared with temperate latitudes, the most recent emerging infectious diseases have occurred in mid-latitudes ($30°$ to $60°$) (Jones et al. 2008). Whether a similar pattern occurs for wildlife diseases is unknown. Climate change alone is likely to have serious implications for wildlife diseases, and recent studies suggest that these effects will be magnified when other environmental stressors such as habitat fragmentation or pollution are present. For example, the combined effects of elevated summer temperatures with increasing nutrient runoff from agricultural areas has led to a rising epidemic of algal blooms, anoxic events, and pathogen outbreaks such as avian cholera, which currently kill tens of thousands of fish and birds every year in North American freshwater ecosystems alone. Given the pervasiveness of current anthropogenic disruption (Daszak et al. 2000) and its exacerbating effect on wildlife disease, it is expected that the ultimate effects of climate change on the health of natural ecosystems are going to be particularly severe.

Although there is an overriding international priority to develop strategies to limit climate change through reductions in anthropogenic carbon dioxide emissions, it is evident that substantial warming over the next 50–100 years is unavoidable (Smith et al. 2009a). It is therefore a priority for conservation managers to develop strategies to mitigate or adapt to the consequences of this climate change, especially when it comes to wildlife epizootics. Given that many of these pathogens can also infect humans, trying to mitigate the effects of a warming climate on infection in wildlife populations is likely to also pay important human health dividends.

3.5.1 Climate change and vector-borne diseases

Most, although not all, of the studies investigating the effects of climate change on infectious human diseases have focused on vector-borne diseases and on the likely influence of rising temperatures and changing precipitation patterns on vector populations (e.g., dengue fever; Yang et al. 2009). For many human vector-borne diseases (e.g., malaria; Paaijmans et al. 2009), the pathogen has to undergo specific obligatory ontogenetic developments inside the vector before the vector becomes infectious. The time needed to complete these developments is strongly dependent on temperature, with higher temperatures shortening this period. Because vectors tend to die at a relatively constant daily rate, accelerated parasite development means that a much higher proportion of infected vectors will survive to become infectious, with a consequent increase in the basic reproductive rate of the pathogen (Mordecai et al. 2019). For malaria at least, this may be a more significant factor in increasing the geographic range of the disease than are actual changes in the distribution of the vector (Paaijmans et al. 2009). Given that many important animal diseases are vector-borne and share substantial similarities with human pathogens, it is likely that insights into the relationship between climate change and human health will also find applications in our understanding of wildlife epizootics.

Perhaps the most dramatic potential effect of climate change on the distribution of a wildlife disease of conservation concern comes from the aforementioned studies of vector-transmitted avian malaria (*Plasmodium relictum*) on the Hawaiian Islands. Malaria together with birdpox, have had a devastating effect on the Hawaiian avifauna (van Riper et al. 1986). Because *Plasmodium* transmission peaks at $17°C$ and then effectively drops to zero below $13°C$, higher elevation areas are disease-free (LaPointe et al. 2010). At present, several species of native Hawaiian birds that have disappeared from low altitudes are surviving in these high altitude refugia. Computer model projections unfortunately suggest

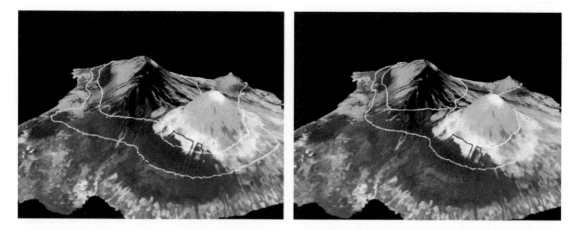

Figure 3.5 Effects of climate change on the distribution of avian malaria on the main island of Hawaii. Projected changes in forest cover in relation to 17°C (yellow, lower line) and 13°C (white, upper line) isotherms, which denote the upper limits in the distribution of avian malaria. Figure on the left shows current conditions, while figure on the right shows future 2°C warming conditions. Dark line denotes the boundary of the Hakalau Refuge, an important sanctuary for many endangered Hawaiian birds. Note how under projected climate change scenarios most of the forest habitat in the refuge will shift into the range of malarial parasites. Images were created by using a 1995 SPOT satellite, false-color composite image draped over 30 m digital elevation models. (Source: Benning et al. 2002.)

that climate change will allow effective parasite transmission at substantially higher altitudes, thus sharply reducing the amount of disease-free habitat available to the surviving bird species (Benning et al. 2002; see Figure 3.5). Data collected from the Hawaiian island of Kauai suggest a progressive expansion of mosquito vectors, as well as *P. relictum* malaria into the last high elevation strongholds of the native bird species (Atkinson et al. 2014).

3.5.1.1 Chytridiomycosis and climate change—A causal link?

The amphibian chytrid fungus *Batrachochytrium dendrobatidis* has been responsible for the extinction of many frog species worldwide, particularly in Australia and the neotropics (Berger et al. 1998; Lips et al. 2006; Skerratt et al. 2007). It has been suggested that climate change has facilitated the spread of chytrid and has therefore been at least partially responsible for a wave of extinctions in Central America (Pounds et al. 2006). This was a rather counterintuitive suggestion, because in vivo and in vitro evidence shows that the fungus grows most rapidly at cool temperatures (17–23°C; Piotrowski et al. 2004) and can be experimentally cleared from infected frogs by exposing them to temperatures above 28°C (Woodhams

et al. 2003). One might therefore expect increased temperatures to reduce the impacts of this particular disease. On the contrary, Pounds et al. argued that increased temperatures have led to increased cloud cover and hence have caused a convergence in the minimum and maximum temperatures in the relatively high altitudes areas in which extinctions have primarily occurred, producing a temperature regime increasingly favorable to chytridiomycosis. Since then, more detailed analyses have revealed that the observed epizootics have spread in a wave-like pattern away from a limited number of foci of introduction, thus providing a more plausible explanation for the observed pattern of amphibian extinctions in Mesoamerica (Lips et al. 2008; Rohr et al. 2008). The possible role of climate change in the emergence of chytridiomycosis therefore remains contentious. It is clear that both the transmission of the pathogen and its impact on the host are strongly dependent on the conditions of the microclimate, such as humidity and temperature, in which the frog and the pathogen exist (Bosch et al. 2007; Rowley and Alford 2007; Puschendorf et al. 2008; Rödder et al. 2008). This case system therefore cautions that the effects of climate change on disease emergence are likely to be subtle and complex and not related simply to increases in average temperature.

References

Aguirre, A.A., Ostfeld, R.S., Tabor, G.M., House, C., and Pearl, M.C. (2002). *Conservation Medicine. Ecological Health in Perspective*. Oxford University Press, Oxford.

Aleccia, J. (2021). A shot in the dark: Endangered ferrets get experimental vaccine. *Undark*, June 1. https://undark.org/2021/01/06/endangered-ferrets-get-experimental-vaccine (accessed July 6, 2021).

Altizer, S., Nunn, C.L., and Lindenfors, P. (2007). Do threatened hosts have fewer parasites? A comparative study in primates. *Journal of Animal Ecology*, 76: 304–14.

Amo, L., Lopez, P., and Martin, J. (2006). Nature-based tourism as a form of predation risk affects body condition and health state of *Podarcis muralis* lizards. *Biological Conservation*, 131:402–09.

Anagnostakis, S.L. (1987). Chestnut blight: The classical problem of an introduced pathogen. *Mycologia*, 79: 23–37.

Anderson, R.C. (1992). *Nematode Parasites of Vertebrates: Their Development and Transmission*. CAB International, Wallingford.

Appleby, B.M., Anwar, M.A., and Petty, S.J. (1999). Short-term and long-term effects of food supply on parasite burdens in tawny owls, *Strix aluco. Functional Ecology*, 13:315–21.

Atkinson, C.T. (2008). Avian malaria. In: Atkinson, C.T., Thomas, N.J., and Hunter, D.B. (eds.), *Parasitic Diseases of Wild Birds*. pp 35-53 Wiley-Blackwell, Ames, IA.

Atkinson, C.T., Lease, J.K., Drake, B.M., and Shema, N.P. (2001). Pathogenicity, serological responses, and diagnosis of experimental and natural malarial infections in native Hawaiian thrushes. *Condor*, 103:209–18.

Atkinson, C.T., Utzurrum R.B., and Lapointe, D.A. (2014). Changing climate and the altitudinal range of avian malaria in the Hawaiian Islands—An ongoing conservation crisis on the island of Kaua'i. *Global Change Biology*, 20:2426–36.

Baer, G.M. (1985). Wildlife control: new problems and strategies. In: Kooprowski, H. and Plotkin, S.A. (eds.), *World Debt to Pasteur*, pp. 235–47. Alan R. Liss Inc., New York.

Barnes, T.S., Goldizen, A.W., Morton, J.M., and Coleman, G.T. (2008). Cystic echinococcosis in a wild population of the brush-tailed rock-wallaby (*Petrogale penicillata*), a threatened macropodid. *Parasitology*, 135:715–23.

Bennett, G.F., Peirce, M.A., and Ashford, R.W. (1993). Avian haematozoa: Mortality and pathogenicity. *Journal of Natural History*, 27:993–1001.

Benning, T.L., LaPointe, D.A., Atkinson, A.C., and Vitousek, P.M. (2002). Interactions of climate change with biological invasions and land use in the Hawaiian Islands, modeling the fate of endemic birds using a geographic information system. *Proceedings of the National Academy of Sciences*, 99 (14):246–49.

Berger, L., Speare, R., Daszak, P., Green, D.E., Cunningham, A.A., Goggin, C.L., et al. (1998). Chytridiomycosis causes amphibian mortality associated with population declines in the rain forests of Australia and Central America. *Proceedings of the National Academy of Sciences*, 95 (9):31–36.

Black, P.F., Murray, J.G., and Nunn, M.J. (2008). Managing animal disease risk in Australia: The impact of climate change. *Scientific and Technical Review, OIE*, 27: 563–80.

Bolker, B.M., Nanda, A., and Shah, D. (2009). Transient virulence of emerging pathogens. *Journal of the Royal Society Interface*, 7:811–22.

Bosch, J., Carrascal, L.M., Durán, L., Walker, S., and Fisher M.C. (2007). Climate change and outbreaks of amphibian chytridiomycosis in a montane area of Central Spain; is there a link? *Proceedings of the Royal Society of London B: Biological Sciences*, 274:253–60.

Bradley, C.A. and Altizer, S. (2006). Urbanization and the ecology of wildlife diseases. *Trends in Ecology & Evolution*, 22:95–102.

Brearley, G., Rhodes, J., Bradley, A., Baxter, G., Seabrook, L., Lunney, D., et al. (2013). Wildlife disease prevalence in human-modified landscapes. *Biological Reviews*, 88:427–42.

Bregman, T.P., Sekercioglu, C.H., and Tobias, J.A. (2014). Global patterns and predictors of bird species responses to forest fragmentation: implications for ecosystem function and conservation. *Biological Conservation*, 169:372–83.

Cable, J., Barber, I., Boag, B., Ellison, A.R., Morgan, E.R., Murray, K. et al. (2017). Global change, parasite transmission and disease control: Lessons from ecology. *Philosophical Transactions of the Royal Society B: Biological Sciences*, 372:20160088.

Cahan, E. (2020). COVID-19 hits U.S. mink farms after ripping through Europe. *Science*, August 18. https://www.sciencemag.org/news/2020/08/covid-19-hits-us-mink-farms-after-ripping-through-europe (accessed July 6, 2021).

Cleaveland, S., Laurenson, M.K., and Taylor, L.H. (2001). Diseases of humans and their domestic mammals: Pathogen characteristics, host range and the risk of emergence. *Philosophical Transactions of the Royal Society of London B: Biological Sciences*, 356:991–99.

Cockle, K.L., Martin K., and Robledo, G. (2012). Linking fungi, trees, and hole-using birds in a neotropical tree-cavity network: Pathways of cavity production and implications for conservation. *Forest Ecology and Management*, 264:210–19.

Conrad, P.A., Miller, M.A., Kreuder, C., James, E.R., Mazet, J., Dabritz, H., et al. (2005). Transmission of toxoplasma: Clues from the study of sea otters as sentinels of *Toxoplasma gondii* flow into the marine environment. *International Journal for Parasitology*, **35**:1155–68.

Cooke, B.D. and Fenner, F. (2002). Rabbit haemorrhagic disease and the biological control of wild rabbits, *Oryctolagus cuniculus*, in Australia and New Zealand. *Wildlife Research*, **29**:689–706.

Crooks, K.R. and Soulé, M.E. (1999). Mesopredator release and avifaunal extinctions in a fragmented system. *Nature*, **400**:563–66.

Daszak, P., Cunningham, A.A., and Hyatt, A.D. (2000). Emerging infectious diseases of wildlife—Threats to biodiversity and human health. *Science*, **287**:443–49.

Davidson, R.K., Romig, T., Jenkins, E., Tryland, M., and Robertson, L.J. (2012). The impact of globalisation on the distribution of *Echinococcus multilocularis*. *Trends in Parasitology*, **28**:239–47.

Day, D. (1981). *The Doomsday Book of Animals*, Viking, New York.

De Castro, F. and Bolker, B. (2005). Mechanisms of disease-induced extinction. *Ecology Letters*, **8**:117–26.

De La Roque, S., Rioux, J.A., and Slingenbergh J. (2008). Climate change: Effects on animal disease systems and implications for surveillance and control. *Scientific and Technical Review, OIE*, **27**:339–54.

Dobson, A.P. and Foufopoulos, J. (2001). Emerging infectious pathogens in wildlife. *Philosophical Transactions of the Royal Society of London B: Biological Sciences*, **356**:1001–12.

Dubey, J.P. (2008). The history of *Toxoplasma gondii*—The first 100 years. *Journal of Eucaryotic Microbiology*, **55**:467–75.

Ellenberg, U., Setiawan, A.N., Cree, A., Houston, D.M., and Seddon, P. (2007). Elevated hormonal stress response and reduced reproductive output in yellow-eyed penguins exposed to unregulated tourism. *General and Comparative Endocrinology*, **152**:54–63.

Epstein, P.R. (2002). Climate change and infectious disease: Stormy weather ahead? *Epidemiology*, **13**:373–75.

Faust, C.L., McCallum, H.I., Bloomfield, L.S.P., Gottdenker, N.L., Gillespie, T.R., Torney, C.J., et al. (2018). Pathogen spillover during land conversion. *Ecology Letters*, **21**:471–83.

Fenner, F. and Fantini, B. (1999). *Biological Control of Vertebrate Pests: The History of Myxomatosis; An Experiment in Evolution*. CAB International, Wallingford.

Foufopoulos, J., Roca, V., White, K.A., Pafilis, P., and Valakos, E.D. (2016). Effects of island characteristics on parasitism in a Mediterranean lizard (*Podarcis erhardii*): A role for population size and island history? *Northwestern Journal of Zoology*, **13**:70–76.

Fournier-Cambrillon, C., Aasted, B., Perrot, A., Pontier, D., Sauvage, F., Artois, M., et al. (2004). Antibodies to Aleutian mink disease parvovirus in free-ranging European mink (*Mustela lutreola*) and other small carnivores from southwestern France. *Journal of Wildlife Diseases*, **40**:394–402.

Friend, M., Franson, J.C., and Ciganovich, E.A. (1999). *Field Manual of Wildlife Diseases: General Field Procedures and Diseases of Birds*. Geological Survey (U.S.). Biological Resources Division, Washington, DC.

Fuglei, E., Stien, A., Yoccoz, N.G., Ims, R.A., Eide, N.E., Prestrud, P. et al. (2008). Spatial distribution of *Echinococcus multilocularis*, Svalbard, Norway. *Emerging Infectious Diseases*, **14**:73–75.

Garnham, P.C.C. (1966). *Malaria Parasites and Other Haemosporidia*. Blackwell Scientific Publications, Oxford.

Gates, J.E. and Evans, D.R. (1998). Cowbirds breeding in the central Appalachians: Spatial and temporal patterns and habitat selection. *Ecological Applications*, **8**:27–40.

Gillespie, T.R. and Chapman, C.A. (2006). Prediction of parasite infection dynamics in primate metapopulations based on attributes of forest fragmentation. *Conservation Biology*, **20**:441–48.

Gillespie, T.R., Chapman, C.A., and Greiner, E.C. (2005). Effects of logging on gastrointestinal parasite infections and infection risk in African primates. *Journal of Applied Ecology*, **42**:699–707.

Gilman S., Blumstein, D.T., and Foufopoulos, J. (2007). The effect of haemosporidian infections on white-crowned sparrow singing behaviour. *Ethology*, **113**:437–45.

Gottdenker, N.L., Streicker, D.G., Faust, C.L., and Carroll, C.R. (2014). Anthropogenic land use change and infectious diseases: A review of the evidence. *EcoHealth*, **11**:619–32.

Groombridge, B. (1992). *Global Biodiversity*. Chapman and Hall, London.

Haines, A., Kovats, R.S., Campbell-Lendrum, D., and Corvalan, C. (2006). Climate change and human health: Impacts, vulnerability and public health. *Public Health*, **120**:585–96.

Hale, K. and Briskie, J. (2007). Decreased immunocompetence in a severely bottlenecked population of an endemic New Zealand bird. *Animal Conservation*, **10**:2–10.

Harvell, D., Altizer, S., Cattadori, I.M., Harrington, L., and Weil, E. (2009). Climate change and wildlife diseases: When does the host matter the most? *Ecology*, **90**:912–20.

Hawley, D.M., Lindström, K., and Wikelski, M. (2005). Experimentally increased social competition compromises humoral immune responses in house finches. *Hormones and Behavior*, **49**:417–24.

Henle, K., Davies, K.F., Kleyer, M., Margules, C., and Settele, J. (2004). Predictors of species sensitivity to fragmentation. *Biodiversity and Conservation*, **13**:207–51.

Henttonen, H., Fuegli, E., Gower, C.N., Haukisalmi, V., Ims, R.A., Niemimaa, J., et al. (2001). *Echinococcus multilocularis* on Svalbard: Introduction of an intermediate host has enabled the life-cycle. *Parasitology*, **123**:547–52.

Hillman, A.E., Lymbery, A.J., and Thompson, R.A. (2016). Is Toxoplasma gondii a threat to the conservation of free-ranging Australian marsupial populations? *International Journal for Parasitology: Parasites and Wildlife*, **5**:17–27.

Hoberg, E.P., Polley, L., Jenkins, E.J., and Kutz S.J. (2008). Pathogens of domestic and free-ranging ungulates: Global climate change in temperate to boreal latitudes across North America. *Scientific and Technical Review, OIE*, **27**:511–28.

Hudson, P.J., Dobson, A.P., and Lafferty K.D. (2006). Is a healthy ecosystem one that is rich in parasites? *Trends in Ecology & Evolution*, **21**:381–85.

Hudson, P.J., Dobson, A.P., and Newborn, D. (1992). Do parasites make prey vulnerable to predation? Red grouse and parasites. *Journal of Animal Ecology*, **61**: 681–92.

Hudson, P.J., Dobson, A.P., and Newborn, D. (1998). Prevention of population cycles by parasite removal. *Science*, **282**:2256–58.

Hudson, P.J., Rizzoli, A.P., Grenfell B.T., Heesterbeek, J.A.P., and Dobson, A.P. (2002). *The Ecology of Wildlife Diseases*. Oxford University Press, Oxford.

Hurston, H., Bonanno, L., Voith, J., Foufopoulos, J., Pafilis, P., Valakos, E., et al. (2009). Effects of fragmentation on genetic diversity in island populations of the Aegean wall lizard *Podarcis erhardii* (Lacertidae, Reptilia). *Molecular Phylogenetics and Evolution*, **52**:395–405.

Huspeni, T.C. and Lafferty, K.D. (2004). Using larval trematodes that parasitize snails to evaluate a saltmarsh restoration project. *Ecological Applications*, **14**:795–804.

Inskeep, W., Gardiner, C.H., Harris, R.K., Dubey J.P., and Goldston R.T. (1990). Toxoplasmosis in Atlantic bottle-nosed dolphins (*Tursiops truncatus*). *Journal of Wildlife Diseases*, **26**:377–82.

Ives, A.R. and Murray, D.L. (1997). Can sublethal parasitism destabilize predator-prey population dynamics? A model of snowshoe hares, predators and parasites. *Journal of Animal Ecology*, **66**:265–78.

Jenkins, D.J. and MacPherson, C.N.L. (2003). Transmission ecology of *Echinococcus* in wildlife in Australia and Africa. *Parasitology*, **127**:S63–72.

Jenkins, D.J. and Morris, B. (2003). *Echinococcus granulosus* in wildlife in and around the Kosciuszko National Park, south-eastern Australia. *Australian Veterinary Journal*, **81**:81–85.

Jones, K.E., Patel, N.E., Levy, M.A., Storeygard, A., Balk, D., Gittleman, J.L., et al. (2008). Global trends in emerging infectious diseases. *Nature*, **451**:990–94.

Kock, R.A., Orynbayev, M., Robinson, S., Zuther, S., Singh, N.J., Beauvais, W., et al. (2018). Saigas on the brink: Multidisciplinary analysis of the factors influencing mass mortality events. *Science Advances*, **4**(1): p.eaao2314.

Kovats, R.S., Campbell-Lendrum, D.H., McMichael A.J., Woodward, A., and Cox, J.S. (2001). Early effects of climate change: Do they include changes in vector-borne disease? *Philosophical Transactions of the Royal Society of London B: Biological Sciences*, **356**:1057–68.

Lafferty, K.D. (2009). The ecology of climate change and infectious diseases. *Ecology*, **90**:888–900.

Lafferty, K.D. (2015). Sea otter health: Challenging a pet hypothesis. *International Journal for Parasitology: Parasites and Wildlife* **4**:291–94.

Laidlaw, W.S. and Wilson, B.A. (2006). Habitat utilisation by small mammals in a coastal heathland exhibiting symptoms of *Phytophthora cinnamomi* infestation. *Wildlife Research*, **33**:639–49.

Lankester, M.W. (2001). Extrapulmonary lungworms of cervids. In: Samuel, W.M., Pybus, M.J., and Kocan, A.A. (eds.), *Parasitic Diseases of Wild Mammals*, 2nd edn, pp. 228–78. Iowa State University Press, Ames, IA.

LaPointe, D.A., Goff, M.L., and Atkinson, C.T. (2010). Thermal constraints to the sporogonic development and altitudinal distribution of avian malaria *Plasmodium relictum* in Hawai'i. *Journal of Parasitology*, **96**: 318–24.

Lebarbenchon, C., Poulin, R., Gauthier-Clerc, M., and Thomas, F. (2007). Parasitological consequences of overcrowding in protected areas. *EcoHealth*, **3**:303–07.

Leberg, P.L. and Vrijenhoek, R.C. (1994). Variation among desert topminnows in their susceptibility to attack by exotic parasites. *Conservation Biology*, **8**:419–24.

Leighton, F.A. (2002). Health risk assessment of the translocation of wild animals. *Revue Scientifique et Technique—OIE*, **21**:187–95.

Levin, I.I., Zwiers, P., Deem, S.L., Geest, E.A., Higashiguchi, J.M., Iezhova, T.A., et al. (2013). Multiple lineages of avian malaria parasites (*Plasmodium*) in the Galapagos Islands and evidence for arrival via migratory birds. *Conservation Biology*, **27**:1366–77.

Lindström, K., Foufopoulos, J., Pärn, H., and Wikelski, M. (2004). Immunological investments reflect parasite abundance in island populations of Darwin's finches. *Proceedings of the Royal Society of London B: Biological Sciences*, **271**:513–19.

Lips, K.R., Brem, F., Brenes, R., Reeve, J.D., Alford, R.A., Voyles, J. et al. (2006). Emerging infectious disease and the loss of biodiversity in a neotropical amphibian community. *Proceedings of the National Academy of Sciences*, **103**:3165–70.

Lips, K.R., Diffendorfer, J., Mendelson, J.R. III, and Sears, M.W. (2008). Riding the wave: Reconciling the roles of disease and climate change in amphibian declines. *PLoS Biology*, **6**(3):e72.

Losin, N., Floyd, C.H., Schweitzer, T.E., and Keller, S.J. (2006). Relationship between aspen heartwood rot and the location of cavity excavation by a primary cavity-nester, the red-naped sapsucker. *Condor*, **108**: 706–10.

Luikart, G., Pilgrim, K., Visty, J., Ezenwa, V.O., and Schwartz, M.K. (2008). Candidate gene microsatellite variation is associated with parasitism in wild bighorn sheep. *Biology Letters*, **23**:228–31.

Malkinson, M., Banet, C., Weisman, Y., Pokamunski, S., King, R., Drouet, M.-T., et al. (2002). Introduction of West Nile virus in the Middle East by migrating white storks. *Emerging Infectious Diseases*, **8**:392–97.

Mallapaty, S. (2020). What's the risk that animals will spread the coronavirus? *Nature*, June 1. https://www.nature.com/articles/d41586-020-01574-4 (accessed July 6, 2021).

Martınez-Mota, R., Valdespino, C., Śanchez-ramos, M.A., and Serio-Silva, J.C. (2007). Effects of forest fragmentation on the physiological stress response of black howler monkeys. *Animal Conservation*, **10**:374–79.

Mbora, N.M.D. and McPeek, M.A. (2009). Host density and human activities mediate increased parasite prevalence and richness in primates threatened by habitat loss and fragmentation. *Journal of Animal Ecology*, **78**:210–18.

McCallum, H. (2008). Tasmanian devil facial tumour disease: Lessons for conservation biology. *Trends in Ecology Evolution*, **23**:631–37.

Meffe, G. and Carroll, R. (1997). *Principles of Conservation Biology*, 2nd edn. Sinauer, Sunderland, MA.

Miller, G.D., Snell, H.H., Hahn, A., and Miller, R.D. (2001). Avian malaria and Marek's disease: Potential threats to Galapagos penguins *Spheniscus mendiculus. Marine Ornithology*, **29**:43–46.

Møller, A.P. and Nielsen, J.T. (2007). Malaria and risk of predation: A comparative study of birds. *Ecology*, **88**:871–81.

Mordecai, E.A., Caldwell, J.M., Grossman, M.K., Lippi, C.A., Johnson, L.R., Neira, M., et al. (2019). Thermal biology of mosquito-borne disease. *Ecology Letters*, **22**:1690–708.

Morshed, M.G., Scott, J.D., Fernando, K., Beati, L., Mazerolle, D.F., Geddes, G., et al. (2005). Migratory songbirds disperse ticks across Canada, and first isolation of the Lyme disease spirochete, *Borrelia burgdorferi*, from the avian tick, *Ixodes auritulus. Journal of Parasitology*, **91**:780–90.

Muellner, A., Linsenmair, K.E., and M. Wikelski. (2004). Exposure to ecotourism reduces survival and affects stress response in hoatzin chicks (*Opisthocomus hoazin*). *Biological Conservation*, **118**:549–58.

Newmark, W.D. (1987). A landbridge island perspective on mammalian extinctions in western North American parks. *Nature*, **325**:430–32.

Norris, K. and Evans, M.R. (2000). Ecological immunology: Life history trade-offs and immune defense in birds. *Behavioral Ecology*, **11**:19–26.

O'Neil, B.D. and Pharo, H.J. (1995). The control of bovine tuberculosis in New Zealand. *New Zealand Veterinary Journal*, **43**:249–55.

Oppliger, A., Clobert, J., Lecomte, J., Lorenzon, P., Boudjemadi, K., and John-Alder, H.B. (1998). Environmental stress increases the prevalence and intensity of blood parasite infection in the common lizard *Lacerta vivipara. Ecology Letters*, **1**:129–38.

Paaijmans, K.P., Read, A.F., and Thomas M.B. (2009). Understanding the link between malaria risk and climate. *Proceedings of the National Academy of Sciences*, **106**:13844–49.

Packer, C., Holt, R.D., Hudson, P.J., Lafferty, K.D., and Dobson A.P. (2003). Keeping the herds healthy and alert: Implications of predator control for infectious disease. *Ecology Letters*, **6**:797–802.

Palacios, M.G., D'Amico, V.L., and Bertellotti, M. (2018). Ecotourism effects on health and immunity of Magellanic penguins at two reproductive colonies with disparate touristic regimes and population trends. *Conservation Physiology*, **6**:p.coy 060.

Parameswaran, N., O'Handley, R.M., Grigg, M.E., Fenwick, S.G., and Thompson, R.C.A. (2009). Seroprevalence of *Toxoplasma gondii* in wild kangaroos using an ELISA. *Parasitology International*, **58**:161–65.

Pedersen, A.B. and Greives, T.J. (2008). The interaction of parasites and resources cause crashes in a wild mouse population. *Journal of Animal Ecology*, **77**:370–77.

Piotrowski, J.S., Annis, S.L., and Longcore, J.E. (2004). Physiology of *Batrachochytrium dendrobatidis*, a chytrid pathogen of amphibians. *Mycologia*, **96**:9–15.

Pounds, A.J., Bustamante, M.R., Coloma, L.A., Consuegra, J.A., Fogden, M.P.L., Foster, P.N., et al. (2006). Widespread amphibian extinctions from epidemic disease driven by global warming. *Nature*, **439**: 161–67.

Power, A.G. and Mitchell, C.E. (2004). Pathogen spillover in disease epidemics. *The American Naturalist*, **164**: S79–89.

Presidente, P. (1986). Tissue worm implications for live deer imports. In: Owen, P. (ed.), *Deer Farming into the Nineties*, pp. 192–202. Owen Art Publishing, Brisbane.

Puettker, T., Meyer-Lucht, Y., and Sommer, S. (2008). Effects of fragmentation on parasite burden (nematodes) of generalist and specialist small mammal species in secondary forest fragments of the coastal Atlantic Forest, Brazil. *Ecological Research*, **23**:207–15.

Puschendorf, R., Carnaval, A.C., VanDerWal, J., Zumbado-Ulate, H., Chaves, G., Bolaños, F., and Alford, R.A. (2008). Distribution models for the amphibian chytrid *Batrachochytrium dendrobatidis* in Costa Rica: Proposing climatic refuges as a conservation tool. *Diversity and Distributions*, **15**:401–08.

Randolph, S.E. (2008). Dynamics of tick-borne disease systems: Minor role of recent climate change. *Revue scientifique et technique—OIE*, **27**:367–81.

Reddacliff, G., Hartley, W., Dubey, J., and Cooper, D. (1993). Pathology of experimentally-induced, acute toxoplasmosis in macropods. *Australian Veterinary Journal*, **70**:4–6.

Reiter, P. (2008). Climate change and mosquito-borne disease: Knowing the horse before hitching the cart. *Revue scientifique et technique—OIE*, **27**:383–98.

Remm, J. and Lohmus A. (2011). Tree cavities in forests—The broad distribution pattern of a keystone structure for biodiversity. *Forest Ecology & Management*, **262**:579–85.

Roca, V., Foufopoulos, J., Valakos, E.D., and Pafilis P. (2009). Parasitic infracommunities of the Aegean wall lizard *Podarcis erhardii* (Lacertidae, Sauria): Isolation and impoverishment in small island populations. *Amphibia-Reptilia*, **30**:493–503.

Rödder, D., Veith, M., and Lötters, S. (2008). Environmental gradients explaining the prevalence and intensity of infection with the amphibian chytrid fungus: The host's perspective. *Animal Conservation*, **11**:513–17.

Rohr, J.R., Dobson, A.P., Johnson, P.T., Kilpatrick, A.M., Paull, S.H., Raffel, T.R., et al. (2011). Frontiers in climate change—disease research. *Trends in Ecology & Evolution*, **26**:270–77.

Rohr, J.R., Raffel, T.R., Romansic, J.M., McCallum, H., and Hudson, P.J. (2008). Evaluating the links between climate, disease spread, and amphibian declines. *Proceedings of the National Academy of Sciences*, **105** (17): 436–41.

Rowley, J.J.L. and Alford, R.A. (2007). Behaviour of Australian rainforest stream frogs may affect the transmission of chytridiomycosis. *Diseases of Aquatic Organisms*, **77**:1–9.

Samuel, M.D., Woodworth, B.L., Atkinson, C.T., Hart, P.J., and LaPointe, D.A. (2015). Avian malaria in Hawaiian forest birds: Infection and population impacts across species and elevations. *Ecosphere*, **6**:1–21.

Skerratt, L.F., Berger, L., Speare, R., Cashins, S., McDonald, K.R., Phillott, A.D., et al. (2007). Spread of chytridiomycosis has caused the rapid global decline and extinction of frogs. *EcoHealth*, **4**:125–34.

Smith, J.B., Schneider, S.H., Oppenheimer, M., Yohe, G.W., Hare, W., Mastrandrea, M.D., et al. (2009a). Assessing dangerous climate change through an update of the Intergovernmental Panel on Climate Change (IPCC) "reasons for concern." *Proceedings of the National Academy of Sciences*, **106**:4133–37.

Smith, K.F., Behrens, M., Schloegel, L.M., Marano, N., Burgiel, S., and Daszak, P. (2009b). Reducing the risks of the wildlife trade. *Science*, **324**:594–95.

Stoskopf, M.K. and Beir, J. (1979). Avian malaria in African black-footed penguins. *Journal of the American Veterinary Medical Association*, **175**:944–46.

Suorsa, P., Huhta, E., Nikula, A., Nikinmaa, M., Jantti, A., Helle, H., et al. (2003). Forest management is associated with physiological stress in an old-growth forest passerine. *Proceedings of the Royal Society of London B: Biological Sciences*, **270**:963–69.

Sutherst, R.W. (2004). Global change and human vulnerability to vector-borne diseases. *Clinical Microbiology Reviews* **17**:136–73.

Tamerius, J.D., Wise, E.K., Uejio, C.K., McCoy, A.L., and Comrie, A.C. (2007). Climate and human health: Synthesizing environmental complexity and uncertainty. *Stochastic Environmental Research and Risk Assessment*, **21**:601–13.

Thiele, D., Jenni-Eiermann, S., Braunisch, V., Palme, R., and Jenni, L. (2008). Ski tourism affects habitat use and evokes a physiological stress response in capercaillie *Tetrao urogallus*: A new methodological approach. *Journal of Applied Ecology*, **45**:845–53.

Timm, S.F., Munson, L., Summers, B.A., Terio, K.A., Dubovi, E.J., Rupprecht, C.E. et al. (2009). A suspected canine distemper epidemic as the cause of a catastrophic decline in Santa Catalina Island foxes (*Urocyon littoralis catalinae*). *Journal of Wildlife Diseases*, **45**: 333–43.

Tompkins, D.M. and Gleeson, D.M. (2006). Relationship between avian malaria distribution and an exotic invasive mosquito in New Zealand. *Journal of the Royal Society of New Zealand*, **36**:51–62.

Tompkins, D.M., White, A.R., and Boots, M. (2003). Ecological replacement of native red squirrels by invasive greys driven by disease. *Ecology Letters*, **6**:189–96.

Torgerson, P.R. and Heath, D.D. (2003). Transmission dynamics and control options for *Echinococcus granulosus*. *Parasitology*, **127**: S143–58.

Torgerson, P.R., Keller, K., Magnotta, M., and Ragland, N., (2010). The global burden of alveolar echinococcosis. *PLoS Neglected Tropical Diseases*, **4**:e722.

Trejo-Macias, G., Estrada, A., and Cabrera, M.A.M. (2007). Survey of helminth parasites in populations of *Alouatta palliata mexicana* and *A. pigra* in continuous and in fragmented habitat in southern Mexico. *International Journal of Primatology*, **28**:931–45.

Turni, C. and Smales, L.R. (2001). Parasites of the bridled nailtail wallaby (*Onychogalea fraenata*) (Marsupialia: Macropodidae). *Wildlife Research* **28**:403–11.

Valkiunas, G. (2005). *Avian Malaria Parasites and Other Haemosporidia*. CRC Press, New York.

Van Meter, P.E., French, J.A., Dloniak, S.M., Watts, H.E., Kolowski, J.M., and Holekamp, K.E. (2009). Fecal glucocorticoids reflect socio-ecological and anthropogenic stressors in the lives of wild spotted hyenas. *Hormones and Behavior*, **55**:329–37.

Van Riper, C., Van Riper, S.G., Lee Goff, M., and Laird, M. (1986). The epizootiology and ecological significance of malaria in Hawaiian land birds. *Ecological Monographs*, **56**:327–44.

Walsh, P.D., Abernethy, K.A., Bermejo, M., Beyers, R., Wachter, P.D., Akou, M.E., et al. (2003). Catastrophic ape decline in western equatorial Africa. *Nature*, **422**:611–14.

Warner, R.E. (1968). The role of introduced diseases in the extinction of the endemic Hawaiian avifauna. *Condor*, **70**:101–20.

Wasser, S.K., Bevis, K., King, G., and Hanson, E. (1997). Non-invasive physiological measures of disturbance in the northern spotted owl. *Conservation Biology*, **11**: 1019–22.

Watson, T.G. and Gill, J.M. (1985). The experimental infection of guinea pigs with the tissue worm of deer—*Elaphostrongylus cervi*. *New Zealand Veterinary Journal*, **33**:81–83. https://doi.org/10.1080/00480169.1985. 35174

Wells, K., Smales, L.R., Kalko, E.K.V., and Pfeiffer, M. (2007). Impact of rain-forest logging on helminth assemblages in small mammals (Muridae, Tupaiidae) from Borneo. *Journal of Tropical Ecology*, **23**:35–43.

Whiteman, N.K., Matson, K.D., Bollmer, J.L., and Parker, P.G. (2006). Disease ecology in the Galapagos hawk (*Buteo galapagoensis*): Host genetic diversity, parasite load and natural antibodies. *Proceedings of the Royal Society of London B: Biological Sciences*, **273**:797–804.

Wikelski, M., Foufopoulos, J., Vargas, H., and Snell, H. (2004). Galápagos birds and diseases: Invasive pathogens as threats for island species. *Ecology and Society* **9** (1):5. http://www.ecologyandsociety.org/vol9/iss1/art5 (accessed July 11, 2018).

Wild, M.A., Hobbs, N.T., Graham, M.S., and Miller, M.W. (2011). The role of predation in disease control: A comparison of selective and nonselective removal on prion disease dynamics in deer. *Journal of Wildlife Diseases*, **47**:78–93.

Woodhams, D.C., Alford, R.A., and Marantelli, G. (2003). Emerging disease of amphibians cured by elevated body temperature. *Diseases of Aquatic Organisms*, **55**: 65–67.

Woodroffe, R. and Ginsberg, J.R. (1998). Edge effects and the extinction of populations inside protected areas. *Science*, **280**:2126–28.

Woolhouse, M.E., Dye, C., Etard, J.F., Smith, T., Charlwood, J.D., Garnett, G.P., et al. (1997). Heterogeneities in the transmission of infectious agents: implications for the design of control programs. *Proceedings of the National Academy of Sciences*, **94**:338–42.

Wyatt, K.B., Campos, P.F., Gilbert, M.T.P., Kolokotronis, S.-O., Hynes, W.H., DeSalle, R., et al. (2008). Historical mammal extinction on Christmas Island (Indian Ocean) correlates with introduced infectious disease. *PLoS ONE*, **3** (11):e3602. DOI:10.1371/journal.pone. 0003602

Yang, H.M., Macoris, M.L.G., Galvani, K.C., Andrighetti, M.T.M., and Wanderley, D.M.V. (2009). Assessing the effects of temperature on dengue transmission. *Epidemiology and Infection*, **137**:1179–87.

PART II

Acquisition of Field Data

CHAPTER 4

Sampling, Experimental Design, and Statistical Analysis

A lady declares that by tasting a cup of tea made with milk, she can discriminate whether the milk or the tea infusion was added first to the cup... Our experiment consists in mixing eight cups of tea, four in one way and four in the other, and presenting them to the subject for judgement in a random order.

R.A. Fisher, *The Design of Experiments*, 1935 p 11 (pagination from Eighth Edition, reprinted 1971, Hafner Publishing Company, New York)

4.1 Introduction

The success of a research project depends critically on careful planning and this in turn hinges on clarity of purpose. Perhaps the most important question to ask at the beginning of any research project is: "What do I want to do with the answers?" Because the focus of this book is infectious disease and conservation biology, we will assume that the fundamental objective here is to improve conservation outcomes in the light of infectious diseases or parasites. While infectious agents typically impact taxa of conservation interest directly, here we will also consider the possibility of indirect impacts to competitors, predators, or prey of the species of interest. Since conservation biology is fundamentally an applied science, researchers need to recognize that the answers provided by their research should ideally be relevant to conservation practices as applied in the field. Accordingly, the initial question is perhaps best refined to: "How could the outcome of my study influence the way in which this species, interaction, or ecosystem should be managed?" Once that question is answered, a subsidiary question might be: "Can a study be designed with a realistic budget and time frame that will answer this question with sufficient precision and generality to be useful in guiding management practices?" In the following sections we will discuss some of the general principles of sampling, study design, and statistical analysis, and consider how they can be applied to the specific case of understanding wildlife disease.

4.2 The three Rs of sampling and study design: Representativeness, replication, and randomness

Three fundamental principles underlie the design of sampling strategies and experimental design – representativeness, replication and randomness.

4.2.1 Representativeness

A central principle of experimental design is that the individuals sampled, or selected as experimental

Infectious Disease Ecology and Conservation. Johannes Foufopoulos, Gary A. Wobeser and Hamish McCallum, Oxford University Press.
© Johannes Foufopoulos, Gary A. Wobeser and Hamish McCallum (2022). DOI: 10.1093/oso/9780199583508.003.0004

subjects, must be *representative* of the population to which the conclusions are to be applied. "Individuals" in most cases will mean individual animals, and "population" will mean the population of animals about which one wishes to draw conclusions. However, it is also important to recognize that one is really talking about the statistical, rather than biological, meaning of a population. For instance, the "individuals" sampled could perhaps be study sites or locations and the population could be the total collection of potential study sites or locations to which one wants the answers to apply.

There will always be trade-offs between designing an experiment that will apply to as broad a population as possible (i.e., generality), and one that has the maximum *power* or *precision*. The concepts of power and precision will be discussed in more detail later in this chapter. For the time being, "power" is the ability of an experiment to detect an effect that does exist, and "precision" is a measure of how much error there is in estimating a quantity of interest.

As an example of the trade-off between generality and power or precision, suppose the objective of an experiment is to determine the influence that infection with a particular pathogen has on the survival of individuals of a given species. If your experimental subjects are as similar as possible (e.g., all are 1-year-old males, all fed *ad libitum* in a standard laboratory environment), then the possibility of precisely measuring a difference in survival between infected and uninfected individuals will be maximized. Unfortunately, no matter how well designed the experiment is, the conclusions will only apply to the population of 1-year-old males fed *ad libitum* in the standard environment. It may therefore not provide good information on the effect of the pathogen on all members of the particular species under natural conditions. If the experimental subjects are selected to be representative of all individuals of the species, across the whole range of all-natural conditions that the species inhabits, not only will the experiment be logistically far more complex and expensive to undertake, but it is also likely that background variability in survival between individuals will be far greater than in the controlled experimental situation. This means that any difference in survival between infected and uninfected animals

will be measured with much lower precision (see Chapter 5).

4.2.2 Replication

This leads directly on to the second R of experimental design—*replication*. Differences in a response variable between treatment groups can only be demonstrated relative to variation in that response variable within treatment groups. It is therefore essential to expose a number of experimental subjects to each experimental treatment, in order to obtain an estimate of the extent of variation in the response variable between individuals treated in the same way. There is often uncertainty among inexperienced scientists concerning what a replicate actually is. Replicate experimental subjects do not have to be identical. In fact, if they were, replication would not be needed because identical individuals would respond identically to the experimental treatment. Rather, replication is needed precisely because individuals are not identical. In the ideal experimental design, replicates within each treatment group will reflect the range of variation in the study population: as defined earlier, they should be representative of the population under investigation. Most importantly, there should be no systematic differences in the replicates of the different treatments, other than in variables explicitly accounted for in the experimental or analytical design. For example, in an experimental infection study, one treatment group should not include disproportionately more male individuals than the others, unless an individual's sex is taken explicitly into account in the statistical analyses.

Consultant statisticians are often asked how many replicates are necessary. There is no simple and general answer to this question, but there are two key considerations. First, what is the magnitude of the effect you need to be able to detect, or equivalently, what is the precision with which you need to be able to estimate the parameter in question? The smaller the effect size that needs to be detected, the larger the sample size that is needed. Deciding on the minimum effect size that needs to be detectable is a biological, not a statistical question. For instance, if there were a 1% difference in survival between infected and uninfected animals, would this have any practical implications for conservation management? Probably not. On the other hand, if survival of infected individuals were half that of uninfected individuals, it is quite likely that this would be an important issue for management. So, in this hypothetical example, one does not need to be able to detect a 1% difference in survival, but one would want to be able to detect a 50% difference. Sometimes simple biological intuition or practical experience is sufficient to determine what size of an effect is "important." In other cases, epidemiological models (see Chapter 9) can provide guidance on the size of effect that needs to be detected. For example, a pathogen can invade a population if R_0 > 1 (see Chapter 2). An appropriate epidemiological model will enable R_0 to be expressed in terms of various parameters such as the transmission rate. If the management question is to determine whether or not the pathogen can invade, then estimating this rate with sufficient precision to determine whether or not R_0 > 1 is important, whereas additional precision that does not change which side of the threshold value of 1 R_0 falls may be less important.

The second consideration in estimating sample size is knowledge of the extent of variation in the response variable between replicates. Usually, this will require prior knowledge, perhaps based on a pilot study. For certain types of problems, however, one can obtain an approximation of the variation without detailed knowledge of the particular biological circumstances. Proportions are a case in point. Many of the problems we deal with in this book are essentially a matter of estimating proportions. For example, survival over a given time period is simply the proportion of individuals from an initial cohort that are still alive at the end of the study period. Similarly, the prevalence of infection in a population is simply the proportion of individuals that are infected. Provided (and this can be a big proviso under some circumstances) that each individual in the experiment can be considered to be independent of every other individual, statistical theory tells us that the number of individuals x surviving out of an initial cohort n should follow a binomial distribution. Thus:

$$P(x) = \binom{n}{x} p^x (1 - p)^{n-x} \qquad (4.1)$$

Where p is the probability that any one individual survives and $P(x)$ is the probability that exactly x individuals of the initial n survive or are infected. Usually, p is unknown and our objective is to derive an estimate of its value $\hat{p}$. Provided n is not too small (above ca. 20) and p is not too close to either 0 or 1 (in the range of ca. $0.2 < p < 0.8$), it turns out that $\hat{p}$ has close to a normal distribution with a mean $\mu = p$ and variance $\sigma^2 = p(1 - p)/n$. So, if p is reasonably close to 0.5, a 95% confidence interval for p, the proportion surviving is approximately:

$$\frac{x}{n} \pm 1.96 \sqrt{\frac{0.5(1 - 0.5)}{n}} \qquad (4.2)$$

Which approximately simplifies to:

$$\frac{x}{n} \pm \frac{1}{\sqrt{n}} \qquad (4.3)$$

This is essentially a long-winded way of showing that the plausible error on an estimate of a proportion based on a sample size of n is approximately $1/\sqrt{n}$. So, to generate an estimate of survival that is within 10% of the true value, one would need to follow the fate of about 100 animals ($0.1 = 1/\sqrt{100}$). Similarly, to estimate the prevalence of infection in a population to within 10% of its true value, a sample size of ca. 100 would be needed. Figure 4.1 shows how well these various approximations to calculate confidence intervals for proportions work for differing sample sizes.

For most other analyses, independent information on the extent of variation in the response variable within treatment groups is often obtained from a pilot study or from previous research on a similar system. Determining the sample size needed for statistical tests is called power analysis and is discussed later in this chapter (Section 4.5). More detail on the binomial and normal distributions can be found in most basic statistics books. Chapter 4 in Bolker (2008) provides a good discussion, written for ecologists.

4.2.3 Randomness

Employing some form of *random* selection of experimental subjects or individuals to be included in a survey is a good way of ensuring that study replicates are indeed representative of the population

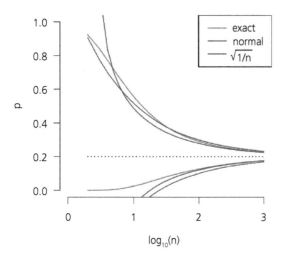

Figure 4.1 The ability to accurately estimate the "true" proportion (e.g., the fraction of infected individual in a population) increases with increasing sample size. Shown here are 95% confidence intervals for that proportion p as a function of sample size, calculated from the exact binomial distribution, the normal approximation, and the $\sqrt{1/n}$ approximation. The "true" proportion in the population is 0.2.

of interest. It is important to recognize that random does not mean haphazard. Substantial effort is often needed to ensure that a sample is indeed random. A random sample is one in which some chance process determines which individual replicates end up in the survey or experiment. A sample will be representative of the population if all individuals in the population have an equal probability of being included in the sample. The best way to ensure this is to assign numbers to every individual in the population in question (often called the "sampling frame" in the statistical literature) and to then use random numbers to select individuals for the sample. This is called a simple random sample. All that is necessary is to allocate a uniformly distributed random number to each individual in the population and then to rank the numbers in ascending order. Then a random sample size of n is simply the first n members of the list. In practice, this can easily be accomplished with a software program like MS Excel: generate the column of random numbers using =*Rand()*, copy the column, and use "Paste Special" as "Values" to generate numbers in an adjacent column before sorting to prevent Excel

from recalculating the random numbers. R has a function **sample()** to draw simple random samples.

In theory, taking a simple random sample is straightforward, but in practice, numbering every member of the population will often not be possible. Nonetheless the general idea provides a gold standard against which to assess alternative ways of ensuring that each individual in the population has the same probability of inclusion in the sample. It is often difficult to ensure unbiased random samples of animals in any study of disease in a free-ranging wildlife population, especially since sampling usually involves capturing or killing animals. Depending on the nature of the sampling method, infection may either make animals harder to catch (e.g., diseased animals may be less active and therefore less likely to encounter a trap), or possibly easier to catch (if diseased animals are less able to escape from a hunter). Prevalence of infection based on a sample will then not be an unbiased estimate of prevalence in the population as a whole (Jennelle et al. 2007).

Simple random sampling is often not the best way to ensure that a sample is representative of the population under investigation (Figure 4.2a). If the population consists of distinct groups that differ in the mean value of the response variable (or are suspected to do so), and the proportion of the population that falls into each of these group is known, stratified random sampling is a preferable approach. Stratified random sampling is most easily explained by considering a survey in which the mean value of a response variable is to be estimated across a study area on the basis of observations at a number of sites within the study area. In Figure 4.2, for example, sampling needs to be done in a landscape consisting of two distinct habitats: an area covered by natural vegetation (light colored) and then road habitat (grayed area). Suppose that the objective is to estimate the overall proportion of groundcover in the landscape as a whole. In this context, the total number of potential study sites within the area is the statistical population and the sites selected for study constitute the sample. Because the study area consists of two different habitat types, there are reasons to suspect that groundcover differs between habitat types. It is sensible to treat each of the habitat types as

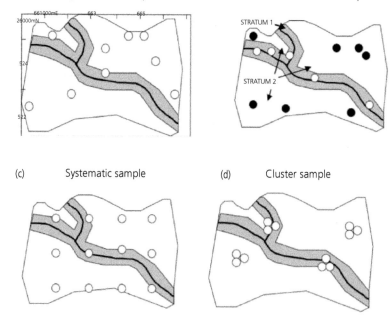

Figure 4.2 Examples of different sampling methods. Inside the study area, two different habitats exist: natural habitat (light colored), and habitat impacted by the presence of a nearby road (dark gray). Within this spatial context, sampling stations (circles) representative of the area need to be placed. (a) Sampling stations are placed in a random manner across the study site irrespective of habitat type. (b) Stratified random sample, with sampling stations placed randomly but within each habitat type. Hence the differences between the two habitat types are taken explicitly into account during the sampling procedure. (c) Stations are placed in a systematic manner, again irrespective of habitat type. (d) Sampling sites placed in clusters. (Simplified from Garton et al. 2005.)

a separate stratum, and because the area of each habitat type has been mapped, it is possible to use a stratified random sample (Figure 4.2b). A decision needs to be made about what proportion of the study sites should be allocated to each of the habitat types and samples are then taken randomly from within each of the habitat types. The formulae necessary to derive an overall estimate and its variance from a stratified random sample are a little complex but can be found in most statistics textbooks (e.g., Thompson 2012).

The overall estimate of the response variable will have the minimum possible variance when samples are allocated in proportion to the variance within each stratum in the response variable. In practice, a good rule of thumb to achieve this is to allocate samples between strata proportional to the number of individuals in the population within each stratum. For example, if the objective is to estimate the overall prevalence of a pathogen in a population and the stratification is done by age class, samples would be allocated proportional to the relative number of individuals in each age class. A suboptimal allocation of sampling effort will not produce a biased estimate, just one with a little

more variation than if the allocation were optimal. Stratification is therefore worthwhile even without unequivocal evidence that the response variable varies between strata. In the particular case of estimating prevalence of a pathogen, one might decide to stratify on the basis of sex and age, selecting random samples from each age and sex class. To calculate overall prevalence, the age and sex structure of the population as a whole would need to be known.

For practical reasons, ecologists and epidemiologists sometimes use sampling methods that are not random. Systematic samples, in which data are collected at fixed, regular intervals along a transect or across a grid, are often much easier and cheaper to collect than random samples so that, for the same amount of effort, sample size can be much greater for a systematic sample than a random sample (see Figure 4.2c). Systematic samples can also help avoid the unconscious bias that can often occur when attempting to take random samples at points in a landscape. For instance, if samples are being collected from 1 m^2 quadrats using a navigation system such as a global positioning system (GPS) (which is often accurate only up to 10–20 m), the investigator will likely be subjective about the placement of the quadrat (e.g., by avoiding placement in an area with thick overgrowth). This could become a significant source of bias regarding the representation of the overall area. In contrast, if the quadrats are placed at fixed intervals along a measuring tape, samples will be taken regardless of convenience of placement. Systematic samples also ensure that the entire study area is sampled, whereas a random sample may occasionally be aggregated by chance. For most practical applications, systematic samples can be treated as if they were random samples. The one important exception is if there is some form of periodicity in the response variable across the sampling area. For example, systematic sampling is a risky strategy in a regular dune field where regularly spaced out sampling points may end up "lining up" in nonrandom ways with the underlying dune pattern resulting in a biased sample. In the special case of estimating prevalence of a pathogen in a population, a systematic sample might consist of sampling from every 10th animal captured. This would be a sensible strategy if detection of infection was resource intensive and it was therefore not possible to determine the infection status of every individual captured. It is unlikely that such a sample would have some underlying periodicity that might bias the result.

Cluster sampling (Figure 4.2d) is an approach that incorporates some of the advantages of both random sampling and systematic sampling. Around randomly selected points, a cluster in a fixed array of samples is taken. Provided that the overall dimensions of the cluster are greater than the precision of a navigation instrument (i.e., if using a GPS the sampling points would be spaced out more than 20 m apart), this method avoids the issue of unconscious bias against "difficult" sampling points. When analyzing data from cluster samples, the mean of the response variable should be taken for each cluster and treated as the individual observation in any further statistical analysis. The sample size n is thus the number of clusters.

There are many texts on ecological methodology that provide more detailed discussions of sampling strategies in specific ecological contexts. Chapter 2 in Sutherland (2006) has a particularly good discussion on sampling written for a general ecological audience.

4.3 Statistical tools

We assume that readers of this book have completed a basic undergraduate course in biostatistics. If not, there are several very good introductory texts worth consulting. Bolker (2008) is an excellent text from an author with interests in disease ecology. Most introductory courses are based around linear models with normal errors. These include t-tests, regression, and analysis of variance (ANOVA). For categorical data, the reader is likely to be familiar with contingency tables analyzed using chi-squared tests. Readers should also be familiar with hypothesis testing (see Section 4.6) and constructing confidence intervals. These approaches are useful, but have their limitations.

There are many excellent statistical packages available, most of which are capable of performing the analyses discussed in this book. We particularly recommend R (R Core Team 2021). R is completely free and available for Windows, Mac, and Linux. It has become the de facto standard statistical tool for ecologists. A particular advantage is that it has an enormous range of user-contributed packages, which are capable of performing almost any statistical analysis described in the scientific literature, with additional packages continually being made available. A somewhat daunting aspect for beginners is that R requires the writing of code rather than the use of a point-and-click interface. However, writing code has the major advantage that any analysis is reproducible and is documented. We also recommend the use of a graphical interface to the program itself. There are several of these, but we suggest RStudio (RStudio Team 2020), which is also free. Many resources to assist with using R are available on the web and are readily accessible through any search engine. Crawley (2012) is an excellent reference text, written by an ecologist.

4.3.1 Likelihood

A fundamental concept in biostatistics is the idea of "likelihood." Given a particular model or explanation for the data at hand—for example, that the average survival between infected and uninfected animals differs by a certain amount and that the number of individuals in the infected and uninfected groups that survive over some specified interval follows a binomial distribution—the likelihood is the chance of observing a particular set of experimental outcomes. The maximum likelihood estimate of the difference in survival between the two groups is the difference value for which the likelihood is greatest. For fairly simple models such as the one just described, likelihood is not difficult to compute.

There are statistical packages (such as R) that will perform maximum likelihood estimation for all of the situations described in this book. Further, in the case of normally distributed errors and additive effects (the assumptions underlying ANOVA and regression), the least squares estimates familiar from basic statistics are in fact maximum likelihood estimates. The data we often deal with in epidemiology, however, do not follow these basic assumptions. If we were to look at survival of a number of individuals, we would expect the error or variation to follow a binomial rather than a normal distribution, because essentially, we are dealing with a number of "trials," with the variable we measure being the number of "successes," or individuals that have survived (see Section 4.2.2). If we wished to model the number of survivors as a function of parasite burden, a simple linear model, (assuming a straight-line decrease in the number of survivors with increasing parasite burden) would be inappropriate because at some parasite burden, the number of survivors predicted would become a negative number, which is impossible.

In this particular case, suppose that x is the number of survivors out of n individuals present originally with a parasite burden p. A sensible way to model the predicted number of survivors, ensuring that the number of survivors cannot be negative is:

$$\ln\left(\frac{x}{n-x}\right) = \alpha + \beta p \qquad (4.4)$$

Where α and β are the two parameters to be estimated. Here, for a given parasite burden, we would expect the actual number of survivors to follow a binomial distribution.

This model is called a *generalized linear model*. The right-hand side of the equation is a linear function (technically, this means that the parameters α and β are separated only by a + sign). The function that links the linear right-hand side of the equation to the predicted response (the number of successes x out of n trials) on the left-hand side is a logit function. This particular generalized linear model is described as a logistic model, having a logit link and a binomial error structure. Most modern statistical packages (including R, which uses the function **glm()** can fit generalized linear models.

4.3.2 The Akaike information criterion

There will always be several different statistical models that can be used to describe any given set of observed data. With conventional linear models, these different models are often compared using F ratios in ANOVA. This is an application of null hypothesis testing (see Section 4.4). Conventionally, a more complex model (e.g., one that in an ANOVA includes in addition to the main variables the interaction term between these two variables) is considered to be justifiable only if the F ratio comparing this with a simpler model (e.g., one without this interaction term) is "significant" (conventionally, at $P = 0.05$). For generalized linear models, this approach can be generalized using likelihood ratio tests. Where there is a good reason for preferring a simple model to a more complex one, such as is often the case in a designed experiment, likelihood ratios and F ratios are a useful approach.

However, many situations in epidemiology involve dealing with a set of observed data from a field population. Several competing models could be used to describe the observed data and the question is to determine how likely it is that each model might have generated the observed results. In this case, there is no necessary reason to prefer any particular "null hypothesis." The question then becomes "Which models are most strongly supported by the observed data?" Rather than the analyst attempting to decide on a single "correct" model, there may be several models with varying levels of support from the observed data.

The Akaike information criterion (AIC), a metric built on the likelihood approach, is widely used to compare the degree of support in the data for a given model (see, for example, Bolker, 2008). It is defined as:

$$AIC = -2L + 2k \qquad (4.5)$$

Where L is the log likelihood for the particular model and k is the number of parameters estimated from the data. The lower the AIC, the better the fit of the model to the observed data. For relatively small sample sizes, one should use a small-sample-corrected AIC_c:

$$AIC_C = AIC + \frac{2k(k+1)}{n-k-1} \qquad (4.6)$$

Where n is the number of observations. Rather than performing hypothesis tests or calculating likelihood ratios, Akaike weights can be calculated for a series of competing models. These can be loosely interpreted as the probability that a given model is the best for the data of the range of models investigated. To calculate Akaike weights for several competing models:

1. Calculate AIC or AIC_C for each of the models; then
2. Calculate the difference in AIC between each of the models and the model with the smallest AIC. Call this Δ.

The AIC weight for model i is then:

$$w_i = \frac{\exp(-\Delta_i/2)}{\sum_{i=1}^{M} \exp(-\Delta_i/2)} \qquad (4.7)$$

Where there are M candidate models.

AIC can also be extracted from most models analyzed in R simply by using the command **AIC(model)**.

It is simple to calculate AIC_C from this using Equation (4.6) and to calculate AIC weights using Equation (4.7). R also has a package to automate this process called MuMIn (Bartoń 2019). Burnham and Anderson (2002) provide a detailed description of model selection methods using these approaches and a more concise summary is provided in Burnham and Anderson (2004).

4.4 Hypothesis testing

Most elementary statistics courses emphasize the hypothesis testing approach, which focuses on distinguishing between a null hypothesis (no effect) and an alternative (research) hypothesis (i.e., that there is some effect), with the help of a statistical hypothesis test used to compare them. The basic paradigm is that the null hypothesis should continue to be accepted unless there is strong evidence to suggest it should be rejected (see, e.g., Sokal and Rohlf 1995). Generations of students have learned the $P = 0.05$ rule: that the null hypothesis is accepted unless there is less than a 5% chance that the observed results could have been obtained if the hypothesis were in fact true. The rule makes an implicit assumption that it is much more serious to make a Type I error (saying that there is an effect when there in fact is not) than to make a Type II error (saying that there is not an effect when in fact there is). The probability of making a Type I error is called the significance level of a test, while one minus the probability of making a Type II error is called the power. A graphical representation of this fundamental concept can be seen in Figure 4.3.

For values of the test statistic larger than the critical value, the shaded area under the probability density for H_0 (solid curve) is the statistical significance α (i.e., if H_0 is true, α is the probability of observing a test statistic bigger than the critical value). For this same critical value, the hatched area under the probability density for H_a (dashed line) is the power $1 - \beta$ (i.e., if the alternative hypothesis is true, there would be a probability $1 - \beta$ of observing a test statistic larger than the critical value). The statistical significance α can also be described as the probability of making a Type I error (accepting the null hypothesis when the alternative hypothesis is in fact true), whereas the probability of making a Type II error (accepting the alternative hypothesis when the null hypothesis is in fact true) is given by β.

Increasingly, however, statisticians are warning against an overemphasis on hypothesis testing and in particular on the rigid application of the $P = 0.05$ rule (Baker 2016; Wasserstein and Lazar 2016; Amrhein et al. 2019). It is a very limiting view of statistical analysis, particularly when the objective

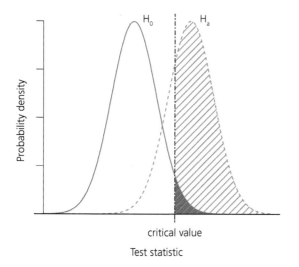

Figure 4.3 Graphical representation of statistical significance α, power $(1 - \beta)$, and the critical value for testing a null hypothesis (H_0) against an alternative hypothesis (H_a). Here, the vertical axis is the probability density, meaning that the area under the curve represents probability, and the horizontal axis is the test statistic (for example, the calculated t-statistic from a t-test comparing the means of two samples). If the value of the test statistic is greater than the critical value, shown as a vertical dot-dashed line, the null hypothesis is rejected and the alternative is accepted. Otherwise, the null hypothesis is accepted.

of the analysis is to assist in decision making for management. It is almost always more useful to have an estimate of the size of an effect and, in conjunction, some idea of the precision with which the estimate has been made (e.g., a 95% confidence interval, or its Bayesian equivalent, a 95% credibility interval). For example, suppose an experiment has been designed and undertaken to investigate the effect of a pathogen on survival of a particular wildlife species (details on how to design such experiments are in Chapter 10) and the mean difference in survival $\bar{x}$ between an infected and a control group is found to be 0.1/year. The standard error of this mean difference $se_{\bar{x}}$ is 0.075 and let us suppose that the degrees of freedom associated with this error are 28. Using a hypothesis testing approach, this information would typically be converted into a t-statistic:

$$t = \frac{\bar{x}}{se_{\bar{x}}} = \frac{0.1}{0.075} = 1.33 \quad (4.8)$$

A t table would show that the probability of getting this t value (ignoring its sign) is 0.194. As this probability is more than 0.05, one would accept the null hypothesis and conclude that there is insufficient evidence to conclude that this particular pathogen had an effect on survival. However, it is important to recognize that this is not evidence that the pathogen has no effect on survival; merely that in this particular scenario there is insufficient evidence to infer that it does affect survival.

Alternatively, the same information can be used to construct a 95% confidence interval for the actual effect of the pathogen on survival based on this experiment. The confidence interval is given by:

$$\bar{x} \pm t_{crit}se_{\bar{x}} = 0.10 \pm 2.05 \times 0.075 = (-0.053, 0.254)$$
$$(4.9)$$

Here, t_{crit} is the critical value of the t-statistic with 28 degrees of freedom and a two-tailed $P = 0.05$. So, while this result is consistent with there being no effect of infection on survival, because the interval includes 0 (and indeed with the possibility that infection might even slightly improve survival), it is also consistent with survival being up to 0.254/year lower in infected individuals on average. The confidence interval thus provides much more potentially useful information on which to base a management decision than the simple hypothesis test. Further, if the objective of the experiment is to use the information in some form of epidemiological model (see Chapter 9), then it is an estimate together with a measure of its error that is required.

4.5 Power analysis

A critical issue in experimental design is to ensure that the experiment has adequate power: there is little point in conducting an experiment if it is unlikely to be able to detect an effect or difference that is important biologically. Power is the ability to detect an effect that does exist. As discussed previously, how big an effect needs to be for it to be of biological importance is a biological, rather than a statistical, question. Some guidance might be provided by sensitivity analysis, which is the process of exploring how the outcomes of a mathematical model depend on changes in its input parameters (see Chapter 9 for

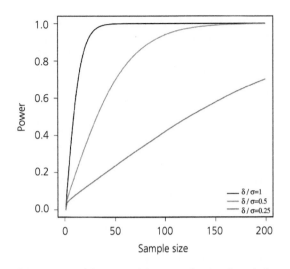

Figure 4.4 Power for a two-tailed *t*-test as a function of sample size within each treatment. Power curves are shown for three different ratios of the minimum difference to be detected δ relative to the within-sample standard deviation σ. Power was calculated using the R function power.t.test.

more details). In the terminology of hypothesis testing, statistical power is the ability to reject the null hypothesis. Since alternative hypotheses are usually expressed as a range of differences (e.g. $\mu_1 \neq \mu_2$ for a simple *t*-test), rather than a specific difference, for numerical calculations of power, it is first necessary to determine the minimum effect size required to be detected. While increased statistical power is desired, it comes at a cost, and there are only a limited number of ways to increase power:

1. Increase the effect size one wants to detect: big effects are easier to detect than small ones. One way of achieving this by comparing groups that are likely to show the largest differences in the response variable.
2. Increase the number of replicates (sample size). The end result of most power analyses is the minimum number of replicates required to detect a certain effect size with a given power. This is specific to the particular statistical analysis and experimental design.
3. Tolerate a higher probability of making a Type I error. All things being equal, there is a direct trade-off between Type I and Type II errors.

4. Decrease the amount of variation in the response between individual experimental subjects receiving the same treatment. This may be achievable by running an experiment with more carefully controlled conditions, or ensuring that subjects are as similar to each other as possible. However, the more tightly controlled the experimental conditions, or the more similar the subjects, the less widely applicable any conclusions one reaches are to the real world. For example, if one is investigating the effects of a pathogen using inbred laboratory mice in laboratory conditions, one may obtain clear statistical results, yet the study may say little about the effect of the pathogen on wild mice in the field.
5. Use an experimental design in which experimental subjects with similar characteristics are blocked into groups and use a mixed model analysis to separate between-block variation from variation between treatments. The traditional way of doing this was with a mixed effects ANOVA (e.g., Sokal and Rohlf 1995), but more recently developed generalized linear mixed models (Bolker et al. 2009; O'Hara 2009) are able to handle a wider range of problems.
6. Use covariates (continuous variables such as size) to extract extraneous variation from within-treatment variation. This is similar in principle to blocking, except that the variables are continuous rather than categorical. The traditional approach was analysis of covariance (ANCOVA), but again generalized linear mixed models can handle a greater range of problems.

For simple experimental designs such as comparing means between two groups (i.e., using a *t*-test), there are relatively simple methods to calculate power, provided the experimenter has some idea of the variability within the groups she or he wishes to compare. Figure 4.4 shows how the power of a *t*-test with equal numbers of replicates per treatment rises with increasing within-treatment sample size. This relationship is shown for three different values of the ratio of the minimum difference between means to be detected δ, and the within-treatment standard deviation *s* (because a *t*-test is simply the difference between two means relative to variability within the samples, this ratio is the appropriate way to scale the effect size). The most obvious conclusion to draw from this figure is that, for a given level of replication and within-sample variability, power decreases dramatically as the minimum difference one wishes to detect decreases. It has become conventional to accept power of 0.8 as adequate (i.e., an 80% chance of detecting a difference given that it exists). This convention should, however, not be adopted uncritically (Di Stefano 2003).

For more complex designs, it will usually be necessary to simulate data with the properties of the experimental data you wish to collect (Johnson et al. 2014). The exercise of simulating data is a good way of thinking clearly about the sources of error and variation in your experiment, and will usually lead to a better experimental design, quite apart from any benefits in calculating power.

References

Amrhein, V., Greenland, S., and McShane, B. (2019). Scientists rise up against statistical significance. *Nature*, **567**:305–07.

Baker, M. (2016). Statisticians issue warning on P values. *Nature*, **531**:151.

Bartoń, K. (2019). MuMIn: Multi-model inference. R package version 1.43.6.

Bolker, B. M. 2008. Ecological Models and Data in R. Princeton University Press, Princeton.

Bolker, B.M., Brooks, M.E., Clark, C.J., Geange, S.W., Poulsen, J.R., Stevens, M.H.H., and White, J.S.S. (2009). Generalized linear mixed models: A practical guide for ecology and evolution. *Trends in Ecology & Evolution*, **24**:127–35.

Burnham, K.P. and Anderson, D.R. (2002). *Model Selection and Multi-model Inference: A Practical Information-theoretic Approach*, 2nd edn. Springer, New York.

Burnham, K.P. and Anderson, D.R. (2004). Multimodel inference: Understanding AIC and BIC in model selection. *Sociological Methods Research*, **33**:261–304.

Crawley, M.J. (2012). *The R Book*. John Wiley & Sons, London.

Di Stefano, J. (2003). How much power is enough? Against the development of an arbitrary convention for statistical power calculations. *Functional Ecology*, **17**:707–09.

Garton, E.O., Ratti, J.T., and Giudice, J.H. (2005). Research and experimental design. In: Braun, C.E. (ed.), *Techniques for Wildlife Investigations and Management*, pp. 43–71. The Wildlife Society, Bethesda, MD.

Jennelle, C.S., Cooch, E.G., Conroy, M.J., and Senar, J.C. (2007). State-specific detection probabilities and disease prevalence. *Ecological Applications*, **17**:154–67.

Johnson, P.C.D., Barry, S.J.E., Ferguson, H.M., and Müller, P. (2014). Power analysis for generalized linear mixed models in ecology and evolution. *Methods in Ecology & Evolution*, **6**:132–42.

O'Hara, R.B. (2009). How to make models add up—A primer on GLMMs. *Annales Zoologici Fennici*, **46**:124–37.

Sokal, R.R. and Rohlf, F.J. (1995). *Biometry*, 3rd edn. W.H. Freeman, New York.

Steidl, R.J. and Thomas, L. (2001). Power analysis and experimental design. In: Scheiner, S.M. and Gurevitch J. (eds.), *Design and Analysis of Ecological Experiments*, 2nd edn, pp.14–36. Oxford University Press, New York.

Sutherland, W.J. 2006. *Ecological Census Techniques: A Handbook*, 2nd edn. Cambridge University Press, Cambridge.

R Core Team (2021). R: A language and environment for statistical computing. R Foundation for Statistical Computing, Vienna, Austria. URL https://www.R-project.org/

RStudio Team (2020). RStudio: Integrated development for R. *RStudio*. PBC, Boston, MA. www.rstudio.com (accessed July 7, 2021).

Thompson, S.K. (2012). *Sampling*, 3rd edn. Wiley & Sons, Hoboken, NJ.

Wasserstein, R.L. and Lazar, N.A. (2016). The ASA's statement on p-values: Context, process, and purpose. *American Statistician*, **70**:129–31.

CHAPTER 5

Capture, Restraint, and Euthanasia of Target Species

The danger and costs to animals should be weighed against the scientific or management benefits to be gained any time that wild animals are handled.

(Williams 1993).

5.1 Introduction

Despite recent advances in remote or noninvasive data collection methods, animal capture remains a fundamental component of wildlife disease study. Modern science, whether through the use of DNA from naturally shed tissue samples, hormone information from feces, or individual identity data from remotely triggered cameras, has simplified data collection, and allows biologists to minimize contact with wildlife. Despite these innovations, the actual physical capture of free-ranging animals provides a wealth of information that other less invasive methods cannot offer—including data on the identity and condition of the host, infection status, and pathogen characteristics. Repeated (longitudinal) samples from the same individuals recaptured through time produce information on disease progression that can be critical to understanding the impact of disease at both the individual and the population level. Catching animals therefore remains the golden standard for wildlife disease studies and obtaining enough individuals is frequently the central challenge in field investigations.

At the most basic level, capture and physical examination of wild animals provide critical information on host age, sex, body size, and condition (Figure 5.1). At the same time, investigators can also determine presence, prevalence, load, and distribution of pathogens and parasites, and DNA samples collected can be used to confirm a pathogen's phylogenetic identity. Biological samples, especially blood for hematology, biochemistry, and immunology measurements, can also shed light on the physiological condition of the host. Additionally, tissue samples can be used for detection of toxins and to identify risk factors associated with disease. In many cases, capture is also required for wildlife management practices including treatment, culling, and relocation.

Animals may be captured for a variety of reasons, and the motivation behind this effort determines not only the number of individuals that need to be caught, but also how, when, and where they should be captured, and whether they should be recaptured. Six basic considerations need to be taken into account in any capture of wild animals for disease sampling:

1. Are captured animals representative of the population?
2. Are biological samples collected representative

Infectious Disease Ecology and Conservation. Johannes Foufopoulos, Gary A. Wobeser and Hamish McCallum, Oxford University Press.
© Johannes Foufopoulos, Gary A. Wobeser and Hamish McCallum (2022). DOI: 10.1093/oso/9780199583508.003.0005

Figure 5.1 Important information about host physiology and condition can only be obtained by capturing animals. Male Aegean wall lizard (*Podarcis erhardii*) being captured by use of a fishing rod and hookless baiting. (Source: J. Foufopoulos.)

of the physiological state of the animal and not unduly influenced by the capture procedure?

3. Will capture affect the study animals' subsequent behavior, activity, or survival?
4. Are sampling frequency and intensity sufficient to accurately represent host population demographics and dynamics of the disease?
5. Will the capture technique endanger the investigators?
6. How can the capturing/sampling process be improved?

5.2 Are captured animals representative of the population?

Sampling is usually done to collect information from captured individuals, which can then be extrapolated to the source population. To ensure that such extrapolations are statistically acceptable, the sampled animals must reflect features of the population, at least for the parameters being studied. This implies that the investigator should have some knowledge of the features of the total population to be certain that the animals sampled are representative of relevant biological, spatial, and temporal variation.

Ideally, the animals chosen should be selected randomly; that is, each animal in the population has an equal opportunity to be captured, and the proportion of various subgroups in the sample is equivalent to the proportion in the wild population (see Chapter 4 for more detail on sampling). However, it is virtually impossible to capture a random sample of wild animals and "sample collection usually represents a compromise associated with availability" (Stallknecht 2007). Whether samples are collected by trapping and other means of capture or through "opportunistic" or "convenience" sampling, such as animals shot by hunters or animals killed on roads, there is always some selection bias. For instance, older animals, especially if captured previously, may be less likely than juveniles to enter traps (Figure 5.2; Davis 2005), and hunters may select for, or against, a particular sex or age of animal. Even mass capture systems are unlikely to result in unbiased samples from a population (Raveling 1966; Sulzbach and Cooke 1978). Selection bias is most serious when the probability of being sampled is related to the parameter being studied. For instance, a sample of road-killed deer would likely result in overestimation of the prevalence of chronic wasting disease (CWD), because deer with clinical CWD are more likely to be killed by automobiles (Krumm et al. 2005). Conversely, the effect of pesticides on birds may be underestimated, because the least affected members of the population (birds with low levels of cholinesterase inhibition) are more likely to survive and be sampled than more severely affected birds in forests sprayed with pesticide (Mineau and Peakall 1987).

A study of feline immunodeficiency virus (FIV) in feral cats (*Felis catus*) illustrates how sampling bias—related to capture—can affect interpretation of disease occurrence in a wild population (Courchamp et al. 2000). It also demonstrates the difficulty in identifying bias without detailed information about the study population. A population of cats was sampled by trapping twice each year for 3 years. About 60% of the population was sampled during each trapping period. Sex ratio, age, and weight distribution were stable in the samples over time, and the sex ratio was similar to that in the population, all of which suggested that the sample was representative. However, the prevalence of FIV in trapped cats declined more than 10-fold (from 33% to 3%) over the 3-year study, without an apparent biological explanation. The study was unusual—among studies of wild animals—in that

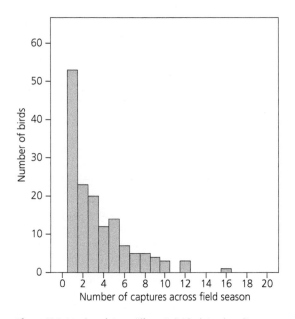

Figure 5.2 Number of times different individuals in a breeding population of free-ranging white-crowned sparrows (*Zonotrichia leucophrys*) were captured over the course of a breeding season. Although trapping effort was held constant over the summer season and across the study plots, frequency of capture varied widely in the resident bird population. Whereas most birds were captured only one ot two times, a few birds were trapped quite frequently, highlighting the differential probability of capture for different individuals and underscoring the difficulty of sampling the population in a representative fashion. (Source: J. Foufopoulos.)

the investigators had intimate knowledge of the population and of individual animals that allowed identification of cats that were present, but not captured. Dominant adult males, which were more likely to be infected with FIV than other members of the population, had a greater probability of being captured initially than more "shy" individuals (inflating the prevalence based on initial capture), but were less likely to be recaptured than other groups (thus leading to a progressive underestimation of FIV prevalence). When the bias was recognized, a new type of trap was used and trapping was intensified, resulting in capture of >90% of the population. Based on this enhanced sampling—"believed to be as free of bias as possible"—it was concluded that the prevalence had actually been stable, at about 14%, during the 3 years (Figure 5.3). This study demonstrated that a large sample size is not, in itself, a guarantee that the sample is representative, and also that examining the

recapture rate of subgroups in the population may aid in detecting bias. If individually marked animals can be recaptured on multiple occasions, modern methods for analyzing mark–recapture data can estimate the probability of recapturing individuals in particular groups (e.g., infected and uninfected animals) known to be present in the population (Cooch and White 2018). This enables capture bias to be detected and potentially corrected for. See Chapter 10 for further details.

The more that is known about the population, in terms of its composition, spatial/temporal distribution, and behavior, the more likely it is that sampling bias will be detected. Information about the population should be collected by methods that are independent of the method used for capturing animals for sample collection. For instance, Hiller et al. (2010) compared the proportion of sex and age classes of white-tailed deer (*Odocoileus virginianus*) captured by trapping with that of the population available for capture (based on a sex–age–kill model). They found that more fawns and fewer adult males were captured than expected. Use of multiple methods of capture (the results of which can be compared) and long-term studies in which the consistency of parameters can be assessed relative to population information will also aid in detecting bias. Knowledge of the ecology of the disease involved is vital for assessing the effect of sampling bias on the interpretation of results. If skewed sampling is detected, more targeted approaches may be required to sample underrepresented groups.

One special case is when the purpose of sampling animals is to determine whether a disease is present within a population or area. If the probability of infection is known to be greater in one particular group than in the remainder of the population, it is sensible to concentrate on that group to maximize the likelihood of detecting an infected individual. Walsh and Miller (2010) developed a weighted surveillance system for detecting cases of CWD in areas where the disease had not been previously detected. In this system, demographic strata of deer that had a higher prevalence of infection and a lower probability of being sampled received more weight in analysis than deer with a lower prevalence of infection but from which samples were more common.

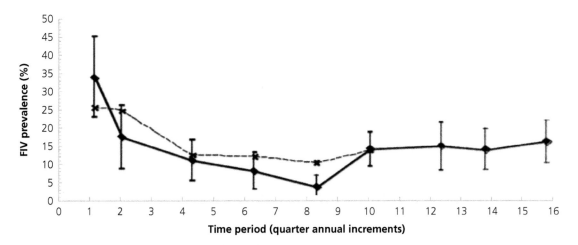

Figure 5.3 Results of a feline immunodeficiency virus (FIV) study in feral cats (*Felis catus*). Shown is prevalence of FIV across quarter-annual increments, together with associated confidence limits (black diamonds and heavy line). Corrected prevalences (represented by crosses and dashed line) were determined after a change to a more representative capture method. This demonstrates how sampling bias during the trapping period can significantly affect the estimation of disease occurrence in the wild. (Figure from Courchamp et al. 2000.)

5.3 Are biological samples collected from study animals representative of the normal physiological state and not unduly influenced by the capture procedure?

"Even the most reliable test, performed in the most reliable facility and interpreted by the most skilled diagnostician, cannot overcome the error introduced by an inappropriate technique used in sample collection or handling."

(Thrall et al. 2004)

A difficulty facing investigators collecting biological samples from captured wild animals is assuring that the samples are free of pre-analytical artifact (PAA) that may interfere with interpretation. PAA has two components:

1. Alterations that occur in the animal prior to sample collection.
2. Alterations that occur in the sample between collection and analysis.

Alterations that occur after the sample is collected can be reduced by prior consultation about sample handling with the laboratory that will do the analysis, followed by attention to detail in how samples

are handled until they reach the laboratory. If multiple people are involved in collecting samples it is important that all participants are trained in proper animal handling and sample collecting methods. Complementing this with detailed written instructions prevents "protocol creep," which occurs in large sampling efforts and over extended periods of time.

Changes that occur in the animal between capture and sample collection are more difficult to assess and correct. Capture, restraint, handling, marking, and release (the usual components of a capture event) are severe stressors that may alter biological samples collected from wild animals. The alterations may compromise the ability to determine "normal" values (the physiological reference interval) for the species and situation, and to detect and assess deviations related to disease.

Effects of capture on individual parameters are species- and situation-specific, but the time interval between capture and sample collection is a variable that must be considered in every circumstance. This time interval is important because a sequence of events, some or all of which may be relevant to the variable measured, occurs once an animal is captured. Emphasis will be placed on two components of the sequence: the release of catecholamines and the release of glucocorticoids (GCs), but other hormones, neurotransmitters, peptides, and cytokines are also involved (Figure 5.4; Sapolsky et al. 2000). The time interval between capture and sampling must be known to those who analyze the samples and interpret the data and should be reported as part of the methods section in descriptions of studies. The time elapsed from capture to sampling can be measured precisely in animals captured by some methods, such as by driving animals into nets or by net gun or immobilizing dart after pursuit from a helicopter; in contrast, it is more difficult to establish the accurate time interval for animals obtained by other passive methods, such as capture by trap or mist-nets.

When an animal is captured, the first event that occurs is the excitement response, resulting in catecholamine (epinephrine and norepinephrine) release from the adrenal medulla. This results in alterations in many physiological parameters, including increased heart rate, blood pressure, blood flow to muscles, numbers of erythrocytes, leukocytes, and concentration of glucose in circulating blood. Because this response is almost instantaneous (within seconds), even "normal" baseline or reference intervals derived from blood samples collected almost immediately after capture from healthy animals, will contain artifactual changes (Le Maho et al. 1992).

Activation of the hypothalamic–pituitary–adrenal axis, with release of a surge of GCs, is the second event in the sequence, resulting in the "emergency life history stage" with "redirection of behavior and physiology towards survival" (Wingfield et al. 1998). Elevation of GCs in plasma occurs within minutes after capture in mammals and birds (Vleck et al. 2000; Romero and Reed 2005; Lynn and Porter 2008; Delehanty and Boonstra 2009). Samples collected less than 2 minutes after capture are thought to reflect baseline ("pre-stress") concentration of GCs, while samples collected 2–3 minutes after capture are near baseline (Romero and Reed 2005).

The concentration of GCs in blood has been measured for many purposes; for example, to determine if an individual animal is stressed, to identify populations vulnerable to, or already

under stress, to assess the importance of natural or anthropogenic stressors (including capture), and to predict survival (Wasser et al. 1997; Wingfield et al. 1997; Wingfield 2005; Romero et al. 2008; Barry et al. 2010; Bian et al. 2011). Busch and Hayward (2009) cautioned "that many factors influence GCs" and to link "GCs with conservation-relevant variables, more basic causes of variation must be controlled for"; the latter including factors like taxon, age, latitude, and habitat. It is critical to know if the GC concentration measured is a true baseline for an undisturbed animal, a response to some chronic stressor, or a stress-induced level in response to the capture event, because "investigator-imposed stressors may swamp other inputs" (Vleck et al. 2000). There is controversy about the use of baseline GC concentrations for assessing the health of individuals and populations, particularly in regard to fitness (Bonier et al. 2009, 2010; Dingemanse et al. 2010). Because baseline and stress-induced concentrations of GCs have "completely different physiological and behavioral effects," both should be measured for drawing conclusions about the relative health of populations (Romero 2004). A standardized capture stress protocol (Wingfield et al. 1997) allows measurement of both parameters.

The degree and rate of change in GC concentration resulting from stressful stimuli vary among species and situations (Wingfield et al. 1995; Romero et al. 1997; Lynn and Porter 2008; Krause et al. 2016) and baseline concentrations of GCs cannot necessarily be inferred from one species to another (McLaren et al. 2007). To interpret the significance of GC concentration, an investigator needs to establish how the species responds to stress in a particular situation. For example, marine iguanas (*Amblyrhynchus cristatus*), an island species that has evolved in a low-predation environment, appear to have a different endocrine stress response than what would be expected from similar, mainland species (Rödl et al. 2007). Baseline levels can also be obtained by measuring the concentration of GCs in samples other than blood, such as the concentration in feces (perhaps collected from within a trap), which reflects the blood concentration several hours prior to capture (Harper and Austad

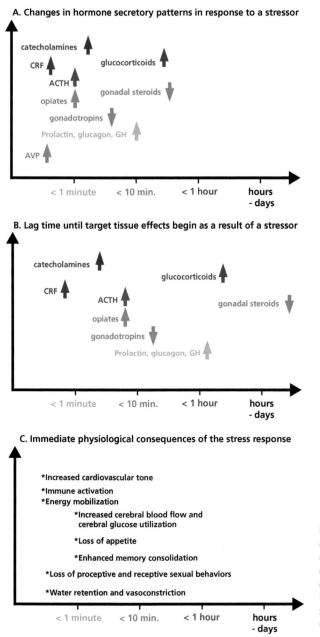

A. Changes in hormone secretory patterns in response to a stressor

catecholamines
CRF
ACTH
opiates
gonadotropins
Prolactin, glucagon, GH
AVP
glucocorticoids
gonadal steroids

< 1 minute   < 10 min.   < 1 hour   hours - days

B. Lag time until target tissue effects begin as a result of a stressor

catecholamines
CRF
ACTH
opiates
gonadotropins
Prolactin, glucagon, GH
glucocorticoids
gonadal steroids

< 1 minute   < 10 min.   < 1 hour   hours - days

C. Immediate physiological consequences of the stress response

*Increased cardiovascular tone
*Immune activation
*Energy mobilization
    *Increased cerebral blood flow and
    cerebral glucose utilization
    *Loss of appetite
    *Enhanced memory consolidation
*Loss of proceptive and receptive sexual behaviors
*Water retention and vasoconstriction

< 1 minute   < 10 min.   < 1 hour   hours - days

Figure 5.4 Capture of an animal is a frequently stressful event that can precipitate a cascade of hormone releases and other physiological changes, which in turn may affect the outcome of a study's measurements. Each of these changes occurs on a distinct time scale that needs to be understood so it can be considered appropriately. These graphs summarize some of these responses as a function of time after capture. (Figure redrawn from Sapolsky et al. 2000.)

2001), or the concentration in hair, which represents conditions during the time the hair was growing (MacBeth et al. 2010; Nejad et al. 2014), but Burnard et al. (2016) cautioned that "there is currently insufficient evidence to conclude that cortisol concentration in hair accurately reflects long-term blood cortisol concentrations." Reliable measurement of stress associated with capture is important, because "if stress can be measured then protocols can be selected to minimize it" (Narayan et al. 2010). A review by Sheriff et al. (2011) provides an overview of available techniques.

The third event in the sequence is a "suite of physiological and behavioral responses" (Wingfield et al. 1998) resulting from the elevation of GCs in plasma. Few of these target tissue effects are exerted until about an hour after the onset of the stressor (Sapolsky et al. 2000), but the actual time interval between the surge of GCs and a measurable response in target tissues is unknown in most species. Hematological and biochemical parameters are measured frequently in wild animals as indices of health, condition, nutrition, immunity, stress, and potential survival, and to assess the effects of capture and handling. Interpretation of analytical results is complicated, because many parameters are altered by an elevation in GCs and are affected by seasonal and life cycle factors (Beldomenico et al. 2008; Franson et al. 2009), and because "the meaningfulness of the changes in particular indices is largely unknown" in many species (Kilgas et al. 2006). To use a parameter as an index to the state of an animal, the investigator ideally needs to know how the time of onset, extent, and duration of capture affect that parameter in healthy undisturbed individuals.

Counts of different types of leukocytes present per unit volume of blood are among the most frequently measured parameters in captured wild animals. The ratio of granulocytes (neutrophils or heterophils, depending on species) to lymphocytes appears to be a useful index of stress in vertebrates (Davis et al. 2008). This ratio varies with the time interval elapsed after an acute stress event, because catecholamines induce increased numbers of both granulocytes and lymphocytes, while GCs induce increased granulocytes and reduced numbers of lymphocytes. In domestic chickens (*Gallus domesticus*), the heterophil/lymphocyte ratio increases about 18 hours after exposure to a stressor, is maximal at about 20 hours, and declines to baseline after 30 hours (Gross 1990). In cynomolgus monkeys (*Macaca fascicularis*) the neutrophil/lymphocyte returned to normal about 1 week after transport (Kim et al. 2005). Unfortunately, this type of data is not available for most wild species. Capture, handling, and the concomitant stress have the potential to confound interpretation of many other immunological parameters. For instance, lymphocyte response to mitogens (a commonly used measure of cell-mediated immunity) is depressed following acute stress in some mammals (Hudson 1973; Griffin 1989; Buddle et al. 1992; Cross et al. 1999), and the intrinsic bacteria-killing ability of blood and plasma is reduced by acute stress in some birds (Matson et al. 2006).

The effects of capture on biological parameters can only be established by sequential sampling at intervals after capture (e.g., Hajduk et al. 1992; Lopez-Olvera et al. 2006). However, interpretation of the serially collected samples may be confounded by the cumulative effects of handling the animals repeatedly for sampling (Hoffmayer and Parsons 2001).

A fourth event that may occur following capture includes hematological and biochemical changes as a result of tissue injury, and alterations in concentration of blood components as a result of dehydration. The hydration state of the animal should be considered in interpretation of biological samples, because blood constituents are measured on a concentration basis. Dehydration in animals confined in traps without access to water may complicate interpretation of data, as well as being an additional stressor for the animal. Alterations in blood acid–base status and leakage of enzymes such as creatine kinase and aspartate aminotransferase from damaged muscle into plasma are indicators of problems in capture technique and can serve as predictors for survival after release (Bollinger et al. 1989; Nicholson et al. 2000; Mandelman and Skomal 2009). Measuring these variables can be an important tool for improving capture methods.

Finally, when dealing with ectoparasites, the parasite burden may itself be affected by the capture of the host. Fleas, for example, may rapidly leave a captured mammal, especially if lethal means of trapping are used (Elzinga 1964). It is very difficult to eliminate all biasing factors, but field investigators should be aware of potential confounding factors and try to keep them the same across all animals. By keeping biases at a similar level across treatment groups, one will hopefully still be able to detect an underlying signal of treatment or disease.

5.4 Will capture affect the study animal's subsequent behavior, activity, or survival?

Studying the effects of capture on wild animals is important for ethical, animal welfare, and scientific reasons. Although emphasis here will be placed on the effect on research, the importance of ethics and animal welfare should not be understated (see also Chapter 16).

An assumption of studies that involve capture and subsequent monitoring of animals is that the capture and marking process has no long-term effect on the animal. Investigators nowadays recognize that capture is not without consequences, and it is common practice to exclude data collected from animals during an initial period following release. The length of this "adjustment" or "conditioning" period is often chosen arbitrarily, without empirical evidence. Holt et al. (2009) reported that adjustment periods ranging from 0 to 14 days had been used in radio-telemetry studies of bobwhite quail (*Colinus virginianus*) and suggested an analytical approach for determining an appropriate adjustment period.

The most serious consequence of capture and handling is that animals die as a result of capture. Because of difficulty in tracking animals after release, it may be challenging to determine the frequency of such deaths. Whenever possible, animals that die after capture should be retrieved for examination by a pathologist to identify the cause of death and to differentiate between "obviously anthropogenic" (Clinchy et al. 2001) and natural mortality. Many such deaths are attributable to capture myopathy (CM), (also called exertional myopathy or exertional rhabdomyolysis), a noninfectious, degenerative disease of skeletal and cardiac muscle (Williams and Thorne 1996). CM is characterized by metabolic acidosis, muscle necrosis, and myoglobinuria, and may be directly fatal, with death occurring within minutes or hours after capture, or days to weeks later. Even in nonfatal forms, it may interfere with mobility, which might bias data collected from captured animals. Although CM has been recognized most commonly in ungulates, macropods, and birds, it can likely occur in

any vertebrate in association with strenuous use of large muscles during a capture event (Vogelnest and Portas 2008). Successful treatment of CM is rare (Rogers et al. 2004; Businga et al. 2007; Paterson 2014), so emphasis must be on preventing the condition (Vogelnest and Portas 2008). Analysis of hematological and biochemical parameters, particularly the activity of enzymes released from damaged muscle, is useful for linking injury to specific components of the capture event (Bollinger et al. 1989; Cattet et al. 2003; Lopez-Olvera et al. 2006), which can then be modified. Arnemo et al. (2006) suggested that wildlife professionals should strive for zero capture-related mortality by: "1) using an experienced professional capture team, 2) developing a capture protocol specific to each species, and 3) requiring a mortality assessment be undertaken after any capture-related death."

The more subtle effects of capture on reproduction and behavior may be even more difficult to measure. Long-term monitoring to detect such effects should be a part of every study, rather than assuming that there is no effect. A study on the effects of capture on bears (*Ursus americanus, U. arctos*) (see Box 5.1) concluded that "failure to recognize and account for long-term effects of capture and handling can potentially confound results leading to erroneous interpretations" (Cattet et al. 2008).

5.5 Are sampling frequency and intensity sufficient to accurately represent host population demographics and the dynamics of the disease?

During the planning stage of a study, it is important to establish how many animals need to be captured to achieve the desired result. If too few are captured, the results may not provide an adequate answer, while capturing more than needed exposes extra animals to needless risk. Methods for determining appropriate sample size for various purposes related to disease, such as determining if a disease is present in a population or measuring prevalence, are available in texts of veterinary epidemiology (Thrushfield 2007; see also Chapter 4).

Box 5.1 Effects of capture on wildlife

Figure 1 Several findings from an analysis of standard types of data originating from two independent studies, one of grizzly bears *Ursus arctos* (pictured) and other of black bears *U. americanus*, point toward the conclusion that failure to recognize and account for long-term effects of capture and handling can potentially confound results leading to erroneous interpretations (Cattet et al. 2008).

Although a need to capture wild animals for scientific or management purposes is well substantiated, the question of long-term effects (effects lasting weeks, months, and even years) from such activities is often overlooked or assumed to be negligible. However, recent long-term wildlife studies challenge this assumption by demonstrating that the effects on individuals may in fact have enduring consequences for populations, and further may introduce biases to research results (Marco et al. 2006; Cattet et al. 2008; McCarthy and Parris 2008; Saraux et al. 2011). In one such investigation (Figure 1), standard types of data originating from two independent studies, one of grizzly bears (*Ursus arctos*) and the other of black bears (*U. americanus*), conducted by different teams of researchers in geographically distinct areas, were evaluated to determine whether long-term effects of capture and handling were detectable and, if so, to ascertain what possible implications this could have for the welfare of released animals and the interpretation of research results (Cattet et al. 2008).

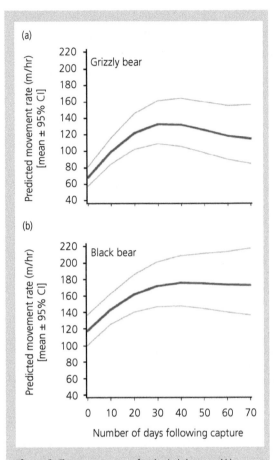

(a) Grizzly bear

(b) Black bear

Number of days following capture

Figure 2 The movement rates for a) grizzly bears and b) American black bears captured for two independent research projects were decreased below normal immediately after capture (grizzly bears: 57% of normal; black bears: 77%) and then returned to approximately normal rates in 3 to 6 weeks (Cattet et al. 2008).

The daily movement rates for radiocollared grizzly bears and radiocollared black bears were analyzed to determine if capture and handling affected their mobility during a 100-day period after capture. In both species, movement rates decreased below mean normal rate immediately after capture (grizzly bears: 57% of normal; black bears: 77%) and then returned to normal in 3–6 weeks (Figure 2). Reduced mobility may, in turn, adversely affect other biological parameters such as tissue maintenance and growth, by lessening food intake, or reproduction, by lessening encounter rates with potential mates. Further, from the standpoint of research results, descriptions of activity patterns or determination of home ranges may be

---

Box 5.1 *Continued*

inaccurate if time elapsed after capture is not considered as a potential factor in analysis of movement rates or locations.

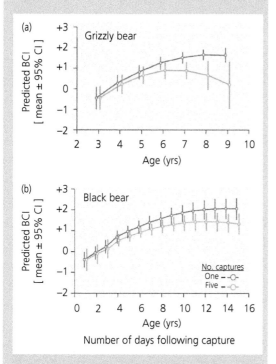

Figure 3 The relationship between the body condition index (BCI) of a) grizzly bears and b) black bears and their age as a function of number of times they were captured (once or 5 times) over the course of their lifetime (Cattet et al. 2008). Because this capture effect is directly proportional to number of times captured, one can interpolate that curves for capture levels from 2 to 4 fall between predicted curves shown in the figure and curves for capture levels > 5 fall below the curve for 'captured 5 times.'

The effect of repeated captures on age-related changes in body condition of grizzly bears and black bears was also examined. In both species, the age-specific body condition of bears captured two or more times tended to be poorer than that of bears captured once only, with the magnitude of effect directly proportional to number of times captured (Figure 3). Importantly, the condition of bears did not affect their probability of capture or recapture. The biological implications of these findings are basic: as body condition fades, so too does an animal's potential for growth, reproduction, and survival.

Overall, these findings should compel investigators engaged in wildlife capture to monitor the effects of their procedures on study animals carefully. Significant capture-related effects may go undetected, providing a false sense of the welfare of released animals. Further, failure to recognize and account for long-term effects of capture and handling on research results can potentially lead to erroneous interpretations.

M.cattet

Canadian Cooperative Wildlife Health Centre, Department of Veterinary Pathology, University of Saskatchewan, Saskatoon, Saskatchewan, Canada

Questions of sample size are particularly pertinent for studies of conditions caused by macroparasites, since these parasites and their effects are often concentrated in a small proportion of the host population. Other special cases include diseases that are endemic at a low prevalence and diseases with a high case fatality rate in which very few survivors are present in the population.

Animals need to be sampled in the right place and at the right time, which depends upon the ecology of the disease. Many parameters related to disease vary strongly across small spatial scales, making careful selection of sampling locations paramount (Schall and Marghoob 1995). Similarly, parasite prevalence and intensity may differ temporally over the course of a year, a season, or even a day, and this must be considered in designing a capture program (Misof 2004; Weatherhead and Bennett 1991). For instance, the number of some protozoa in peripheral blood varies diurnally (Roller and Desser 1973), so that capturing birds at the wrong time of day results in distorted estimates of frequency and intensity of infection. Similarly, intestinal *Isospora* parasites are shed in avian feces only during afternoon hours, perhaps as a mechanism to reduce duration of exposure to the sterilizing effects of the midday sun (Martinaud et al. 2009); morning sampling hence leads to severe underestimation of the true rates of parasitism (Figure 5.5). On a larger scale, prevalence of malarial parasites in migrating birds increases later in the season as infection appears to be associated with slower migration (Valkiunas 2005).

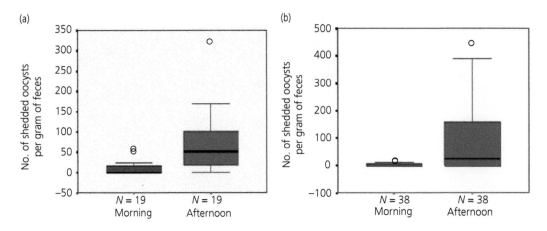

Figure 5.5 Time of capture can influence significantly the results of parasitological measurements and needs to be considered in wildlife disease investigations. The average number of shedded *Isospora* sp. oocyst per gram of feces from adult (a) and nestling (b) Eurasian blackbirds (*Turdus merula*) follows a strong diurnal cycle. Box-and-whisker plots exhibit the median and 25th and 75th percentiles. (Figure from Misof 2004.)

Disease agents may be present (or detectable) in or on hosts only during a limited time period, for example, fleas affect reproductive success of lesser snow and Ross's geese (*Anser caerulescens, A. rossi*) in the Canadian arctic (Harriman et al. 2008; Harriman and Alisauskas 2010), yet fleas are only present on the birds for a short period of time during nesting, because they spend most of their life cycle in the nest rather than on the bird.

To understand the dynamics of a pathogen–host relationship, it is important to monitor both demographic features of the host population (abundance, survival, recruitment, dispersal) and the frequency of infection over time. See Chapter 10 for information on how to estimate these parameters. The frequency with which samples need to be collected from the host population to measure demographic impacts of infection may differ from the frequency required for measuring characteristics of the disease agent. In a study of hantavirus infection in deer mice (*Peromyscus maniculatus*), Carver et al. (2010) found that reducing the frequency of sampling resulted in underestimation of host abundance and reduced detection of annual peaks of host abundance; it also led to overestimation of the prevalence of infection when prevalence was high.

5.6 Will the capture technique endanger the investigator?

Researchers involved in capture/handling of wild animals are exposed to a variety of health risks including:

1. Injury associated with use of equipment, including aircraft, boats, snowmobiles, all-terrain vehicles, traps, firearms, and anesthetic/immobilizing drugs.
2. Injury by captured animals. While the risk of bite wounds by carnivores is obvious, other species, such as white-tailed deer (*O. virginianus*), are not usually considered a "dangerous animal," except by those who have received lightning-fast strikes by a deer's forefeet.
3. Exposure to an infectious agent carried by the animal.

When large animals are being restrained and handled, or when techniques such as the use of low-flying helicopters are employed, the potential for physical injury is particularly severe. When drugs are used for immobilization, tranquilization, or anesthesia there exists risk of accidental exposure. Only personnel with specific training in safe drug usage should be allowed to handle drugs and associated equipment, but all members of the capture team must be trained in emergency procedures in the event of accidental human exposure, since the person handling the drug is the one most likely to be exposed. All persons in close contact with captured wild animals should be aware of the potential presence of disease agents transmissible to humans and must be trained in how to prevent exposure. Generally, anyone working with wild animals should be familiar with the zoonotic agents that are known or are likely to occur in the species. For instance, persons working with mammals, particularly carnivores or bats, must be aware of the importance of these host species for rabies virus transmission, and special measures, including prophylactic immunization, must be taken where appropriate. It is important to avoid contact with saliva and to reduce the chance of being bitten—all bite wounds must be treated as a medical emergency that may require postexposure treatment for rabies. Indeed exposure to rabies virus via bite wounds from carnivores or bats is generally recognized as a grave risk by biologists. However, other routes of exposure, such as inhalation of hantavirus from rodent droppings or chlamydia from birds, oral exposure to eggs of helminths, such as *Echinococcus* spp. or *Baylisascaris* spp. from carnivore feces, infection of skin wounds by bacteria, for example, *Francisella tularensis* while handling infected rodents, and transfer of vector-transmitted pathogens by ectoparasites present on the captured animal, are less well-known risks. In general, however, the closer related the target wildlife species is to humans, the higher the probability that cross-infection can happen. As a result, work with primates and mammals poses the highest risk of infection, whereas amphibians and fish represent relatively lower-risk study organisms.

Investigators must also be conscious of the possibility that the animals that they capture may be infected with a "new" agent that has never been reported in the species, or has not been recognized to be infectious to humans, or is new to science. Most recently, the emergence of novel coronaviruses (severe acute respiratory syndrome (SARS), SARS-CoV-2) in humans, which can be traced at least partially back to Asian *Rhinolophus* bats, points to the risks of working with

Chiroptera. Discovery of human disease with high mortality, caused by a previously unknown hantavirus infecting deer mice (*P. maniculatus*) (Nichol et al. 1993), was a startling reminder that little is known about the range of disease agents present in most wild species, even those that have been the subject of many biological studies. Following the discovery of hantavirus pulmonary syndrome associated with deer mice, many other rodent-borne viruses with the potential for human transmission have been identified (Childs et al. 1995). Guidelines developed for working with rodents (Mills et al. 1995), including appropriate personal protection equipment (Figure 5.6), provide a model that can be applied in many situations. These should be reviewed prior to contact with animals, modified as appropriate for disease agents that might be encountered in the particular species, and then applied whenever animals are handled.

Risks to the investigators may be further exacerbated if aid in the event of injury or infection is delayed, because of lack of reporting or recording incidents, or when working in remote or inaccessible locations. Investigators who become ill after contact with wild species should seek medical assistance. It is vital that the attending medical staff be made aware of the possibility of a zoonotic disease, "because the diagnosis may not be considered by physicians without some prompting from the patient" (Wobeser and Brand 1982).

Prior to capture and handling animals, there should be a thorough review of potential risks (including consultation with an infectious disease specialist), followed by creation of a standard set of operating guidelines, training of personnel regarding risks, use of personal protective equipment, and at least initial confirmation that said procedures are indeed being followed in the field. Appropriate protective equipment will depend upon the situation, but in most cases should include protective gloves and clothing, eye protection, a surgical mask or respirator, and equipment for first aid treatment of wounds. In some situations, such as rabies, protective immunization of staff may be required. While the use of protective gear, such as gloves or facemasks, serves primarily to protect research investigators, it has also the added benefit

Figure 5.6 In Bangladesh, Dr Jonathan Epstein and a colleague feed a bat mango juice before it is released. Note the personal protective equipment donned by both individuals to prevent infectious disease transmission. Image reproduced from the website gallery for the PBS special documentary *Spillover: Zika, Ebola & Beyond*.(http://www.pbs.org/spillover-zika-ebola-beyond/image-gallery, accessed 13 May 2021.)

of preventing captured wildlife from being exposed to potentially harmful human pathogens.

5.7 How can the capturing/sampling process be improved?

When contemplating a capture program, it is important to assess whether the intended result can be attained by other, less invasive, means. If capture is required for a management purpose, such as translocation of animals, the goal must be to reduce the consequences for captured animals as much as possible. If animals are captured for scientific purposes, such as to mark them for movement studies or to collect biological samples, the same goal applies, in addition to the need to ensure that scientific data collected from the animals are as free of sampling artifact as possible.

Because little information is available on the effectiveness and consequences of various capture techniques for many wild species, investigators must often begin by using methods developed for a similar species. For instance, information on the effect of various capture methods on mallards (Bollinger et al. 1989; Dabbert and Powell 1993) might provide a starting point for developing capture methods for other waterfowl. Regardless of what technique is used to capture wild animals, the method should be subject to continual evaluation and improvement.

This evaluation needs to focus both on the efficacy and efficiency of obtaining samples and meeting the study objectives, as well as reducing the effects on animals, the environment, and the investigator. Long-term monitoring could, for example, include measuring indices of acute stress and injury (such as GC concentrations and other hematological and biochemical parameters; Figure 5.7), as well as long-term effects on animal behavior, reproduction, and survival. This information can then be used to modify capturing and sampling protocols.

To assess effects of the capture process, animals need to be followed in some manner after release. Several authors have suggested that a mortality rate of >2% in capture programs needs careful reevaluation (Spraker 1993; Arnemo et al. 2006; Ponjoaan et al. 2008). General recommendations to prevent CM include that: (i) operators recognize the environmental limitations posed by adverse weather conditions and difficult terrain when planning capture; (ii) pursuit, struggling, and handling times are minimized, and these activities are performed by experienced personnel; (iii) a protocol specific to the species and situation is developed and followed; (iv) drugs used for immobilization produce rapid induction, recovery, and physiological stability; and (v) the animal is released into less stressful conditions as quickly as possible (Williams and Thorne 1996; Arnemo et al. 2006; Paterson 2014).

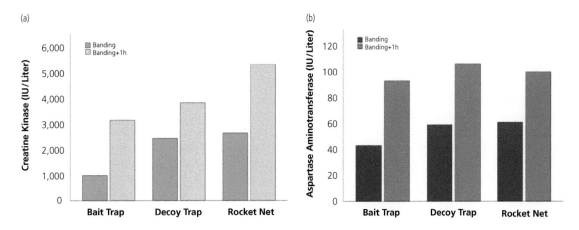

Figure 5.7 Concentrations of (a) creatine kinase (CK) and (b) aspartate aminotransferase (AST) in the blood serum of wild adult male mallards (*Anas platyrhynchos*) captured during a banding study. Both CK and AST concentrations—which are indicative of capture stress—vary depending on bird capture method. Shown are levels of these enzymes at banding time, as well as 1 hour after. This type of information can be used to improve capture and sampling protocols. (Figure based on data from Bollinger et al. 1989.)

5.8 Euthanasia (humane killing)

On occasion, it may be necessary to kill wild animals either because of the type of information/tissue that needs to be collected, or because they were injured during the capturing process. Euthanasia is the term used to describe the act of providing a humane death; the goal is that the "animal must be rendered irreversibly unconscious as rapidly as possible with the least possible pain, fear, and anxiety. The preferred methods used to achieve this are those that affect the brain first, followed quickly by cessation of cardiac and respiratory function" (Canadian Veterinary Medical Association 2006). Although the need to kill animals may seem unlikely at the onset of a study, experience shows that euthanasia eventually becomes necessary once large enough numbers of animals are handled. As a result, a rapidly increasing number of countries are requiring provisions for euthanasia for wildlife studies and institutions are now establishing animal care and use committees to oversee this process. The first step in preparing for this eventuality is to become thoroughly familiar with accepted methods for euthanizing the species (or taxon) being captured. A publication by the American Association of Zoo Veterinarians, *Guidelines for Euthanasia of Nondomestic Animals* (Baer 2006), provides an excellent starting point for interested researchers.

It also may be important to consider other species that can be captured by mistake. These other species ("bycatch") may be more likely than the target animals to be injured, since the capture system has not been optimized to avoid their injury. Methods to be used for euthanasia should be described as part of the study or management protocol and approved by the appropriate animal care and use committee. Before fieldwork begins, there should be clear guidelines developed that describe the circumstances in which euthanasia is required, as well as who is responsible for euthanizing the animals.

Individuals responsible for euthanasia must be fully trained with the equipment and technique that will be used, and the required equipment and drugs must be readily available and accessible in the field. Euthanasia is highly stressful for all individuals involved, including the animal, the person responsible, and others that are present. This is not the time to be trying to recall where a suitable vein can be found, or to be carrying out one's first intravenous injection, or to be trying to remember how a gun is loaded. In these sensitive situations, the "experience, training, sensitivity, and compassion of the individual carrying out the procedure are critical" (Canadian Veterinary Medical Association 2006).

Methods used for euthanasia are, by design, dangerous to life, so the safety of all individuals present must be the primary concern. All members of a capture team must be trained to deal with emergencies, particularly if drugs are used for euthanasia, since the person responsible for euthanasia is the individual most likely to be injured or accidentally exposed to the drugs used to kill the animal. Therefore, another member of the team must always be well equipped to deal with that situation if deemed necessary.

There are other concerns when an animal must be euthanized. Whenever possible, maximum benefit should be derived from the information and tissues obtained from the animal. For instance, if biological samples are to be collected from the animal after death, the effect of the euthanasia method on the integrity of the samples should be also considered—although this should be secondary to the need to ensure that the method is humane. If the animal is killed in a remote location, the carcass must be disposed of in a manner that precludes secondary poisoning of other animals by agents (including lead bullets) used to kill the animal.

5.9 Capture considerations for various categories of vertebrate hosts

Because ecological preferences and life histories patterns vary so much between different groups of wildlife, very different methodologies are needed for their capture and study. Although much of the relevant detail goes beyond the scope of this book and can be found in specialized literature (e.g., Sutherland 2006), some general considerations relevant for the capture of each major group of terrestrial vertebrates are given in the next few sections.

5.9.1 Amphibians

Although temperate amphibians tend to be closely tied to water, and the group as a whole is clearly dependent on moisture, amphibian habitat use is surprisingly varied. Amphibians can be found in habitats ranging from rainforest canopy to sandy deserts, and appropriate capture methods vary accordingly. Aquatic taxa like frogs and some salamanders can often be captured using nets or seines. Many terrestrial species are best acquired through focused searches of appropriate habitats and by turning logs, stones, and other hiding places. Especially in more arid regions, amphibians tend to be active at night, so nocturnal searches using spotlights can be particularly effective. Capture success in many frog species can be further enhanced by focusing on chorusing aggregations when animals congregate to particular breeding sites that can be heard from afar. At times, the bulk of an otherwise widely dispersed population can be encountered for a period of a few nights around specific breeding ponds, making population studies (and parasite transmission) easier. Individuals migrating to breeding sites can be intercepted through the strategic use of drift fences and pitfall traps (Heyer et al. 1994). The various capturing methods used for amphibians are reviewed in a number of standard references including Heyer et al. (1994), Ferner (2007), and Dodd (2010).

5.9.2 Reptiles

Reptiles constitute another diverse vertebrate group with varied ecologies that inhabit a wide spectrum of—mostly warm—habitats. A correspondingly broad range of active search and passive capturing methods have been developed over the years for their study (Ferner 2007; McDiarmid et al. 2011). Many species are caught most effectively through opportunistic searches of their hiding places (refugia) in preferred habitats. Because this targeted search approach is most successful if the investigator has a detailed understanding of the habitat preferences and the ecology of the particular species, it is often essential to obtain local information on the species' habits before engaging in these types of searches. During field collection it is very important to replace all investigated hiding places (such as rocks, logs) in exactly the same position as before. Displacing them can alter their thermal properties thus making them unsuitable for further use by wildlife. Many species of lizards can be captured using nooses, whereas climbing geckos can be often shot down from their perches using size-appropriate rubber bands. For lower-density or

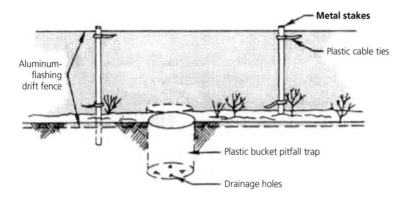

Figure 5.8 Pitfall trap deployed in combination with a drift fence to increase capture rates. (Figure from Gibbons and Semlitsch 1981.)

wide-ranging species, it is sometimes most effective to set up arrays of pitfall traps, which, in conjunction with drift fences, can result in substantial capture rates (Figure 5.8).

This method can also work well for snakes, particularly smaller-bodied ones, although larger species still need to be located through targeted habitat searches. Lastly, when working with certain snakes or other potentially dangerous taxa it is important to use appropriate tools and protective clothing such as snake hooks, snake tubes, gloves, boots, etc., and to practice safe handling techniques ahead of time, to avoid envenomation.

5.9.3 Birds

In contrast to reptiles and amphibians, birds are very active, fast-moving organisms, and capturing them almost always involves the use of nets or traps. From a practical perspective, epidemiological studies often focus on small- to medium-sized (e.g., passerine) birds, which are typically captured with mist-nets. Use of these fine-meshed nets requires specialized training that can often be obtained by volunteer work at bird-banding stations. Mist-nets are most effective when used in good weather with low wind speeds, and when carefully placed within the existing vegetation to maximize concealment and to take advantage of bird flight corridors. Alternatively, other types of traps, varying in size and shape, are used to attract and capture different species. Handling captured birds and obtaining blood samples, especially under adverse field conditions, is a nontrivial task that

Figure 5.9 Blood sample collection from white-crowned sparrow (*Zonotrichia leucophrys oriantha*). Obtaining samples from small wildlife requires a substantial amount of physical dexterity and care. (Source: J. Foufopoulos.)

requires manual dexterity and needs to be practiced ahead of time (Figure 5.9). While detailed descriptions are beyond the scope of this work, much additional information can be found in the specialized literature (e.g., Rose et al. 2006).

5.9.4 Mammals

General guidelines for the use of wild mammals in research (Sikes et al. 2016) provide a foundation, which can be augmented by additional information related to particular species. Because of their close phylogenetic proximity to humans, and the elevated risk of cross-species pathogen transmission, mammals constitute the most important source of zoonotic diseases to humans. Consequently, mammals are a frequent object of public health investigations, but they also require diligent use of prophylactic measures to prevent investigator exposure. Mammal disease studies often focus on two types of hosts: small-bodied ground dwellers like rodents, and bats, as both of these groups harbor diverse and epidemiologically important pathogen communities. For the first group, the most important capturing method involves the use of live traps (Sherman or similar) deployed in capture grids (Wolff and Sherman 2008; Ryan 2011). Pitfall traps can also be very effective (Umetsu et al. 2006). Capture success can be low and is influenced by careful selection of trapping locale, ambient conditions, positioning of the individual traps, and use of the appropriate bait (Wilson et al. 1996). Bats are often captured using harp traps or mist-nets, and both methods require substantial investigator training to avoid bites, scratches, or other exposure. Specialized trapping methods ranging from the use of ungulate traps to cannon-nets and tranquilizing darts are available for many large-bodied mammals (Silvy 2012). Capturing large mammals also entails additional special animal welfare considerations, such as prevention of capture myopathy (see section 5.4).

Captured mammals, whether lions or mice, represent both a significant physical injury and infection risk that requires careful investigator preparation and training. Depending on the host and pathogen studied, gloves and even face masks may be required. For many mammal studies, especially when hosts are expected to carry rabies (e.g., bats, small carnivores), it may be necessary for investigators to receive the full pre-exposure series of rabies vaccinations. Other vaccines may be required depending on the mammal species studied.

References

Arnemo, J.M., Ahlqvist, P., Andersen, R., Berntsen, F., Ericsson, G., Odden, J., et al. (2006). Risk of capture-related mortality in large free-ranging mammals: Experiences from Scandinavia. *Wildlife Biology*, **12**:109–13.

Baer, C.K. (2006). *Guidelines for Euthanasia of Nondomestic Animals*. American Association of Zoo Veterinarians (AAZV), Lawrence, KS.

Barry, M., Cockrem, J.F., and Brunton, D.H. (2010). Seasonal variation in plasma corticosterone concentrations in wild and captive adult Duvaucel's geckos (*Hoplodactylus duvaucelii*) in New Zealand. *Australian Journal of Zoology*, **58**:234–42.

Beldomonico, P.M., Telfer, S., Gebert, S., Lukomski, L., Bennett M. and Begon, M. (2008). The dynamics of health in wild field vole populations: A haematological perspective. *Journal of Animal Ecology*, **77**:984–97.

Bian, J., Wu, Y., Getz, L.L., Cao, Y.-F., Chen, F., and Yang, L. (2011). Does maternal stress influence winter survival of offspring in root voles *Microtus oeconomus*? A field experiment. *Oikos*, **120**:47–56.

Bollinger, T., Wobeser G., Clark, R.G., Nieman, D.J., and Smith, J.R. (1989). Concentration of creatine kinase and aspartate aminotransferase in the blood of wild mallards following capture by three methods for banding. *Journal of Wildlife Disease*, **25**:225–31.

Bonier, F., Martin, P.R., Moore, I.T., and Wingfield, J.C. (2009). Do baseline glucocorticoids predict fitness? *Trends in Ecology & Evolution*, **24**:634–42.

Bonier, F., Martin, P.R., Moore, I.T., and Wingfield, J.C. (2010). Clarifying the cort-fitness hypothesis: A response to Dingemanse et al. *Trends in Ecology & Evolution*, **25**:262–63.

Breed, D., L. C. R. Meyer, J. C. A. Steyl, A. Goddard, R. Burroughs, and T. A. Kohn. 2019. Conserving wildlife in a changing world: Understanding capture myopathy—a malignant outcome of stress during capture and translocation. *Conservation Physiology* 7. 10.1093/conphys/coz027

Buddle, B.M., Aldwell, F.E., Jowett, G., Thomson, A., Jackson, R., and Paterson, B.M. (1992). Influence of stress of capture on haematological values and cellular immune responses in the Australian brushtail possum (*Trichosurus vulpecula*). *New Zealand Veterinary Journal*, **40**:155–59.

Burnard, C., Ralph, C., Hynd, P., Hocking Edwards, J., and Tilbrook, A. (2016). Hair cortisol and its potential value as a physiological measure of stress response in human and non-human animals. *Animal Production Science*, **57**:401–14.

Busch, D.S. and Hayward, L.S. (2009). Stress in a conservation context: A discussion of glucocorticoid actions and how levels change with conservation-relevant variables. *Biological Conservation*, **142**:2844–53.

Businga, N.K., Langenberg, J., and Carlson. L. (2007). Successful treatment of capture myopathy in three wild Greater Sandhill cranes (*Grus canadensis tabida). Journal of Avian Medicine and Surgery*, **21**:294–98.

Canadian Veterinary Medical Association. (2006). Position statement, euthanasia. http://canadianveterinarians.net?ShowText.aspx?ResourceID=34 (accessed August 1, 2018).

Carver, S., Mils, J.N., Kuenzi, A., Flietstra, T., and Douglass, R. (2010). Sampling frequency differentially influences interpretation of zoonotic pathogen and host dynamics: Sin Nombre virus and deer mice. *Vector-Borne and Zoonotic Diseases*, **10**:575–83.

Cattet, M., Boulanger, J., Stenhouse, G., Powell, R.A., and Reynolds-Hogland, M.J. (2008). An evaluation of long-term capture effects in ursids: Implications for wildlife welfare and research. *Journal of Mammalogy*, **89**:973–90.

Cattet, M.R.L., Christison, K., Caulkett, N.A., and Stenhouse, G.B. (2003). Physiologic responses of grizzly bears to different methods of capture. *Journal of Wildlife Diseases*, **39**:649–54.

Childs, J.E., Mills, J.N., and Glass, G.E. (1995). Rodent-borne hemorrhagic fever viruses: A special risk for mammalogists? *Journal of Mammalogy*, **76**:664–80.

Clinchy, M., Krebs, C.J., and Jarman, P.J. (2001). Dispersal sinks and handling effects: Interpreting the role of immigration in common brushtail possum populations. *Journal of Animal Ecology*, **70**:515–26.

Cooch, E. and White, G. (2018). Using MARK—A gentle introduction, 17th edn. http://www.phidot.org/software/mark/docs/book (accessed August 1, 2018).

Courchamp, F., Say, L., and D. Pontier (2000). Detection, identification, and correction of bias in an epidemiological study. *Journal of Wildlife Diseases*, **36**:71–78.

Cross, M.L., Swale, E., Young, G., and Mackintosh, C. (1999). Effect of field capture on the measurement of cellular immune responses in wild ferrets (*Mustela furo*), vectors of bovine tuberculosis in New Zealand. *Veterinary Research*, **30**:401–10.

Dabbert, C.B. and Powell, K.C. (1993). Serum enzymes as indicators of capture myopathy in mallards (*Anas platyrhynchos). Journal of Wildlife Disease*,**29**:304–09.

Davis, A.K. (2005). A comparison of age, size, and health of house finches captured with two trapping methods. *Journal of Field Ornithology*, **76**:339–44.

Davis, A.K., Maney, D.L., and Maerz, J.C. (2008) The use of leukocyte profiles to measure stress in vertebrates: A review for ecologists. *Functional Ecology*, **22**:760–72.

Delehanty, B. and Boonstra, R. (2009). Impact of live trapping on stress profiles of Richardson's ground squirrel (*Spermophilus richardsonii). General and Comparative Endocrinology*, **160**:176–82.

Dingemanse, N.J., Edelaar, P., and Kempenaers, B. (2010). Why is there variation in baseline glucocorticoid levels? *Trends in Ecology & Evolution*, **25**:261–62.

Dodd C.K. (2010). *Amphibian Ecology and Conservation: A Handbook of Techniques*. Oxford University Press, Oxford.

Elzinga, R.J. (1964). The importance of the Berlese technique in studying ectoparasite populations upon rodent hosts. *Journal of the Kansas Entomological Society*, **37**:52–56.

Ferner, J.W. (2007). A review of marking and individual recognition techniques for amphibians and reptiles. In *Herpetological Circular*, 35. Society for the Study of Amphibians and Reptiles, Atlanta, GA.

Franson, J.C., Hoffman, D.J., and Schmutz, J.A. (2009). Plasma biochemistry values in emperor geese (*Chen canagica*) in Alaska: Comparisons among age, sex, incubation, and molt. *Journal of Zoo and Wildlife Medicine*, **40**:321–27.

Gibbons, J.W. and Semlitsch, R.D. (1981). Terrestrial drift fences with pitfall traps: An effective technique for quantitative sampling of animal populations. *Brimleyana*, **7**:1–16.

Griffin, J.F.T. (1989). Stress and immunity: A unifying concept. *Veterinary Immunology and Immunopathology*, **20**:263–312.

Gross, W.B. (1990). Effect of exposure to a short-duration sound on the stress response of chickens. *Avian Diseases*, **34**:759–61.

Hajduk, P., Copland, M.D., and Schultz, D.A. (1992). Effects of capture on hematological values and plasma cortisol levels of free-range koalas (*Phascolarctos cinerus*). *Journal of Wildlife Diseases*, **28**:502–06.

Harper, J.M. and Austad, A.N. (2001). Effects of capture and season on fecal glucocorticoid levels in deer mice (*Peromyscus maniculatus*) and red-backed voles (*Clethrionomys gapperi). General and Comparative Endocrinology*, **123**:337–44.

Harriman, V.B. and Alisauskas, R.T. (2010). Of fleas and geese: The impact of an increasing nest ectoparasite on reproductive success. *Journal of Avian Biology*, **41**:573–79.

Harriman, V.B., Alisauskas, R.T., and Wobeser, G.A. (2008). The case of the blood-covered egg: Ectoparasite abundance in an arctic goose colony. *Canadian Journal of Zoology*, **86**:883–92.

Heyer, W.R., Donnelly, M.A., McDiarmid, R.W., Hayek, L.A.C., and Foster, M.S. (1994). *Measuring and Monitoring Biological Diversity: Standard Methods for Amphibians.* Smithsonian Institution Press, Washington, DC.

Hiller, T.L., Burroughs, J.P., III, Cama, H., Cosgrove, M.K., Rudolph, B.A., and Tyre, A.I. (2010). Age-sex selectivity and correlates of capture for winter-trapped white-tailed deer. *Journal of Wildlife Management*, **74**:564–72.

Hoffmayer, E.R. and Parsons, G.R. (2001). The physiological response to capture and handling stress in the Atlantic sharpnose shark, Rhizopriodon terraenovae. *Fish Physiology and Biochemistry*, **25**:277–85.

Holt, R.D., Burger, L.W. Jr., Dinsmore, S.J., Smith, M.D., Szukaitis, S.J., and Godwin, K.D. (2009). Estimating duration of short-term acute effects of capture handling and radiomarking. *Journal of Wildlife Management*, **73**:989–95.

Hudson, R.J. (1973). Stress and in vitro lymphocyte stimulation by phytohemagglutinin in Rocky Mountain bighorn sheep. *Canadian Journal of Zoology*, **51**:479–82.

Kilgas, P., Tilgar, V., and Mänd, R. (2006). Hematological health state indices predict local survival in a small passerine bird, the great tit (*Parus major*). *Physiological and Biochemical Zoology*, **79**:565–72.

Kim, C.Y., Han, J.S., Suzuki, T., and Han, S.S. (2005). Indirect indicator of transport stress in hematological values in newly acquired cynomolgus monkeys. *Journal of Medical Primatology*, **34**:188–92.

Krause, J.S., Perez, J.H., Chmura, H.E., Meddle, S.L., Hunt, K.E., Gough, L., et al. (2016). The stress response is attenuated during inclement weather in parental, but not in pre-parental Lapland longspurs (*Calcarius lapponicus*) breeding in the Low Arctic. *Hormones and Behaviour*, **83**:68–74.

Krumm, C.E., Connor, M.M., and Miller, M.W. (2005). Relative vulnerability of chronic wasting disease infected mule deer to vehicle collisions. *Journal of Wildlife Diseases*, **41**:503–11.

Le Maho, Y., Karmann, H., Briot, D., Handrich, Y., Robin, J.P., Mioskowski, E., et al. (1992). Stress in birds due to routine handling and a technique to avoid it. American Journal of Physiology, **263** (4, Pt 2):R775–81.

López-Olvera, J.R., Marco, I., Montané, J., and Lavín, S. (2006). Transport stress in southern chamois (*Rupicapra pyrenaica*) and its modulation by acepromazine. *Veterinary Journal*, **172**:347–55.

Lynn, S.E. and Porter, A.J. (2008). Trapping initiates stress response in breeding and non-breeding house sparrows *Passer domesticus*: Implications for using unmonitored traps in field studies. *Journal of Avian Biology*, **39**:87–94.

Macbeth, B.J., Cattet, M.R.L., Stenhouse, G.B., Gibeau, M.L., and Janz, D.M. (2010). Hair cortisol concentration as a noninvasive measure of long-term stress in free-ranging grizzly bears (*Ursus arctos*): Considerations with implications for other wildlife. *Canadian Journal of Zoology*, **88**:935–49.

Mandelman, J.W. and Skomal, G.B. (2009). Differential sensitivity to capture stress assessed by blood acid-base status in five carcharhinid sharks. *Journal of Comparative Physiology*, **179**:267–77.

Marco, I., Mentaberre, G., Ponjoan, A., Bota, G., Manosa, S., and Lavin, S. (2006). Capture myopathy in little bustards after trapping and marking. *Journal of Wildlife Diseases*, **42**:889–91.

Martinaud, G., Billaudelle, M., and Moreau, J. (2009). Circadian variation in shedding of the oocysts of *Isospora turdi* (Apicomplexa) in blackbirds (*Turdus merula*): An adaptive trait against desiccation and ultraviolet radiation. *International Journal for Parasitology*, **39**:735–39.

Matson, K.D., Tieleman, B.I., and Klasing, K.C. (2006). Capture stress and the bactericidal competence of blood and plasma in five species of tropical birds. *Physiological and Biochemical Zoology*, **79**:556–64.

McCarthy, M.A. and Parris, K.M. (2008). Optimal marking of threatened species to balance benefits of information with impacts of marking. *Conservation Biology*, **22**:1506–12.

McDiarmid, R.W., Foster, M. S., Guyer, C., Gibbons, J. W., and Chernoff, N. (eds.). (2011). *Reptile Biodiversity: Standard Methods for Inventory and Monitoring.* University of California Press, Oakland, CA.

McLaren, G., Bonacic, C., and Rowan, A. (2007). Animal welfare and conservation: Measuring stress in the wild. In: Macdonald, D.W and Service, K. (eds.), *Key Topics in Conservation Biology*, pp. 120–33. Blackwell Publishing, Oxford.

Mills, J.N., Yates, T.L., Childs, J.E., Parmenter, R.R., Ksiazek, T.G., Rollin, P.E., et al. (1995). Guidelines for working with rodents potentially infected with hantavirus. *Journal of Mammalogy*, **76**:716–22.

Mineau, P. and Peakall, D.B. (1987). An evaluation of avian impact assessment techniques following broad-scale forest insecticide sprays. *Environmental Toxicology and Chemistry*, **6**:781–91.

Misof, K. (2004). Diurnal cycle of *Isospora* spp. oocyst shedding in Eurasian blackbirds (*Turdus merula*). *Canadian Journal of Zoology*, **82**:764–68.

Narayan, E., Molinia, F., Christi, K., Morley, C., and Cockrem, J. (2010). Urinary corticosterone metabolic responses to capture, and annual patterns of urinary corticosterone concentration in wild and captive endangered Fijian ground frogs (*Platymantis vitiana*). *Australian Journal of Zoology*, **58**:189–97.

Nejad, J.G., Lohakare, J.D., Son, J.K., Kwon, E.G., West, J.W., and Sung, K.I. (2014). Wool cortisol is a better indicator of stress than blood cortisol in ewes exposed to heat stress and water restriction. *Animal*, **8**: 128–32.

Nichol, S.T., Spiropoulou, C.F., Morzunov, S., Rollin, P.E., Ksiazek, T.G., Feldmann, H., et al. (1993). Genetic identification of a hantavirus associated with an outbreak of acute respiratory illness. *Science*, **262**:914–17.

Nicholson, D.S., Lochmiller, R.L., Stewart, M.D., Masters, R.E., and Leslie Jr., D.M. (2000). Risk factors associated with capture-related deaths in eastern wild turkey hens. *Journal of Wildlife Diseases*, **36**:308–15.

Paterson, J. (2014). Capture myopathy. In: West, G., Heard, D., and Caulkett, N. (eds.), *Zoo Animal and Wildlife Immobilization and Anesthesia*, 2nd edn, pp. 171–79. John Wiley & Sons Inc., NY.

Ponjoan, A., Bota, G., Garcia de la Morena, E.L., Morales, M.B., Wolff, A., Marco, I., et al. (2008). Adverse effects of capture and handling little bustard. *Journal of Wildlife Management*, **72**:315–19.

Raveling, D.G. (1966). Factors affecting age ratios of samples of Canada geese caught with cannon nets. *Journal of Wildlife Management*, **30**:682–91.

Rödl, T., Berger, S., Romero, L.M., and Wikelski, M. (2007). Tameness and stress physiology in a predator-naive island species confronted with novel predation threat. *Proceedings of the Royal Society of London B: Biological Sciences*, **274**:577–82.

Rogers, D.I., Battley, P.F., Sparrow, J., Koolhaus, A., and Hassell, C.J. (2004). Treatment of capture myopathy in shorebirds: A successful trial in northwestern Australia. *Journal of Field Ornithology*, **75**:157–64.

Roller, N.F. and Desser, S.S. (1973). Diurnal periodicity in peripheral parasitemias in ducklings (*Anas boschas*) infected with *Leucocytozoon simondi* Mathis and Leger. *Canadian Journal of Zoology*, **51**:1–9.

Romero, L.M. (2004). Physiological stress in ecology: Lessons from biomedical research. *Trends in Ecology & Evolution*, **19**:249–55.

Romero, L.M., Meister, C.J., Cyr, N.E., Kenagy, G.J., and Wingfield, J.C. (2008). Seasonal glucocorticoid responses to capture in wild free-living mammals. *American Journal of Physiology. Regulatory, Integrative and Comparative Physiology*, **294**:R614–22.

Romero, L.M. and Reed, J.M. (2005). Collecting baseline corticosterone samples in the field: Is under 3 min good enough? *Comparative Biochemistry and Physiology A*, **140**:73–79.

Romero, L.M., Ramenosky, M., and Wingfield, J. (1997). Season and migration alters the corticosterone response to capture and handling in an arctic migrant, the white-crowned sparrow (*Zonotrichia leucophrys gambeli*). *Comparative Biochemistry and Physiology*, **116**: 171–77.

Rose, K., Newman, S., Uhart, M., and Lubroth, J. (2006). Wild Bird HPAI Surveillance: Sample collection from healthy, sick and dead birds. FAO animal production and health manual. No. 4. FAO, Rome.

Ryan, J. (2011) *Mammalogy Techniques Manual*, 2nd edn. Lulu Press, Morrisville, NC.

Sapolsky, R.M., Romero, M.L., and Munck, A.U. (2000). How do glucocorticoids influence stress responses? Integrating permissive, suppressive, stimulatory, and preparative actions. *Endocrine Reviews*, **21**: 55–89.

Saraux, C., Le Bohec, C., Durant, J.M., Viblanc, V.A., Gauthier-Clerc, M., Beaune, D., et al. (2011). Reliability of flipper-banded penguins as indicators of climate change. *Nature*, **469**:203–08.

Schall, J.J. and Marghoob, A.B. (1995). Prevalence of a malarial parasite over time and space: *Plasmodium mexicanum* in its vertebrate host, the western fence lizard *Sceloporus occidentalis*. *Journal of Animal Ecology*, **64**: 177–85.

Sheriff, M.J., Danzer, B., Delahanty, B., Palme, R., and Boonstra, R. (2011). Measuring stress in wildlife: Techniques for quantifying glucocorticoids. *Oecologia* **166**:869–87.

Sikes, R.S. and the Animal Care and Use Committee of the American Society of Mammalogists. (2016). 2016 guidelines of the American Society of Mammalogists for the use of wild mammals in research and education. *Journal of Mammalogy*, **97**:663–88.

Silvy, N.J. (ed.). (2012). *The Wildlife Techniques Manual*. 2 vols. Johns Hopkins University Press, Baltimore, MD.

Spraker, T.R. (1993). Stress and capture myopathy in artiodactylids. In: Fowler, M.E. (ed.), *Zoo and Wild Animal Medicine Current Therapy*, 3rd edn, pp. 481–88, W.B. Saunders, Philadelphia, PA.

Stallknecht, D.E. (2007). Impediments to wildlife disease surveillance, research, and diagnostics. *Current Topics in Microbiology and Immunology*, **315**:445–61.

Sulzbach, D. and Cooke, F. (1978). Elements of non-randomness in mass-captured samples of snow geese. *Journal of Wildlife Management*, **42**:437–41.

Sutherland, W.J. (2006). *Ecological Census Techniques: A Handbook*, 2nd edn. Cambridge University Press, Cambridge.

Thrall, M.A., Baker, D.C., Campbell, T.W., Nicola, D., Fettman, M.J., Lassen, E.D., et al. (2004). *Veterinary Hematology and Clinical Chemistry*. Lippincott Williams & Wilkins, Philadelphia, PA.

Thrushfield, M.V. (2007). *Veterinary Epidemiology*, 3rd edn. Blackwell Publishing, Oxford.

Umetsu, F., Naxara, L., and Pardini, R. (2006). Evaluating the efficiency of pitfall traps for sampling small mammals in the neotropics. *Journal of Mammalogy*, **87**: 757–65.

Valkiunas, G. (2005). *Avian Malaria Parasites and Other Haemosporidia*. CRC Press, Boca Raton, FL.

Vleck, C.M., Vertalino, N., Vleck, D., and Bucher, T.L. (2000). Stress, corticosterone, and heterophil to lymphocyte ratios in free-living Adélie penguins. *The Condor*, **102**:392–400.

Vogelnest, L. and Portas, T. (2008). Macropods. In: Vogelnest, L. and Woods, R. (eds.), *Medicine of Australian Mammals*, pp. 133–225. CSIRO Publishing, Collingwood, Victoria.

Walsh, D.P. and Miller, M.W. (2010). A weighted surveillance approach for detecting chronic wasting disease foci. *Journal of Wildlife Diseases*, **46**:118–35.

Wasser, S. K., Bevis, K., King, G., and Hanson, E. (1997). Noninvasive physiological measures of disturbance in the northern spotted owl. *Conservation Biology*, **1**: 1019–22.

Weatherhead, P.J. and Bennett G.F. (1991). Ecology of redwinged blackbird parasitism by haematozoa. *Canadian Journal of Zoology*, **69**:2352–59.

Williams, E.S. (1993). Humane considerations in immobilization and study of free-ranging wildlife. In: Fowler, M.E. (ed.), *Zoo and Wildlife Medicine Current Therapy*, 3rd edn, pp 64–67,W.B. Saunders Company, Philadelphia, PA.

Williams, E.S. and Thorne, E.T. (1996). Exertional myopathy (capture myopathy). In: Fairbrother, A., Locke, L.N., and Hoff, G.L. (eds.), *Noninfectious Diseases of Wildlife*, 2nd edn, pp. 181–93. Iowa State University Press, Ames, IA.

Wilson, D.E., Cole, F.R., Nichols, J.D., Rudran, R., and Foster, M.S. (1996). *Measuring and Monitoring Biological Diversity: Standard Methods for Mammals*. Smithsonian Institution Press, Washington, DC.

Wingfield, J.C. (2005). The concept of allostasis: Coping with a capricious environment. *Journal of Mammalogy*, **86**:248–54.

Wingfield, J.C., Hunt, K., Breuner, C., Dunlap, K., Fowler, G.S., Freed, L., et al. (1997). Environmental stress, field endocrinology, and conservation biology. In: Clemons, J.R. and Buchholz, R. (eds.), *Behavioural Approaches to Conservation in the Wild*, pp. 95–131. Cambridge University Press, Cambridge.

Wingfield, J.C., Maney, D. L., Breuner, C.W., Jacobs, J.D., Lynn, S., Ramenofsky, M., et al. (1998). Ecological basis of hormone-behavior interactions: The "emergency life history stage." *American Zoologist*, **38**: 191–206.

Wingfield, J.C., O'Reilly, K.M., and Astheimer, L.B. (1995). Modulation of the adrenocortical responses to acute stress in arctic birds: A possible ecological basis. *American Zoologist*, **35**:285–94.

Wobeser, G. and Brand C.J. (1982). Chlamydiosis in 2 biologists investigating disease occurrences in wild waterfowl. *Wildlife Society Bulletin*, **10**:170–72.

Wolff, J.O. and Sherman, P.W. (eds.). (2008). *Rodent Societies: An Ecological and Evolutionary Perspective*. University of Chicago Press, Chicago, IL.

CHAPTER 6

Disease and Agent Detection

> Thousands upon thousands of persons have studied disease. Almost no one has studied health.
>
> **Adelle Davis**

6.1 Introduction

Infectious disease did not receive much attention in conservation biology until recently, although even Aldo Leopold (1939) had noted that the role of disease in wildlife conservation had been badly underestimated. There are many reasons for the delayed recognition of the importance of infectious disease (see also Chapter 1, Section 1.4). Two of the most important are, first, the difficulty of detecting infection and disease in wild animals, and, second, establishing a causal relationship between exposure to an agent and an outcome related to health.

Investigation of infection and disease consists of a series of steps that may be initiated in different ways, depending on the situation. The steps include:

1. Detection of infectious agents in an individual or population of host animals.
2. Detection of disease in an individual or population of host animals.
3. Confirmation of a cause–effect relationship between infection and disease in an individual or population of host animals. In some cases, disease may have been recognized before a cause is known; in others, an infection may have been recognized before its effect is known.

6.2 Detecting infectious agents in individuals

Detection of infection in an individual animal would appear to be a relatively simple matter, but, unfortunately, there is no single or universal method for identifying individuals infected with various disease agents. Specific methods are required to identify individuals infected with different agents, so the choice of how to search (i.e., what techniques to use) must be based on a priori knowledge of the type of agent that is sought. For instance, if the type of agent being sought includes intestinal worms and coccidia, examination of feces for eggs or oocysts may be adequate for detecting infection. If blood parasites are of interest, examination of blood smears by optical microscopy or by molecular means is probably appropriate. Infection by microparasites, such as viruses or bacteria, may be determined by directly examining excretions, secretions, blood, or tissues for the agent by various means, but often infection or exposure to these types of agents is detected indirectly, by testing for the presence of antibodies or by some other response to the specific agent. In practice, all detection methods have shortcomings; for example, in the case of avian malaria in Hawaiian honeycreepers, both optical microscopy and molecular approaches miss

Infectious Disease Ecology and Conservation. Johannes Foufopoulos, Gary A. Wobeser and Hamish McCallum, Oxford University Press.
© Johannes Foufopoulos, Gary A. Wobeser and Hamish McCallum (2022). DOI: 10.1093/oso/9780199583508.003.0006

Table 6.1 The relationship between actual infection state and test results.

	Infection present	Infection absent

Test status: positive	**True positive**
	
	$\text{Sensitivity} = \dfrac{\text{True positives}}{\text{True positives} + \text{False negatives}}$
		False positive
		
		Reasons:
		  Antibodies from mother
		  Individual was infected in the past but has now recovered
		  Cross-reactivity with antibodies produced to infection with another pathogen
		  Cross contamination of samples

Test status: negative	**False negative**
	
	Reasons:
	  Unsuitable test
	  Inappropriate timing (too early or too late)
	  Test blocked or inhibited
	  Test insufficiently sensitive
		True negative
		
		$\text{Specificity} = \dfrac{\text{True negatives}}{\text{True negatives} + \text{False positives}}$

detecting infections, although serological methods have been more sensitive than either polymerase chain reaction (PCR)-based approaches or optical microscopy (Jarvi et al. 2002). Typically, testing is done on the basis of either a particular suspicion for a certain pathogen that has been implicated in outbreaks in the past, or by using the nature and the location of symptoms to determine the type of test used. For example, if skin ulcers are present, a search for the causative agents can be conducted by preparing smears of the exudate for microscopic examination.

Discussion of analytical methods (diagnostic tests) for detecting various agents is beyond the scope of this volume and can be found in a variety of standard veterinary textbooks. Most methods require collaboration with specialists; however, anyone concerned with interpretation of results of diagnostic tests should be familiar with common features of the tests that are used, including validity, accuracy, reliability, sensitivity, specificity, and predictive value (for discussion of these features see Dohoo et al. (2003) or Thrusfield (2007)). Each of these features is specific to the situation in which the test is used (Šimundič 2009).

Detection of an agent directly in tissues, excretions, or secretions of an individual animal is good evidence of infection. Therefore, a "true positive"

test result reflects the presence of an actual infection, while a "true negative" test result indicates absence of infection. Failure to detect the agent, although the animal is infected (a "false negative" result), might result from using an unsuitable test or a test with insufficient sensitivity, or because of improper timing of testing (e.g., an animal recently infected with an intestinal worm that has not yet begun shedding eggs into its feces). Table 6.1 summarizes the relationships between actual infection and test results, and also outlines two important concepts mentioned earlier, test sensitivity and test specificity (Fawcett 2006).

Some tests, such as enzyme-linked immunosorbent assays (ELISAs), rely on detecting specific antibodies produced by the host to battle a pathogen infection, as opposed to directly detecting the actual pathogen (see also Chapter 8, Section 8.3). However, interpreting antibody presence is more complex than interpreting the simple presence or absence of an agent. For instance, detection of antibody to an agent could mean that: (i) the animal is presently infected; (ii) it was infected previously and has now recovered; (iii) the animal received antibodies passively from its mother (e.g., through milk); or (iv) there is cross-reaction with some other, related, infectious agent that has triggered production of similar antigens. The latter two are termed "false positive" reactions, because the animal was never infected with the pathogen of interest. Conversely, if no antibodies are detected, this could mean that the animal has indeed never been exposed to the agent (a "true negative"), or that the animal was previously exposed and the test returned a "false negative" result. Reasons for false negatives include: (i) unsuitability of the test (e.g., using a serum neutralization test for an agent that does not cause production of neutralizing antibody); (ii) the test was not sufficiently sensitive; (iii) the test was applied at an inappropriate time (e.g., the animal was exposed recently and the appropriate immune response has not yet developed; (iv) the animal was exposed at some time previously but the immune response has waned and is now no longer detectable; (v) the test was blocked (e.g., by non-specific inhibitors or blocking antibodies); or, lastly, (vi) the animal was exposed or infected but for some reason did not mount a lasting response.

6.3 Detecting infectious agents at the population level

Simply detecting an infection in an individual does not necessarily mean that the infecting agent is of epidemiological or conservation importance: pathogens, for example, will occasionally "spill" from one species to another, without necessarily achieving sustained transmission in their new host (Wolfe et al. 2007). Understanding the epidemiological significance of a parasite or pathogen requires scaling up detection to the population level. Assessing infection within a population, however, has an added level of complexity related to the choice of which, and how many, individuals should be sampled. Such sampling is generally done for one of two reasons: to determine if an infection is, or is not, present in the population, or to determine the prevalence and distribution of infection within the population. It is likely impossible in most situations to obtain a truly random sample of wild animals (i.e., a sample that is chosen directly from the population by a method in which every individual in the population has an equal opportunity of being sampled; see Chapter 4, Section 4.2). Most samples from wild populations are biased in some way, often through selection bias produced by

the method used to capture the animals. Infection and disease are seldom distributed randomly or uniformly within a population. This is most obvious in the case of macroparasites, which characteristically are overdispersed within the population, so that most individuals are uninfected or lightly infected, and a few individuals are heavily infected (Shaw et al. 1998; see also Chapter 2, Section 2.1, Figure 2.2), but infections of all types often are concentrated in a particular age or sex group. In the initial search for an infection within a population, it is appropriate to use some form of stratified sampling (see Chapter 4) to ensure that the sample contains representatives from each subgroup, and that the proportion of each subgroup within the total sample is similar to the proportion of that subgroup within the population. This initial sampling may provide an estimate of prevalence and distribution that can be used later in determining appropriate sample size. Subsequent sampling can then be directed at (intentionally biased toward) a particular subgroup, depending upon the reason for sampling.

The appropriate sample size is determined by the research objective, the level of precision desired, the confidence one wishes to have in the conclusion, and the size of the population from which the sample is drawn. In wild populations, because of the difficulty and cost of obtaining samples, the concern is usually to identify the minimum number of animals that will yield acceptable information. General methods for choosing the minimum sample size for estimating prevalence of infection in a population have been published (DiGiacomo and Koepsell 1986; Thrushfield 2007) and relevant software is available through several sources (e.g., Win Episcope 2.0; see Thrusfield et al. 2001). As already outlined in Chapter 4, a good rule of thumb for the sample size needed to estimate any proportion between ca. 30% and 70% (and prevalence being of course the proportion of individuals infected) is that the sampling error is ca. $\pm 1/\sqrt{n}$, where n is the sample size. So if you want to estimate prevalence to $\pm 10\%$, you will need to sample 100 individuals, but if you want to improve the precision to $\pm 1\%$, you would need to increase the sample size to 10,000.

Determining whether or not an infection is present within a population constitutes a special case of the earlier question. It is particularly important in conservation biology in relation to plans to translocate or introduce animals, and in the latter stages of disease control programs. Finding a single infected animal proves the existence of the infection in the population, but failure to find an infected individual *does not* prove that the infection is not present, unless every individual in the population is examined using a test that is infallible (sensitivity = 100%). On the basis of sampling, one can only be confident, to a defined level of certainty, that the prevalence of infection is not greater than a specified level in the population.

An excellent rule of thumb in this situation is the "rule of three" (Hanley and Lippman-Hand 1983). If you have sampled a reasonably large number of individuals *and* that sample is a random sample of the entire population *and* specificity and sensitivity are 100% (these are crucial provisos!), then if you have not detected infection, the maximum plausible prevalence in the overall population is $3/n$, where n is the sample size. Strictly, by "maximum plausible prevalence" we mean the upper 95% confidence bound for the actual prevalence in the population as a whole. How well this approximation works for different sample sizes is shown in Figure 6.1. The "rule of three" can also be used to calculate the minimum sample size n necessary to detect infection in a population with 95% confidence, given the actual prevalence in the population p. This is simply $n > 3/p$. These rules of thumb assume that the population is large compared with the sample size. More precise minimum sample sizes for finite populations are available in standard epidemiological texts (e.g., Thrushfield 2007). Table 6.2 summarizes these rules of thumb for determining sample sizes to estimate prevalence.

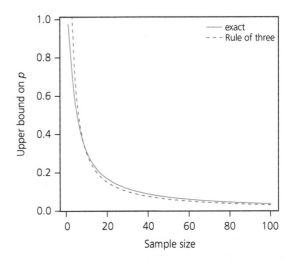

Figure 6.1 Upper 95% confidence intervals for the prevalence p of an infection in a population, given that it has not been found in a sample of n individuals tested. Therefore, the larger the sample size n that failed to document a single infection, the lower the possible prevalence of the pathogen in the population, should it be present. The solid green line shows the upper confidence limit calculated from a binomial distribution and the dashed orange line shows the result obtained by using the "rule of three" approximation.

Table 6.2 Rules of thumb for determining sample sizes to estimate prevalence. For each of the rules below, n is the sample size and p is the actual prevalence in the population as a whole. These rules apply only if the population is substantially greater than the sample size n. Each rule also assumes that both sensitivity and specificity are 100% and that samples are genuine random samples of the population in question.

Quantity	Rule	Limitations
95% confidence interval for p as a function of n, based on an estimate $\hat{p} = x/n$ (x is the number of positive individuals sampled)	$p \approx \hat{p} \pm \dfrac{1}{\sqrt{n}}$	Smaller of x and $n - x > 10$
95% confidence limit for the upper bound of p, given zero observed cases in a sample of n individuals	$p_{\max} \approx \dfrac{3}{n}$	$n > 10$
Minimum sample size necessary for 95% confidence of detecting at least one infected case, given p	$n > \dfrac{3}{p}$	$p < 0.3$

6.4 Detecting disease at the individual or population level

As stated earlier, disease refers to a condition or dysfunction that can be observed or measured. The significance of a disease depends both upon its effect on the individual and its frequency of occurrence in the population—it therefore may be necessary to detect disease at both levels. For conservation purposes, it is important to understand if disease affects fitness (i.e., the likelihood of survival or reproduction of individuals). Diseases differ greatly in the ease with which they can be detected, and often it is easier to detect infection than disease. Two infectious diseases of native birds in Hawaii provide an example of this disparity. The decline of many species of endemic Hawaiian birds has been noted since the late 1800s (Warner 1968) and, while the cause of the decline is multifactorial (Van Riper et al. 1986), there is little doubt that the spread of exotic pathogens following the introduction in the 1800s of the mosquito vector *Culex quinquefasciatus* has been a major factor (Scott et al. 2001; see also Chapter 3, Section 3.5). The presence of disease caused by infection with pox virus was identified very early "because large numbers of diseased and badly debilitated birds" were observed (Warner 1968, p106) and because the skin lesions caused by the virus are conspicuous and distinctive (Figure 6.2). In contrast, "there is no direct historical evidence of the effects of avian malaria" (Warner 1968, p106), likely because infection with *Plasmodium falciparum* does not result in conspicuous changes. For instance, during 2 years of intensive study, Van Riper et al. (1986) captured and examined 2,365 wild birds of which 7.8% (ca. 185 birds) were infected with *Plasmodium*, but only 10 obviously ill birds were recognized.

A critical first step in detecting a disease is for the investigator to define what they are looking for as clearly as possible, so that it is possible to distinguish animals with the particular condition from "normal" animals, and from animals with similar, but unrelated, conditions. This is termed "developing a case definition." The case definition may include various types of information, such as who is involved (e.g., sex and age of affected animals), when and where the condition occurs, as well as clinical and pathological features of affected

Figure 6.2 Avian pox lesion on Hawaii 'amakihi (*Hemignathus virens*). Infection with this mosquito-vectored virus can cause tumor-like growths on unfeathered parts of the avian body. Introduction of this virus is thought to have contributed to the decline of the Hawaiian avifauna. (Source: Carter Atkinson, United States Geological Survey.)

individuals. The objective when developing a case definition is to establish criteria that allow cases of the disease to be distinguished from noncases. Developing a case definition requires knowledge of what is normal, since abnormal can only be defined in terms of the degree to which an observed situation differs from what is expected. However, defining normal when working with wild animals is often difficult. For instance, it would be reasonable, in most situations, to assume that death of a large proportion of a population over a short period of time is an abnormal situation. But in some species of the marsupial genera *Antechinus*, *Phascogale*, *Dasyurus*, and *Parantechinus*, all of the males die during their first mating season (Dickman and Braithwaite 1992; Humphries and Stevens 2001). In these species, mass mortality of males is normal; survival of males during this period would be abnormal. Normal is often derived from measurements of "those who do not stand out from their local contemporaries", with the presumption that what "is common is all right" (Rose 1985, p32). This may be appropriate for conditions that occur in only a small proportion of the population; however, if a disease occurs at high prevalence in a population, it may be difficult to find individuals that are not affected. For example, >89% of blue tits (*Cyanistes caeruleus*) studied by Tomás et al. (2007) were infected by blood parasites. In such a situation, it will be difficult to establish "normal" values (values not influenced by infection) without examining a very large sample to find enough of the rare individuals that are not infected, or without treating animals to remove infection. In any case, substantial effort may be needed to accrue the sufficiently large sample necessary to establish an adequate normal or reference range. In human medicine, Bishop et al. (2005) recommend testing at least 120 individuals in each age and sex category to establish reliable reference intervals (i.e., normal values) for laboratory tests (Bishop et al. 2005). Reference information of this type is seldom available for wild animals.

Historically, interest in wildlife health has centered on conditions in which animals died or were visibly harmed, that is, epidemic pathogens causing high levels of mortality. However, the effect of a disease that is highly pathogenic but occurs very rarely may be trivial compared with the population effect of a disease that is much less pathogenic but affects a large proportion of the population.

When an agent is so pathogenic that many animals die over a short period of time, and their carcasses are available for examination by pathologists, it may be obvious that disease is occurring and the cause may be apparent. For instance, epidemics of plague (*Yersinia pestis* infection) in prairie dogs (*Cynomys* spp.) are "so devastating that their effects are dramatic and obvious" (Biggins et al. 2010, p24) (Figure 6.3). In diseases where sick or dead individuals may be available, these should be submitted to a diagnostic pathology laboratory, where pathologists "are in a strategic position to direct additional diagnostic testing, based on their experience of the likely etiological significance of lesions" (O'Toole 2010, p41). But the effect of even very pathogenic

agents may not be obvious if the effect is dispersed in time and space (Romain et al. 2013; Salkeld et al. 2016). While identifying plague mortality in a prairie dog colony is relatively easy during the epidemic phase, this is often preceded by a very long "smoldering" phase where the pathogen is present but difficult to detect (Figure 6.3a). The fact that even during this hidden endemic phase, plague still takes a toll on both prairie dogs, as well as on black-footed ferrets (*Mustela nigripes*), their specialized, critically endangered predators, was not evident to researchers until infection had been prevented by vector control or vaccination (Biggins et al. 2010; Matchett et al. 2010).

Very few infectious agents in wild animals are as pathogenic as *Y. pestis*, and most do not cause epidemic mortality. Most infectious agents are sublethal, endemic, persistent, and have relatively small fluctuations in prevalence (Anderson and May 1979). However, sublethal effects can have greater regulatory effects than mortality, even though the impairment that the agents cause may be subtle (Dobson and Hudson 1992; Foufopoulos 1998) and difficult to identify or quantify in free-living populations. Often the effects are only evident when the causative agent has been removed or neutralized experimentally, or when both the agent and some other factor, such as food supply, are manipulated simultaneously (Pedersen and Grieves 2008).

The discussion of sampling in regard to detection of infection is equally applicable to the detection of disease at the population level. However, while infection can often be identified by a single sample taken at one time, detection of disease may require long-term observation and repeated sampling to measure subtle effects that are dispersed over time. Alternatively, researchers have used specialized assays to evaluate the effects of infection on individual performance. For example, in *Sceloporus* lizards, malaria-caused anemia precipitated strong declines in running performance as determined by locomotor challenges (Schall 1982). In songbirds, territory intrusion challenges in conjunction with behavioral assays demonstrated that different species of hemosporidian parasites affect male song in distinct ways (Gilman et al. 2007). More remarkably, even simple activation of the immune system, without the presence of parasites, is enough to dramatically undermine male song production (Munoz et al. 2010).

6.5 Confirming a cause–effect relationship between infection and disease

6.5.1 Basic principles and methodological challenges

There are many examples of disease in wild species for which a cause remains unidentified, and there are also long lists of infectious agents for which effects at the individual animal and population level are undetermined (also termed "orphan agents in search of a disease"). Establishing a causal relationship between an infectious agent and a disease is critically important but potentially complicated, because the effects of the agent are often dependent on the effects of other factors with which they interact in complex ways. Among the obvious interactions are those with other infectious agents (Ezenwa and Jolles 2015), host competitors and predators (Hoyer et al. 2017), nutrition (Beldomenico et al. 2008, 2010), habitat area effects (Foufopoulos et al. 2016; Resasco et al. 2019), climate change (Burge et al. 2014; Liao et al. 2015), and various environmental contaminants (Van Loveren et al. 2000). The effect of infectious agents is often indirect rather than direct, further complicating assessment of their role. For instance, when a disease agent increases the vulnerability of a host to predation (e.g., Murray et al. 1997), the proximate and visible outcome is caused by the predator, while the indirect effect of an underlying infection is hidden. Furthermore, the methods used to assess the effect of infection may not be appropriate or sufficiently sensitive to detect outcomes: when Bennett et al. (1993) reviewed the impact of blood parasites on free-ranging avian populations, they found that "the matter of whether the birds survived or died" was the criterion generally used to measure impact, meaning that important sublethal

effects such as impairment of reproduction likely went unrecognized. Even lethal effects may not be detected because of difficulty in finding carcasses: dying animals often hide in inaccessible places and/or their carcasses are quickly removed by predators (Wobeser and Wobeser 1992; Bennett et al. 1993).

"Cause" has been defined in many ways. The definition used here is: "any factor that produces a change in the severity or frequency of the outcome" (Dohoo, et al. 2003). Very few factors are sufficient in and of themselves to be "the" cause of a disease, but a group of factors (sometimes called "component causes") may interact to become a sufficient cause. Some authors separate biological causes (those that operate within an individual animal) from population causes (those that operate at or beyond the level of the individual) (Dohoo et al. 2003). For example, a bacterium might be the direct biological cause of disease in an individual animal, while a number of other causative factors ("determinants") may influence whether or not individuals are exposed or are susceptible to the bacterium. Some determinants may be necessary (a factor without which the disease cannot occur) and others may be sufficient (disease inevitably occurs if the factor is present). For instance, *Mycobacterium tuberculosis* is a necessary cause of human tuberculosis, but it is not a sufficient cause of the disease, because only 5–10% of infected individuals are likely to develop symptomatic tuberculosis (Harries and Dye 2006). The early stages of investigation of a disease often consist of description of the temporal, spatial, clinical, and demographic features of the disease, or of the infectious agents found within a group of animals. Descriptive studies are useful for defining the type of outcome (disease) that is occurring, and for identifying potential causes (e.g., a listing of parasites present in a population), but they are inadequate for inferring a cause–effect relationship. Guidelines have been developed to assist in making judgments about causation. The most widely known of these are Koch's postulates, which were developed more than a century ago for determining if an infection is the cause of a disease (Koch 1884; Löffler 1884; see Box 6. 1).

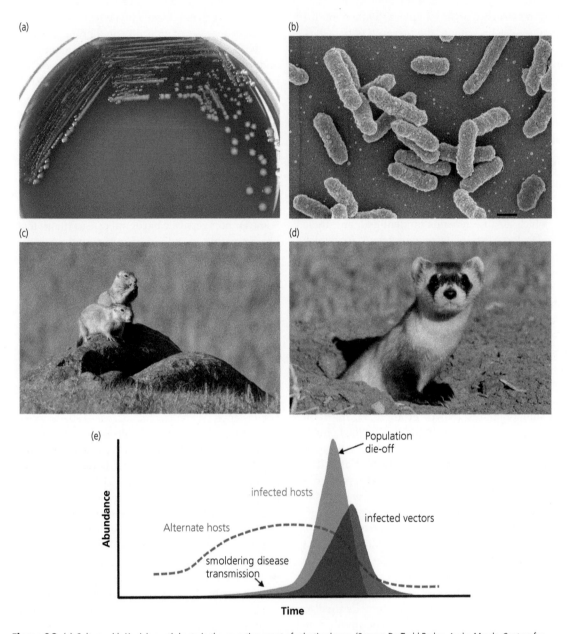

Figure 6.3 (a) Culture with *Yersinia pestis* bacteria the causative agent of sylvatic plague. (Source: Dr. Todd Parker, Audra Marsh, Centers for Disease Control, United States.) (b) Electron microphotograph of *Y. pestis* (Source: Muhsin Özel, Gudrun Holland, Rolf Reissbrodt; Robert Koch Institute, Germany.) (c) Pair of black-tailed prairie dogs (*Cynomys ludovicianus*), Wichita Mountains Oklahoma, United States (Photo credit: Larry Smith.) Colonies of this species can be severely impacted by *Y. pestis* epizootics. (d) Close-up photo of black-footed ferret (*Mustela nigripes*), an endangered prairie dog predator that can also become infected with *Y. pestis*. (Source: United States Geological Service.) (e) A conceptual representation of the dynamics of a plague outbreak in a prairie dog colony. Disease transmission may be ongoing at low levels (smoldering phase) both in prairie dogs and in alternative hosts (e.g., grasshopper mice (*Onychomys leucogaster*)) before an acute prairie dog die-off (Salkeld et al. 2016).

Box 6.1 Koch's Postulates

1. The organism must be present in every case of the disease
2. The organism must be isolated from the tissues and grown in pure culture
3. The agent must reproduce the specific disease when introduced into susceptible animals
4. The agent must be recoverable from experimental animals

These rules are useful for defining a cause–effect relationship in diseases in which the agent is required for the disease to occur, and the agent acting by itself results in disease. But they are not adequate for understanding many diseases of wild animals (Hanson 1969), because:

1. Relatively few agents are both necessary and sufficient to cause disease over a wide range of environmental conditions.
2. A single disease may require interaction of several causes.
3. A single disease may be caused by different factors.
4. A single agent may cause several different diseases.

Other criteria use information on the way that the disease occurs and is distributed in the population, together with information from experimental exposure to infer a cause–effect relationship. Bradford Hill (1965) provided framework criteria for judging causal relationships, five of which appear to be generally accepted:

Time sequence of events: A cause must precede the effect. The sequence of events is often evident in situations of an acute disease caused by a specific agent, but is more difficult to establish in conditions that are chronic or insidious in onset. Establishing a sequence may be particularly difficult if the presence of antibody or some other response is used as a surrogate measure of infection, because the response indicates the animal was infected but does not indicate when infection occurred relative to disease.

Strength of association: The association should not be weak—for example, if a pathogen is causing disease,

there should be strong fitness differences between infected and uninfected individuals.

Biological gradient: If there is a dose–response relationship between the agent and the disease, the plausibility of a cause–effect relationship is increased. For instance, this type of relationship might be evident between the intensity of infection by macroparasites (e.g., the number of worms) and the degree of injury in the host.

Consistency: If the association occurs in a variety of different circumstances, a causal relationship is probable.

Consistency with existing knowledge: If a plausible biological mechanism is evident, a causal relationship is more likely than if one is not evident.

Analytical studies that involve comparisons between those animals with the disease (cases) and those without the disease (noncases) are required to make inferences about the relationship between exposure to potential causes and outcome. These analytical studies can be either observational or experimental, and both types of study may be needed during the course of an investigation to prove a cause–effect relationship beyond a reasonable doubt.

6.5.2 Experimental studies

The traditional approach to identifying causal factors of disease is to perform an experiment in which a set of randomly chosen individuals are exposed to the factor, while others are not. In such a controlled experiment, any confounding factors can be kept constant or regulated, and study subjects differ only with regard to exposure to the suspected causative agent.

Experimental studies are generally of two types: laboratory-based, in which the investigator has almost complete control over the environmental conditions; and controlled trials, in which the investigator controls the allocation of subjects to the study groups but the experiment is carried out under real-world conditions. Laboratory trials are particularly useful for determining whether an agent can cause a specific effect in an individual animal; however, they do this at the cost of reduced relevance to real-world situations. Controlled trials extend the rigorousness of the experimental approach to the more variable conditions of the natural environment. However, such studies may face ethical and environmental challenges stemming from the need to introduce an infectious agent into a wild population, simply to understand its effect. A good alternative to such introductions are controlled trials in which the naturally occurring agent is being removed from a randomly selected subset of wild hosts through medical treatment. For example, administration of antimalarial drugs has been used to evaluate the effects of *Plasmodium* infections on *Sceloporus* lizards (Foufopoulos 1998). Such studies have often produced good evidence of the effect at a population level, and will be discussed in greater detail later (see Chapter 10, Section 10.4.1).

6.5.3 Observational studies

While experimental manipulations provide the most rigorous evidence of causation, it is often impossible "to carry out experiments under conditions that even remotely resemble 'real-world' conditions" (Dohoo et al. 2003). In these situations, disease investigators will employ observational studies, which compare natural disease occurrence and exposure to possible risk factors between groups of individuals in free-ranging populations. Because the subject animals have been exposed naturally to the potential causative factor and little investigator manipulation is needed, such studies tend to be logistically easier. Observational studies are useful for identifying risk factors and quantifying the strength of association between factors and disease, but while an association—which is unlikely to have arisen by chance alone—can be explicit in an observational study, it is not possible to prove using this design that a cause–effect relationship exists between the agent and the disease. There are three types of observational study: cross-sectional, cohort, and case–control; in each of which animals are classified, first into those with the disease and those without, and second into those exposed and those not exposed to a risk factor; this generates a 2 × 2 contingency table for each disease–factor combination.

Cross-sectional studies are often the easiest type to perform. A sample of animals is obtained from the population and their exposure and disease status at that point in time is determined (the type and size of sample being determined by the situation). At the beginning of a cross-sectional study, the numbers of animals with and without disease and the number exposed to various risk factors are unknown; all that is predetermined is the sample size. Many different diagnostic tests may be applied to compare the groups; the choice of which analyses need to be done will depend on the investigator's knowledge of factors known to cause similar disease, as well as of agents known to be present in the population. Some analyses, such as hematology and clinical chemistry, are nonspecific indicators of health, whereas serological tests measure exposure to specific agents. In deciding which analyses should be done, the investigator needs to consider the full range of possibilities in consultation with specialists, and then select those variables that seem most appropriate to investigate first.

Whenever possible, samples should be retained in the event that initial testing fails to produce satisfactory data. For example, aliquots of serum might be preserved and frozen for possible later use, should other tests be required. Advantages of a cross-sectional study design are that multiple potential causes can be studied simultaneously, and that it provides an estimate of the prevalence of disease and exposure in the population. A disadvantage is that the temporal relationship between cause and effect cannot be determined, because the outcome and the suspected causes are measured at the same time.

As an example, let us assume that we sample 40 animals from a population, and that we determine that 15 have the disease of interest (cases), and 25 are noncases. Based on testing, a number of "determinants" that differ in frequency of occurrence between cases and noncases can be identified. Some might be directly causative, while others may be agent, host, or environmental factors that facilitate or interact with a causative agent to result in disease. Some, such as poor body condition, might be either a cause of, or a result of the disease (Beldomenico et al. 2008). Frequency of exposure to agent A (as may be indicated by a significant antibody titer) is measured and the relationship between disease and

Table 6.3 Contingency table for estimating relative risk (RR). In the following table, the formulae for calculating RR and the odds ratio are given in algebraic form with a, b, c, and d representing the numbers in each cell of the table. We also provide a worked example with $a = 12$, $b = 3$, $c = 3$ and $d = 22$.

	Exposure	No exposure	Total
Diseased (cases)	12 (a)	3 (b)	15 (a + b)
Not diseased (noncases)	3 (c)	22 (d)	25 (c + d)
Total	15 (a + c)	25 (b + d)	40 (n)

The RR, which is the probability of being diseased if exposed, relative to the probability of being diseased if not exposed is:

$$RR = \frac{a/(a+c)}{b/(b+d)} = \frac{12/(12+3)}{3/(3+22)} = 6.667$$

Whereas the odds ratio, which is the odds (relative probability) of being diseased if exposed relative to the odds of being diseased if not exposed is:

$$OR = \frac{ad}{bc} = \frac{12 \times 22}{3 \times 3} = 29.33$$

In this case, the null hypothesis that exposure is not associated with disease can be rejected with high confidence ($P = 10^{-5}$; Fisher's exact test).

exposure is shown in a 2×2 contingency table (see Table 6.3).

In this example, it is obvious that exposure to A is more common among cases than in noncases, but not all cases were exposed, and some noncases were exposed to the agent. The strength of association between disease and exposure to A can be measured by several methods. Because they are influenced by sample size, measures of statistical significance (P levels) cannot be used to measure the strength of association (Dohoo et al. 2003). A commonly used measure of strength of association is risk ratio or relative risk (RR), which is the ratio of the rate of disease among the exposed group to the rate of disease in the nonexposed group. If there is no association between exposure and disease, the RR should = 1. If the RR > 1, then the size of the value of the RR is directly related to the strength of association. For the example in table 6.3:

$$RR = \frac{\text{prevalence of disease in those exposed}}{\text{prevalence of disease in those not exposed}}$$

$$= \frac{12/15}{3/25}$$

The risk of having disease is about 6.7 times greater among animals exposed to A than in those not exposed, indicating a relatively strong association. If instead the exposed and unexposed animals did not differ substantially with regard to other identified risk factors, then we could conclude that exposure to A is a risk factor for the disease, but from this study we cannot determine when the animals were exposed to A. The hypothesis that the RR is significantly different than one can be tested statistically by using a Chi-squared test, log linear model, or Fisher's exact test. For small sample sizes, Fisher's exact test will be most accurate (Sokal and Rohlf 2012). Rather than the RR, most standard statistics packages (e.g., SPSS, SAS, or R) will calculate the odds ratio, which is the ratio of the odds (relative probability) of disease if the individual is exposed, divided by the odds of being diseased if not exposed (see Table 6.3). R has a function **riskratio()** in the package **epitools** (Aragon 2020) to calculate risk ratios and associated confidence intervals and significance tests directly. For a given set of results, the probability that the RR differs from one and the probability that the odds ratio differs from one will be identical. However, it is important to recognize that the odds ratio and RR are not the same thing, unless the prevalence of disease is very small. RR is a more intuitive concept than the odds ratio and many epidemiologists recommend it should be used in preference to odds ratios (Davies et al. 1998).

Case–control studies may be more efficient than a cross-sectional study, particularly for rare diseases, because individuals with a specific characteristic, such as presence of disease, are chosen for inclusion (the cases) and then a group of individuals who do not have the characteristic, but are otherwise matched (the noncases) are selected for comparison (Mann 2003). As with a cross-sectional study, multiple potential causes can be studied, but a case–control study does not provide an estimate of the prevalence of disease and exposure in the population. Selection of an appropriate control (noncase group) is often difficult, and the possibility of confounding variables is always present. Case–control studies are best viewed as hypothesis-generating, with more rigorous methods then being used to then test those hypotheses.

Cohort studies are longitudinal studies concerned with the development of disease (incidence) in exposed and unexposed individuals. One method is to choose a group of nonexposed individuals at the outset, and then monitor and test them at regular intervals for evidence of infection (exposure) and measure the outcome (disease or no disease) at some predetermined time. Another method is to choose two groups (exposed and nonexposed) and monitor the occurrence of disease over a predetermined period of time. Cohort studies focus on a specific identified risk factor, do not provide an estimate of the prevalence of disease and exposure in the population, and are difficult because of the need to follow individuals through time, but they provide a higher degree of "proof" of causal association than either a case–control or a cross-sectional study (Dohoo et al. 2003). For example, a long-term cohort study was used to evaluate the effects of bot fly (*Cuterebra* sp.) infestation on white-footed mice (*Peromyscus leucopus*) in the north-eastern United States (Burns et al. 2005). This particular study found that, counterintuitively, bot fly parasitism was associated with longer mouse life expectancy, in part because parasitized mice invested fewer resources in reproduction with increased resources being allocated instead into their own survivorship.

One particular form of experimental study, the controlled trial, has many features of a cohort study in which exposure is manipulated experimentally and then animals are followed longitudinally. This type of study has been used with many types of agent ranging from bacteria to ectoparasites and from protozoans to nematodes (Brown et al. 1995; Hudson et al. 1998; Merino et al. 2000; Potti et al. 2002; Stien et al. 2002; Neuhaus 2003; Redpath et al. 2006; Tomás et al. 2007) and may provide strong evidence of the effects of an agent at both the individual animal and the population level (see also Chapter 10).

References

Anderson, R.M. and May, R.M. (1979). Population biology of infectious disease: Part 1. *Nature*, **280**:361–67.

Aragon, T.J. (2020). epitools: Epidemiology tools. R package version 0.5-10.1. https://CRAN.R-project.org/package=epitools (accessed August 18, 2020).]

Beldomenico, P.M. and Begon, M. (2010). Disease spread, susceptibility and infection intensity: vicious circles? *Trends in Ecology & Evolution*, **25**:21–27.

Beldomenico, P.M., Telfer, S., Gebert, S., Lukomski, L., Bennett, M., and Begon, M. (2008). Poor condition and infection: a vicious circle in natural populations. *Proceedings of the Royal Society B: Biological Sciences*, **275**:1753–59.

Bennett, G.F., Pierce, M.A., and Ashford, R.W. (1993). Avian haematozoa: Mortality and pathogenicity. *Journal of Natural History*, **27**:993–1001.

Biggins, D.E., Godbey, J.L., Gage, K.L., Carter, L.G., and Montenieri, J.A. (2010). Vector control improves survival of three species of prairie dogs (*Cynomys*) in areas considered enzootic for plague. *Vector Borne and Zoonotic Diseases*, **10**:17–26.

Bishop, M.L., Fody, E.P., and Shoeff, L. (2005). *Clinical Chemistry. Principles, Procedures, Correlations*, 5th edn. Lippincott Williams & Wilkins, Philadelphia, PA.

Brown, C.R., Bomberger Brown, M., and Rannala, B. (1995). Ectoparasites reduce the long-term survival of their avian host. *Proceedings of the Royal Society of London: Biological Sciences*, **262**:313–19.

Burge, C.A., Mark Eakin, C., Friedman, C.S., Froelich, B., Hershberger, P.K., Hofmann, E.E., et al. (2014). Climate change influences on marine infectious diseases: Implications for management and society. *Annual Review of Marine Science*, **6**:249–77.

Burns, C.E., Goodwin, B.J., and Ostfeld, R.S. (2005). A prescription for longer life? Bot fly parasitism of the white-footed mouse. *Ecology*, **86**:753–61.

Davies, H.T., Crombie, I.K., and Tavakoli, M. (1998). When can odds ratios mislead? *British Medical Journal*, **316**: 989–91.

Dickman, C.R. and Braithwaite, R.W. (1992). Post-mating mortality of males in the dasyurid marsupials, *Dasyurus* and *Parantechinus*. *Journal of Mammalogy*, **73**: 143–47.

DiGiacomo, R.F. and Koepsell, T.D. (1986). Sampling for detection of infection or disease in animal populations. *Journal of the American Veterinary Medical Association*, **189**:22–23.

Dobson, A.P. and Hudson, P.J. (1992). Regulation and stability of a free-living host-parasite system: *Trichostrongylus tenuis* in red grouse. II. Population models. *Journal of Animal Ecology*, **61**:487–98.

Dohoo, I., Martin, W., and Stryhn, H. (2003). *Veterinary Epidemiologic Research*. AVC Inc., pp. 10–12, Charlottetown.

Ezenwa, V.O. and Jolles, A.E. (2015). Opposite effects of anthelmintic treatment on microbial infection at individual versus population scales. *Science*, **347**: 175–77.

Fawcett, T. (2006). An introduction to ROC analysis. *Pattern Recognition Letters*, **27**:861–74.

Foufopoulos, J. (1998). Host–parasite interactions in the mountain spiny lizard *Sceloporus jarrovii*. PhD dissertation, University of Wisconsin–Madison.

Foufopoulos, J., Roca V., White, K.A., Pafilis, P., and Valakos, E.D. (2016). Effects of island characteristics on parasitism in a Mediterranean lizard (*Podarcis erhardii*): A role of population size and island history. *North-Western Journal of Zoology* **13**:70–76.

Gilman S., Blumstein, D.T., and Foufopoulos, J. (2007). The effect of haemosporidian infections on white-crowned sparrow singing behaviour. *Ethology*, **113**:437–45.

Hanley, J.A. and Lippman-Hand, A. (1983). If nothing goes wrong, is everything all right? Interpreting zero numerators. *JAMA*, **249**:1743–45.

Hanson, R.P. (1969). Koch is dead. *Bulletin of the Wildlife Disease Association*, **5**:150–56.

Harries, A.D. and Dye, C. (2006). Tuberculosis. *Annals of Tropical Medicine & Parasitology*, **100**:415–31.

Hill, A.B. (1965). The environment and disease: Association or causation? *Proceedings of the Royal Society of London B: Biological Sciences*, **58**:295–300.

Hoyer, I.J., Blosser, E.M., Acevedo, C., Thompson, A.C., Reeves, L.E., and Burkett-Cadena, N.D. (2017). Mammal decline, linked to invasive Burmese python, shifts host use of vector mosquito towards reservoir hosts of a zoonotic disease. *Biology Letters*, **13**:p.20170353.

Hudson, P.J., Dobson, A.P., and Newborn, D. (1998). Prevention of population cycles by parasite removal. *Science*, **282**:2256–58.

Humphries, S. and Stevens, D.J. (2001). Reproductive biology: Out with a bang. *Nature*, **410**:758–59.

Jarvi, S.I., Schultz, J.J., and Atkinson, C.T. (2002). PCR diagnostics underestimate the prevalence of avian malaria (*Plasmodium relictum*) in experimentally-infected passerines. *Journal of Parasitology*, **88**:153–58.

Koch, R. (1884). Die Ätiologie der Tuberkulose. *Mittheilungen aus dem kaiserlichen Gesundheitsamte*, **2**:1–88.

Leopold, A. (1939). *Game Management*. Charles Scribner's Sons, New York.

Liao, W., Elison Timm, O., Zhang, C., Atkinson, C.T., LaPointe, D.A., and Samuel, M.D. (2015). Will a warmer and wetter future cause extinction of native Hawaiian forest birds? *Global Change Biology*, **21**: 4342–52.

Löffler, F. (1884). Untersuchung über die Bedeutung der Mikroorganismen für die Entstehung der Diphtherie beim Menschen, bei der Taube und beim Kalbe. *Mittheilungen aus dem kaiserlichen Gesundheitsamte*, **2**: 421–99.

Mann, C.J. (2003). Observational research methods. Research design II: Cohort, cross sectional, and case-control studies. *Emergency Medicine Journal*, **20**: 54–60.

Matchett, M.R., Biggins, D.E., Carlson, V., Powell, B., and Rocke, T. (2010). Enzootic plague reduces black-footed ferret (*Mustela nigripes*) survival in Montana. *Vector-Borne and Zoonotic Diseases*, **10**:27–35.

Merino, S., Moreno, J., Sanz, J.J., and Arriero, E. (2000). Are avian blood parasites pathogenic in the wild? A medication experiment in blue tits (*Parus caeruleus*). *Proceedings of the Royal Society of London B: Biological Sciences*, **267**:2507–10.

Munoz, N.E., Blumstein, D.T., and Foufopoulos, J. (2010). Immune system activation affects song and territorial defense. *Behavioral Ecology*, **21**:788–93.

Murray, D.L., Cary, J.R., and Keith, L.B. (1997). Interactive effects of sub-lethal nematodes and nutritional status on snowshoe hare vulnerability to predation. *Journal of Animal Ecology*, **66**:250–66.

Neuhaus, P. (2003). Parasite removal and its impact on litter size and body condition in Columbian ground squirrels (*Spermophilus columbianus*). *Proceedings of the Royal Society of London B: Biological Sciences*, **270**:S213–15.

O'Toole, D. (2010). Monitoring and investigating natural disease by veterinary pathologists in diagnostic laboratories. *Veterinary Pathology*, **47**:40–44.

Pedersen, A.B. and Greives, T.J. (2008). The interaction of parasites and resources cause crashes in a wild mouse population. *Journal of Animal Ecology*, **77**:370–77.

Potti, J., Moreno, J., Yorio, P., Briones, V., Garcia-Borboroglu, P., Villar, S., et al. (2002). Bacteria divert resources from growth for Magellanic penguin chicks. *Ecology Letters*, **5**:709–14.

Redpath, S.M., Mougeot, F., Leckie, F.M., Elston, D.A., and Hudson, P.J. (2006). Testing the role of parasites in driving the cyclic population dynamics of a gamebird. *Ecology Letters*, **9**:410–18.

Resasco, J., Bitters, M.E., Cunningham, S.A., Jones, H.I., McKenzie, V.J., and Davies, K.F. (2019). Experimental habitat fragmentation disrupts nematode infections in Australian skinks. *Ecology*, **100**:p.e02547.

Romain, K.S., Tripp, D.W., Salkeld, D.J., and Antolin, M.F. (2013). Duration of plague (*Yersinia pestis*) outbreaks in black-tailed prairie dog (*Cynomys ludovicianus*) colonies of northern Colorado. *EcoHealth*, **10**:241–45.

Rose, G. (1985). Sick individuals and sick populations. *International Journal of Epidemiology*, **14**:32–38.

Salkeld, D.J., Stapp, P., Tripp, D.W., Gage, K.L., Lowell, J., Webb, C.T., et. al. (2016). Ecological traits driving the outbreaks and emergence of zoonotic pathogens. *BioScience*, **66**:118–29.

Schall, J.J. (1982). Lizards infected with malaria: Physiological and behavioral consequences. *Science*, **217**:1057–59.

Scott, J.M., Van Riper, C., and Conant, S. (2001). *Evolution, Ecology, Conservation, and Management of Hawaiian Birds: A Vanishing Avifauna*. Cooper Ornithological Society, Camarillo, CA.

Shaw, D.J., Grenfell, B.T., and Dobson, A.P. (1998). Patterns of macroparasite aggregation in wildlife host populations. *Parasitology*, **117**:589–610.

Šimundič, A.M. (2009). Measures of diagnostic accuracy: Basic definitions. *Journal of the International Federation of Clinical Chemistry*, **19**:203.

Sokal, R.R. and Rohlf, F.J. (2012). *Biometry: the Principles and Practice of Statistics in Biological Research*, 4th edn. WH Freeman and Co., New York.

Stien, A., Irvine, R.J., Ropstad, E., Halvorsen, O., Langvatn, R., and Alton, S.D. (2002). The impact of gastrointestinal nematodes on wild reindeer: experimental and cross-sectional studies. *Journal of Animal Ecology*, **71**: 937–45.

Thrusfield, M. (2007). *Veterinary epidemiology*, 3rd edn. Blackwell Publishing, Oxford.

Thrusfield, M., Ortega, C., De Blas, I., Noordhuizen, J.P., and Frankena, K. (2001). WIN EPISCOPE 2.0: Improved epidemiological software for veterinary medicine. *Veterinary Record*, **148**:567–72. https://stevelrandocom.wixsite.com/ensecesub/post/win-episcope-2-0-download (accessed July 20, 2020).

Tomás, G., Merino, S., Moreno, J., Morales, J., and Martinez-De La Puente, J. (2007). Impact of blood parasites on immunoglobulin level and parental effort: A medication field experiment on a wild passerine. *Functional Ecology*, **21**:125–33.

Van Loveren, H., Ross, P.S., Osterhaus, A.D.M.E., and Vos, J.G. (2000). Contaminant-induced immunosuppression and mass mortalities among harbor seals. *Toxicology Letters*, **112**:319–24.

Van Riper, III, C., Van Riper, S.G., Goff, S.G., and Laird, M. (1986). The epizootiology and ecological significance of malaria in Hawaiian land birds. *Ecological Monographs*, **56**:327–44.

Warner, R.E. (1968). The role of introduced diseases in the extinction of the endemic Hawaiian avifauna. *Condor*, **70**:101–20.

Wobeser, G. and Wobeser, A.G. (1992). Carcass disappearance and estimation of mortality in a simulated die-off of small birds. *Journal of Wildlife Diseases*, **28**:548–64.

Wolfe, N.D., Dunavan, C.P., and Diamond, J. (2007). Origins of major human infectious diseases. *Nature*, **447**:279–83.

CHAPTER 7

The Environmental Context of Wildlife Disease

> To do science is to search for repeated patterns, not simply to accumulate facts.
>
> **(Robert MacArthur 1972)**

7.1 Introduction

In the study of infectious disease, environmental features refer to determinants that are neither a feature of the causative agent nor a characteristic of the host population, and include both abiotic and biotic factors (Elith and Leathwick 2009). The relationships among agent, host, and environmental factors in disease have been illustrated in various manners as the "disease triangle" (see Section 2.3.3). The occurrence of disease is usually not homogeneous or random in space or time, and is associated with specific environmental conditions, with disease occurring only where and when there is the appropriate concurrence of suitable host, infectious agent, and environmental factors. The observation that disease occurrence is associated with particular environmental conditions has led to several concepts, including that of the "nidality" (microscale foci of disease, see also Herbold 2005), the landscape epidemiology of disease (Audy 1958; Pavlovsky and Pious 1966) and, more recently, the *ecological niche* of disease. The idea of ecological niche—"the set of conditions under which the species can maintain populations without immigration of individuals from other areas" (Peterson 2006)—was originally applied to free-ranging species (Grinnell

1917; Hutchison 1957) and has been extended more recently to include infectious disease. Hence, the concept of an ecological niche—originally used for a species—can also be applied to an infectious agent; this concept can also be restated as the set of environmental conditions under which an infectious disease can persist without introductions from other areas.

Why would a conservation biologist need to study the environmental context of wildlife disease? The most common reason for sampling environmental variables is to determine how and why disease occurs when and where it does, that is, to identify the environmental features suitable for disease occurrence. The environmental factors most relevant to defining the ecologic niche of a disease are usually related to either transmission of the disease agent among host animals or to the persistence of the agent outside the focal host.

Environmental samples also may be collected as a method of disease surveillance, particularly in situations in which it is easier to detect the presence of the infectious agent outside, rather than within, the target host. Sampling in such situations often takes advantage of a trophic relationship between the host and some other species. For example, detection of a disease that has a high

Infectious Disease Ecology and Conservation. Johannes Foufopoulos, Gary A. Wobeser and Hamish McCallum, Oxford University Press.
© Johannes Foufopoulos, Gary A. Wobeser and Hamish McCallum (2022). DOI: 10.1093/oso/9780199583508.003.0007

case-fatality rate in a wild host may be problematic, because of difficulty in finding dead animals, and because there are few survivors that can be tested for evidence of exposure. An alternative method of sampling in such situations is to search for evidence of exposure to the disease agent in animals that prey upon, or scavenge, the target host. For instance, serological surveys of carnivores have been used to detect the occurrence of zoonoses in wild rodents (Gese et al. 1997; Leighton et al. 2001), and anthrax (*Bacillus anthracis*) in African wildlife (Lembo et al. 2011). The carnivores "sample" the prey species, including sick or dead individuals that are unavailable for sampling directly. In a similar manner, it may be easier to search for an agent within blood ingested by mosquitoes, than to search for the agent in the mammals that are host to both the disease agent and the mosquito (Trienbach et al. 2010).

Recently, environmental DNA has been used to detect some wildlife pathogens. For example, DNA from water samples has been used to detect a range of amphibian pathogens and parasites, including ranaviruses (Hall et al. 2016), trematodes (Huver et al. 2015), and the amphibian chytrid fungus *Batrachochytrium dendrobatidis* (Kamoroff and Goldberg 2017). Knowledge of the specific environmental conditions associated with disease occurrence may be used to guide a search for the agent. For example, searches for animals infected with filoviruses failed to identify a reservoir of Ebola hemorrhagic fever and Marburg disease, but characterizing features of areas where the diseases occur may "provide a framework for a more informed search" for the reservoir(s) (Peterson et al. 2004).

Collection and use of information on environmental features that define the ecological niche of a disease can be divided into three phases:

1. Identifying and quantifying the association between environmental features and disease occurrence.
2. Using knowledge of the ecologic niche to predict where or when disease will occur.
3. Using knowledge of the ecologic niche for disease management.

7.2 Identifying and quantifying the association between environmental features and disease occurrence

The choice of which environmental factors might be involved, and, hence, should be investigated in any disease, should be based upon a hypothesis about the disease process. It is not necessary to understand all of the many steps that may be involved in the disease transmission process in order to produce a hypothesis that can be used to guide sampling.

Coccidioidomycosis, caused by the fungus *Coccidioides immitis*, will be used as an example of how a basic hypothesis can guide environmental sampling. The natural reservoir for *C. immitis* is soil, where it grows as a saprophyte (Fisher et al. 2007). Disease in humans and animals occurs when airborne fungal spores are inhaled. Using this basic information, it was posited that the seasonal occurrence of disease in humans was related to a moist period, favorable for growth of the fungus in soil, followed by a dry period, in which spores form and become airborne, leading to increased risk of inhalation of spores (Kolivras and Comrie 2003). Comrie (2005) used this "grow and blow" hypothesis as the rationale for measuring the temporal distribution of two environmental variables: monthly precipitation (an indicator of soil moisture for fungal growth), and concentration of atmospheric particulate matter <10 μm in diameter (a surrogate for measuring wind-blown fungal spores), in relation to disease occurrence.

If little is known about a particular disease, an initial hypothesis may be formed using what is known about similar infections, following the advice that the "only way we have of studying the unknown is by pretending that it is like the known" (Levins 1995). For instance, if a disease is known to be caused by a trematode, a reasonable hypothesis would be that gastropods are involved in the disease, because other trematodes require a gastropod alternate host. This hypothesis might guide sampling to determine if: (i) gastropods are present where the disease occurs; (ii) the gastropods are infected with trematode larvae; and (iii) the distribution of disease is related to environmental conditions favorable for gastropods.

Figure 7.1 Dead saiga antelope (*Saiga tatarica*) female lying beside her dead calves in Kazakhstan following the pasteurellosis outbreak of 2015. (Source: Sergei Khomenko, Saiga Resource Centre.)

Environmental factors can be critical in determining whether a potential pathogen causes asymptomatic infections, as opposed to overt disease with population-level consequences. For example, the saiga antelope (*Saiga tatarica*) of central Asia suffers occasional mass mortality events—in May 2015 more than 200,000 died in calving aggregations at several sites (Figure 7.1; also Section 2.3.3), a loss of about 62% of the entire species population (Kock et al. 2018). The proximate agent of mortality was hemorrhagic septicemia caused by the bacterium *Pasteurella multocida*. This was not, however, a novel pathogen or even an epidemic. The antelopes frequently harbor this particular bacterium as a commensal and it was unusual environmental conditions that most likely caused this usually subclinical infection to cause mass mortality. Substantial mass mortality events have also occurred at multiple sites in previous years, including 1981 and 1988. Relative to years and sites at which mass mortalities have not occurred, mass mortalities are associated with unusually high humidity immediately prior to the event, and (particularly in 2015) high temperatures (Figure 7.2).

The range of abiotic and biotic factors that might be important for a particular disease is almost unlimited, and it is not practical to provide a meaningful checklist of what should be sampled, or a list of sampling techniques that would cover all situations. Sampling often requires a multidisciplinary approach and one should seek advice from experts in sampling particular factors. On the one hand, among abiotic factors, climate, weather, and water influence almost all diseases. More specifically, "climate restricts the range of infectious diseases, whereas weather affects the timing and intensity of outbreaks" (Epstein 2004). Water in all forms (standing or flowing, precipitation, atmospheric humidity, soil moisture) influences the distribution of animals, plants, infectious agents, and land use by humans. It may be a reservoir of infection, a transporter of disease agents, a factor in survival of agents outside the host, and habitat for many vectors and alternate hosts. "Consider water" should be a motto for disease investigators, in the same way that *"cherchez la femme"* was a guide for detectives hoping to explain unusual male behavior (Dumas 1871).

On the other hand, among biotic features, the most important one to consider when investigating many infectious diseases is the position of the target host species, the infectious agent, as well as other species within a food web, because transmission of many infectious agents is dependent upon trophic relationships (see Section 2.2.2). Hence detecting a parasite in other host species that constitute either prey or predators or ectoparasites of the target species may be an easier and more efficient approach.

The most common method for collecting environmental data is through observational studies that relate the distribution—spatially or temporally—of

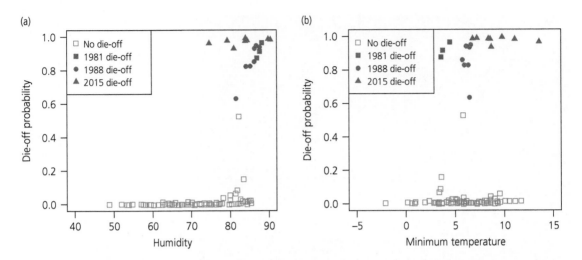

Figure 7.2 Modeled probability of die-off of saiga antelopes (*Saiga tatarica*) related to selected environmental variables in the preceding 10-day period. Open squares indicate site/year combinations in which no die-off took place; solid squares indicate sites where die-off occurred in 1981; solid circles, sites where die-offs occurred in 1988; and solid triangles, sites where die-offs occurred in 2015. (a) Fitted values for the probability of a die-off event against mean maximum daily relative humidity in the previous 10 days. (b) Fitted values for the probability of a die-off event against mean minimum daily temperature in the previous 10 days. There was a strong threshold effect of humidity, as well as a weak but significant effect of minimum temperature on the probability of a die-off. (Redrawn from Kock et al., 2018.)

environmental features to disease occurrence. Commonly, such studies begin with knowledge of where or when disease occurs, followed by identification of the environmental features associated with "foci" of disease occurrence. For example, persistent foci of tularemia (*Francisella tularensis* infection) in the Czech Republic are associated with alluvial forest habitat, at elevations up to 200 m above sea level, and with specific conditions of mean annual temperature, precipitation, and sunshine duration (Pikula et al. 2003). Studies may also proceed in the opposite direction, that is, by searching for evidence of disease in locations with differing environmental conditions. For instance, Limmathurotsakul et al. (2010) searched in uncultivated and cultivated land for *Burkholderia pseudomallei* (a soil-dwelling pathogen), and Swei et al. (2011) sampled plots with differing environmental features for ticks infected with *Borrelia burgdorferi*.

Comparison of environmental features in sites where disease occurs versus those where it does not occur is an important method for characterizing environmental determinants of a disease. In making such comparisons, it is important to recall that absence of evidence of disease is not evidence of absence of disease. For example, a disease may be present but appear to be absent, if sampling has been inadequate (see Chapter 4). "Presence–absence" studies attempt to differentiate environmental conditions where disease is present (presence points) from conditions where disease is *known* to be absent (absence points). The spatial scale of such studies may range from comparison of broad geographic area to the location of individual animals. As an example of the latter, Staubach et al. (2001) determined environmental features associated with *Echinococcus multilocularis* infection in red foxes (*Vulpes vulpes*) in an endemic area of Germany. The location where each fox was killed by a hunter was entered into a geographic information system database, together with data on topographic features, soil humidity, and land use at the site. Infection was associated with pasture areas with high soil humidity near water, while infected foxes were underrepresented in forested areas. A potential problem in presence–absence studies is that absence records might introduce confounding information "because they can indicate either habitat that is unsuitable or habitat that is suitable but unoccupied, perhaps because of inaccessibility" (Elith and Leathwick 2009). "Presence-only" studies attempt to differentiate conditions at presence points from those at "pseudoabsence" points. The latter method has been used in the study of several diseases (Soberon and Peterson 2005; Peterson et al. 2006a; Giles et al. 2011). For a review of the advantages and disadvantages of the numerous approaches used to model species distributions, see Guillera-Arroita et al. (2015).

7.2.1 The importance of scale

There is no single spatial scale at which the relationship between environmental features and disease should be examined; "systems generally show characteristic variability on a range of spatial, temporal, and organizational scales" (Levin 1992). For instance, the geographic area in which coccidioidomycosis occurs in humans is measured in hundreds of kilometers, while growth sites of C. *immitis* in soil are measured in meters and centimeters, and the actual niche within the soil is defined on a scale of millimeters or nanometers (Fisher et al. 2007). In this disease, climatic conditions over a large area may be generally suitable for the fungus, but critical features for fungal growth are determined in the soil microhabitat. Similarly, the biological drivers of Ross River virus, an Australian zoonotic pathogen, are most important at finer spatial scales, whereas abiotic ones are more important at larger scales (Flies et al. 2018). Hence to obtain the best understanding of the epidemiology of the pathogen, eco-epidemiological analyses need to be conducted across multiple spatial scales.

The relationship between environmental features and disease must also be considered on a variety of temporal scales. For instance, the time frame for considering a relationship between bedrock type and occurrence of the disease leptospirosis (Kingscote 1970) may be millennia, while the appropriate scale for other temporal relationships may be hours, days, months, or years. The relationship between an environmental factor and the occurrence of disease may even be different at different time scales (Davis et al. 2005). The effect of environmental factors is seldom instantaneous, and often there is a lag between the occurrence of some critical factor and disease occurrence. Conditions when disease occurs (or when it is detected) may not be as important as those at some distant time. For instance, the incidence of coccidioidomycosis

in humans is related to precipitation that occurred 1.5–2 years prior to exposure (Comrie and Glueck 2007), and the greatest risk of acquiring Lyme disease from nymphal ticks in oak forests occurs 2 years after an abundant acorn crop (Jones et al. 1998).

Establishing a relationship between environmental factors and disease occurrence may be particularly difficult in highly mobile or migratory hosts, because the environmental factors may be separated from disease occurrence in both time and space.

7.3 Modeling the environmental occurrence of disease

Information about environmental conditions may be collected through direct sampling by the investigator, or by using information available in various databases. Databases are used commonly for information on weather conditions, either collected at local weather stations, or from websites such as WorldClim (www.worldclim.org). Indeed, a variety of other publicly available, geo-referenced, satellite-derived information on factors such as topography, vegetation type and cover, land use, and soil parameters is available and can be used for the study of disease.

Numerical models using geo-referenced information of environmental conditions are used widely to describe patterns of species distribution (Elith and Leathwick 2009). Models that include information (both ground-based and remotely sensed) on disease occurrence are used to define the ecological niche of a disease and the geographic distribution of environmental conditions suitable for disease occurrence (e.g., Costa et al. 2002; Baptista-Rosas et al. 2007; Pullan et al. 2011; Giles et al. 2011). This type of modeling may "involve a machine-learning algorithm for discovering associations between point-occurrence data and sets of electronic maps summarizing environmental/ecologic dimensions" (Peterson et al. 2002), and may "center on estimating the dimensions of species' ecologic niches" (Costa et al. 2002). The technique has been used for a variety of purposes including: to identify a reservoir for Chagas disease (*Trypanosoma cruzi* infection) (Costa et al. 2002; Peterson et al. 2002); to examine the distribution of mosquito vectors of malaria (Levine et al.

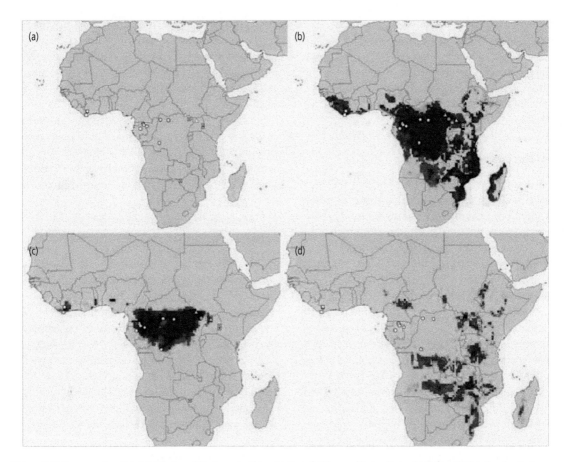

Figure 7.3 Known and predicted distribution of filoviruses in Africa. (a) Map of all known filovirus hemorrhagic fever (HF) outbreaks. Open squares, Ebola Ivory Coast; circles, Ebola Zaire; triangles, Ebola Sudan; dotted squares, Marburg HF occurrences. (b) Predicted geographic distribution of all filoviruses in Africa based on an ecological niche model. (c) Predicted distribution for Ebola HF only, based on an ecological niche model. (d) Predicted distribution for Marburg HF only, based on an ecological niche model. Darker shades of gray indicated higher predicted probability of occurrence. (Source: Peterson et al. 2004.)

2004); to characterize the geographic and ecologic distribution of Ebola and Marburg viruses in Africa (Peterson et al. 2004; see Figures 7.3 and 7.4); to determine the ecology and invasion potential of a rodent host of monkeypox (Peterson et al. 2006b); and to predict the geographic distribution of flea vectors of plague (Adjemian et al. 2006), anthrax (Blackburn et al. 2007), and hotspots for the presence of *Coccidioides* spp. (Baptista-Rosas et al. 2007).

The ecological niche of a disease, when defined by correlative studies or modeling, is often broad, such as a type of forest, certain soil characteristics, or particular climatic conditions. Often additional factors that were not considered in the initial analysis may limit the distribution of the disease

further (Peterson et al. 2004). The environmental features used in the analysis may not be functionally important by themselves, but may only be correlated with the actual environmental factors that drive disease transmission or persistence; in fact, the actual mechanisms involved are often unknown. For example, infection of foxes with *E. multilocularis* has been associated with locations near water and with high soil humidity (Staubach et al. 2001); however, it was not known if this was related to better survival of oncospheres of *E. multilocularis* in moist areas, or to the presence of muskrats (*Ondatra zibethica*—a suitable intermediate host for the parasite), or to some other unidentified factors in these areas. Lack of understanding of the mechanisms involved may not preclude usefulness of the data, and ecologic niche modeling and related techniques appear to have great potential for the study of disease, particularly as methods for predicting disease distribution and occurrence. If a pattern can be identified, "the determinants of pattern, and the mechanisms that generate and maintain those patterns" (Levin 1992) can then be investigated, either through further sampling or through laboratory or field experiments.

7.3.1 Using features of the ecological niche to predict where or when disease will occur

If the environmental features of a disease can be identified, it may be possible to predict where

and when disease may occur in the future, and how environmental change will influence the distribution and frequency of occurrence of disease. Ecological niche modeling and other similar techniques may "identify areas that fit the ecologic bill" (Peterson 2006) for a disease, even if the disease is not already present there. The information may be used to predict infection occurrence in unsurveyed areas, to identify areas for future sampling, and to prioritize data collection (Peterson et al. 2004; Pullan et al. 2011), as well as to guide management. For instance, rivers have been shown act as "semipermeable barriers" that delay spread of rabies epidemics in raccoons (*Procyon lotor*); this information could be used in planning surveillance

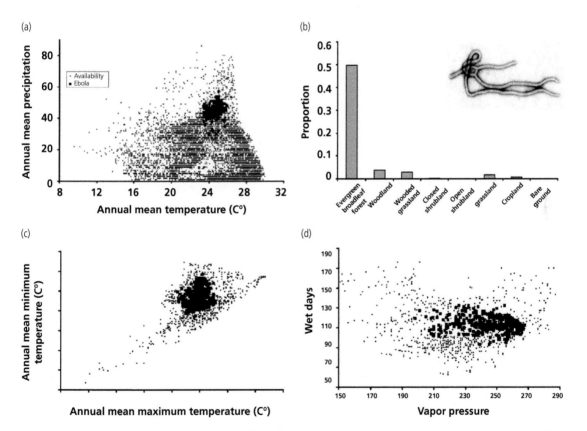

Figure 7.4 Predicted potential distributional areas for Ebola hemorrhagic fever (HF) occurrences, displayed along axes of key climatic variables. (a) Locations of Ebola occurrence (larger black squares) against a backdrop of possible sites (all of Africa, smaller black squares); the pathogen is distributed in areas of warm, wet conditions. (b) Distribution of Ebola HF across different habitat types, shown at the fraction of the total area of each habitat in which the disease is predicted to occur. (c and d) Principal environmental characteristics of sites predicted to harbor Ebola HF (larger black squares) versus all sampled (possible) sites (smaller black squares) (c, minimum/maximum mean temperature; d, vapor pressure/wet days). (Adapted from: Peterson et al. 2004, Ebola virus image: Cynthia Goldsmith/Centers for Disease Control.)

programs (Smith et al. 2002). In some situations, a single environmental feature that defines the distribution of a disease may be identified. For example, the distribution of some ticks may be restricted because they require relative humidity of ≥80% in the microhabitat where they occur while off the host (Gray et al. 2009), and soil-transmitted nematodes may not occur in areas with land surface temperatures >40°C, because development of free-living larval stages ceases at this temperature (Pullan et al. 2011).

There is intense interest in predicting the effect of climate change on infectious diseases, because of the potential effects on host populations, community structure, and ecosystems. Climate change will result in changes in the geographic distribution, phenology, and intensity of many infectious diseases. Rising temperature and changing precipitation patterns are expected to have the greatest effect on those diseases in which a life stage of the agent is free in the environment, and is transmitted by invertebrate vectors or intermediate hosts, or through contaminated water (Kutz et al. 2009; Shuman 2010).

The impact of avian malaria and bird pox on the Hawaiian avifauna is one of the best studied cases of the influence of climate change on wildlife disease. As discussed earlier (Sections 2.3.1, 3.2.3, and 3.5.1) most of the endemic Hawaiian low-altitude bird species were driven to extinction following the nineteenth-century introduction of the mosquito *Culex pipiens*, which is the vector of avian malaria and bird pox (Warner 1968). Several species such as the 'i'iwi (*Drepanis coccinea*) (Figure 7.5) persist at high altitudes, where the current temperature is too low for the vectors to effectively transmit these pathogens. Already, there is evidence that increasing temperature has led to increased prevalence of avian malaria at higher altitudes (Freed et al. 2005), and modeling suggests that further climate change will lead to dramatic declines in the abundance of several endangered Hawaiian birds (Benning et al. 2002; Atkinson et al. 2014; Guillaumet et al. 2017; Figure 7.6, also Figure 3.5).

Because climate change "will send changes propagating through ecosystems" (Poulin 2006), including alterations in the dynamics of disease transmission and host switching (Brooks and Hoberg 2007), and novel behavior of ecosystems (Yakob and Mumby 2011), it is notoriously difficult to predict what the net effect on any particular disease will be (Hueffer et al. 2011). Predictions need to be validated by complete analysis of the agent–host–environment system (Mouriston et al. 2005).

7.3.2 Using knowledge of the ecological niche in disease management

All forms of disease management are based upon reducing exposure of the target host to the disease agent or reducing the effect of the agent on the host. If a specific set of environmental conditions—associated with disease occurrence—can be identified, the disease might be managed by ensuring that those conditions do not occur or occur less frequently. For example, recent investigations have demonstrated associations between anthropogenic environmental degradation and outbreaks of both Ebola virus disease (Olivero et al. 2017) and Buruli ulcer (*Mycobacterium ulcerans*) (Morris et al. 2016) in sub-Saharan Africa, providing powerful arguments for sustainable environmental management. Ultimately, however, for any disease management technique to be effective, it must be applied at the correct place and time. Techniques that define the spatial or temporal distribution of a disease, such as ecologic niche modeling, can be used to direct control or management programs. If hotspots of disease occurrence can be identified, these areas (or times) can be subject to enhanced surveillance to detect disease occurrence; preparations can be made to deal with anticipated disease occurrences; and resources can be directed to them in the same way that biodiversity hotspots are used in setting other conservation priorities (Mittermeier et al. 1998). Examples of intervention techniques directed at small foci include dispersing waterfowl from known foci of duck plague (Pearson 1973), and whooping cranes (*Grus americana*) from sites of avian cholera occurrence (Zinkl et al. 1977), and restricting access by baboons (*Papio ursinus*) to sites where they were exposed to bovine tuberculosis (Keet et al. 2000). Alternatively, if based on the understanding of the ecological niche of a disease, specific focal disease areas are identified, then

Figure 7.5 An 'I'iwi (*Drepanis coccinea*) foraging in its native habitat. (Source: Robby Kohley, from https://abcbirds.org/bird/iiwi.)

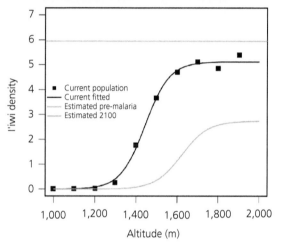

Figure 7.6 Predicted 'I'iwi (*Drepanis coccinea*) density (no. pairs/ha) on the western side of the island of Hawaii, plotted as a function of altitude for pre-malaria, present, and estimated malaria transmission risk in 2100. The observed current 'I'iwi population size is shown by solid squares. (Redrawn from Guillaumet et al. 2017.)

transfer of susceptible animals into that area might be prevented (Wobeser 2007; see also Section 13.6).

7.4 Conclusions and summary

The environmental features of a disease refer to determinants that are neither a feature of the causative agent nor a characteristic of the host population, and which define the ecologic niche of the disease. The choice of which abiotic and biotic features should be investigated must be based on a hypothesis of the disease process; factors that influence transmission among hosts and survival of the agent outside the host are usually most important. Among abiotic factors, climate, weather, and water are important in all diseases. The most important biotic factors are often related to trophic relationships and the location of the host within a food web.

Temporal and spatial scales are important in considering the association between environmental factors and disease occurrence. Techniques such as ecologic niche modeling are very useful for identifying environmental features associated with disease, for predicting the effect of environmental change on disease, and for planning and implementing disease management.

References

Adjemian, J.C.Z., Girvetz, E.H., Beckett, L., and Foley, J.E. (2006). Analysis of genetic algorithm for rule-set production (GARP) modeling approach for predicting distributions of fleas implicated as vectors of plague, *Yersinia pestis*, in California. *Journal of Medical Entomology*, **43**:93–103.

Atkinson, C.T., Utzurrum R.B., and Lapointe, D.A. (2014). Changing climate and the altitudinal range of avian malaria in the Hawaiian Islands—an ongoing conservation crisis on the island of Kaua'i. *Global Change Biology*, **20**:2426–36.

Audy, J.R. (1958). The localization of disease with special reference to the zoonoses. *Transactions of the Royal Society of Tropical Medicine and Hygiene*, **52**:308–28.

Baptista-Rosas, R.C., Hinojosa, A., and Riquelme, M. (2007). Ecological niche modeling of *Coccidioides* spp. in western North American deserts. *Annals of the New York Academy of Sciences*, **1111**:35–46.

Benning, T.L., LaPointe, D.A., Atkinson, A.C., and Vitousek, P.M. (2002). Interactions of climate change with biological invasions and land use in the Hawaiian islands, modeling the fate of endemic birds using a geographic information system. *Proceedings of the National Academy of Sciences*, **99**:14246–49.

Blackburn, J.K., McNyset, K.M., Curtis, A., and Hugh-Jones, M.E. (2007). Modeling the geographic distribution of *Bacillus anthracis*, the causative agent of anthrax disease, for the contiguous United States using predictive ecologic niche modeling. *The American Journal of Tropical Medicine and Hygiene*, **77**:1103–10.

Brooks, D.R. and Hoberg, E.P. (2007). How will global climate change affect parasite-host assemblages? *Trends in Parasitology*, **23**:571–74.

Comrie, A.C. (2005). Climate features influencing coccidioidomycosis seasonality and outbreaks. *Environmental Health Perspectives*, **113**:688–92.

Comrie, A.C. and Glueck, M.F. (2007). Assessment of climate-coccidioidomycosis model: Model sensitivity for assessing climatologic effects on the risk of acquiring coccidioidomycosis. *Annals of the New York Academy of Sciences*, **1111**:83–95.

Costa, J., Peterson, A.T., and Beard, C.B. (2002). Ecological niche modeling and differentiation of populations of *Triatoma brasiliensis neiva*, 1911, the most important Chagas' disease vector in northeastern Brazil (Hemiptera, Reduviidae, Triatominae). *The American Journal of Tropical Medicine and Hygiene*, **67**:516–20.

Davis, S., Calvet, E., and Leirs, H. (2005). Fluctuating rodent populations and risk to humans from rodent-borne zoonoses. *Vector-Borne and Zoonotic Diseases*, **5**:305–14.

Dumas, A. (1871). *Les Mohicans de Paris*. Michel Lévy Frères, Paris.

Elith, J. and Leathwick, J.R. (2009). Species distribution models: Ecological explanation and prediction across space and time. *Annual Review of Ecology, Evolution and Systematics*, **40**:677–97.

Epstein, P.R. (2004). Climate change and public health: Emerging infectious diseases. pp 381–392 In: Cleveland, C.J. (ed.), *Encyclopedia of Energy* Elsevier Academic Press, Amsterdam.

Fisher, F.S., Bultman, M.W., Johnson, S.M, Pappagianis, D., and Zabrosky, E. (2007). Coccidioides niches and habitat parameters in the southwestern United States: A matter of scale. Annals of the New York Academy of Sciences, 1111:47–72.

Flies, E.J., Weinstein, P., Anderson, S.J., Koolhof, I., Foufopoulos, J., and Williams, C.R. (2018). Ross River virus and the necessity of multiscale, eco-epidemiological analyses. *The Journal of Infectious Diseases*, **217**:807–15.

Freed, L.A., Cann, R.L., Goff, M.L., Kuntz, W.A., and Bodner, G.R. (2005). Increase in avian malaria at upper elevation in Hawai'i. *The Condor*, **107**:753–64.

Gese, E.M., Schultz, R.D., Johnson, M.R., Williams, E.S., Crabtree, R.L., and Ruff, R.L. (1997). Serological survey for diseases in free-ranging coyotes (*Canis latrans*) in Yellowstone National Park, Wyoming. *Journal of Wildlife Diseases*, **33**:47–56.

Giles, J., Peterson, A.T., and Almeida, A. (2011). Ecology and geography of plague transmission areas in northeastern Brazil. *PLoS: Neglected Tropical Diseases*, **5**:e925.

Gray, J.S., Dautel, H., Estrada-Peña, A., Kahl, O., and Lindgren, E. (2009). Effects of climate change on ticks and tick-borne diseases in Europe. *Interdisciplinary Perspectives on Infectious Diseases*, 2009:593232. doi:10.1155/2009/593232

Grinnell, J. (1917). Field tests of theories concerning distributional control. *The American Naturalist*, **51**: 115–28.

Guillaumet, A., Kuntz, W.A., Samuel, M.D., and Paxton, E.H. (2017). Altitudinal migration and the future of an iconic Hawaiian honeycreeper in response to climate change and management. *Ecological Monographs*, **87**:410–28.

Guillera-Arroita, G., Lahoz-Monfort, J.J., Elith, J., Gordon, A., Kujala, H., Lentini, P.E., et al. (2015). Is my species distribution model fit for purpose? Matching data and models to applications. *Global Ecology and Biogeography*, **24**:276–92.

Hall, E.M., Crespi, E.J., Goldberg, C.S., and Brunner, J.L. (2016). Evaluating environmental DNA-based quantification of ranavirus infection in wood frog populations. *Molecular Ecology Resources*, **16**:423–33.

Herbold, J.R. (2005). Emerging zoonotic diseases: An opportunity to apply the concepts of nidality and one-medicine. *Environmental Health and Preventive Medicine*, **10**: 260–62.

Hueffer, K., O'Hara, T.M., and Follmann, E.H. (2011). Adaptation of mammalian host-pathogen interactions in a changing arctic environment. *Acta Veterinaria Scandinavica*, **53**:17.

Hutchison, G.E. (1957). Concluding remarks. *Cold Spring Harbor Symposia on Quantitative Biology*, **22**:415–27.

Huver, J.R., Koprivnikar, J., Johnson, P.T.J., and Whyard, S. (2015). Development and application of an eDNA method to detect and quantify a pathogenic parasite in aquatic ecosystems. *Ecological Applications*, **25**:991–1002.

Jones, C.G., Ostfeld, R.S., Richard, M.P., Schauber, E.M., and Wolff, J.O. (1998). Chain reactions linking acorns to gypsy moth outbreaks and Lyme disease risk. *Science*, **279**:1023–26.

Kamoroff, C. and Goldberg, C.S. (2017). Using environmental DNA for early detection of amphibian chytrid fungus *Batrachochytrium dendrobatidis* prior to a ranid die-off. *Diseases of Aquatic Organisms*, **127**:75–79.

Keet, D.F., Kriek, N.P.J., Bengis, R.E., Grobler, D.G., and Michel, A. (2000). The rise and fall of tuberculosis in a free-ranging baboon troop in the Kruger National Park. *Onderstepoort Journal of Veterinary Research*, **67**:115–22.

Kingscote, B.F. (1970). Correlation of bedrock type with geography of leptospirosis. *Canadian Journal of Comparative Medicine*, **34**:31–37.

Kock, R.A., Orynbayev, M., Robinson, S., Zuther, S., Singh, N.J., Beavuvais, W., et al. (2018). Saigas on the brink: Multidisciplinary analysis of the factors influencing mass mortality events. *Science Advances*, **4**:eaao2314.

Kolivras, K.N. and Comrie, A.C. (2003). Modeling valley fever (coccidioidomycosis) incidence based on climatic conditions. *International Journal of Biometeorology*, **47**:87–101.

Kutz, S.J., Jenkins, E.J., Veitch, A.M., Ducrocq, J., Polley, L., Elkin, B., et al. (2009). The Arctic as a model for anticipating, preventing, and mitigating climate change impacts on host-parasite interactions. *Veterinary Parasitology*, **163**:217–28.

Leighton, F.A., Artsob, H.A., Chu, M.C., and Olson, J.G. (2001). A serological survey of rural dogs and cats on the southwestern Canadian prairie for zoonotic pathogens. *Canadian Journal of Public Health*, **92**:67–71.

Lembo, T., Hampson, K., Auty, H., Beesley, C.A., Bessell, P., Packer, C., et al. (2011). Serologic surveillance of anthrax in the Serengeti ecosystem, Tanzania, 1996–2009. *Emerging Infectious Diseases*, **17**:387–94.

Levin, S.A. (1992). The problem of pattern and scale in ecology: The Robert H. MacArthur Award lecture. *Ecology*, **73**:1943–67.

Levine, R.S., Peterson, A.T., and Benedict, M.Q. (2004). Geographic and ecologic distributions of the *Anopheles gambiae* complex predicted using a genetic algorithm. *The American Journal of Tropical Medicine and Hygiene*, **70**:105–09.

Levins, R. (1995). Preparing for uncertainty. *Ecosystem Health*, **1**:47–57.

Limmathurotsakul, D., Wuthiekanun, V., Chantratita, N., Wongsuvan, G., Amornchai, P., Day, N.P.J., et al. (2010). *Burkholderia pseudomallei* is spatially distributed in soil in northeast Thailand. *PLoS: Neglected Tropical Diseases*, https://doi.org/10.1371/journal.pntd.0000694

MacArthur, R.H. (1972). *Geographical Ecology: Patterns in the Distribution of Species*. Princeton University Press, Princeton, NJ.

Mittermeier, R.A., Myers, N., Thomsen, J.B., Da Fonseca, G.A.B., and Olivieri, S. (1998). Biodiversity hotspots and major tropical wilderness areas: Approaches to setting conservation priorities. *Conservation Biology*, **12**:516–20.

Morris, A.L., Guégan, J.F., Andreou, D., Marsollier, L., Carolan, K., Le Croller, M., et al. (2016). Deforestation-driven food-web collapse linked to emerging tropical infectious disease, *Mycobacterium ulcerans*. *Science Advances*, **2**:p.e1600387.

Mouriston, K.N., Tompkins, D.M., and Poulin, R. (2005). Climate warming may cause a parasite-induced collapse in coastal amphipod populations. *Oecologia*, **146**:476–83.

Olivero, J., Fa, J.E., Real, R., Márquez, A.L., Farfán, M.A., Vargas, J.M., et al. (2017). Recent loss of closed forests is associated with Ebola virus disease outbreaks. *Scientific Reports*, **7**:1–9.

Pavlovsky, E.N. and Pious, F.K., Jr. (1966). *Natural Nidality of Transmissible Diseases: With Special Reference to the Landscape Epidemiology of the Zooanthroponoses*. University of Illinois Press, Urbana, IL.

Pearson, G. (1973). The Duck Virus Enteritis Outbreak at Lake Andes National Wildlife Refuge. US Fish and Wildlife Service Mimeo Report, Washington, DC.

Peterson, A.T. (2006). Ecologic niche modeling and spatial patterns of disease transmission. *Emerging Infectious Diseases*, **12**:1822–26.

Peterson, A.T., Bauer, J.T., and Mills, J.N. (2004). Ecologic and geographic distribution of filovirus disease. *Emerging Infectious Diseases*, **10**:40–47.

Peterson, A.T., Lash, R.R., Carroll, D.S., and Johnson, K.M. (2006a). Geographic potential for outbreaks of Marburg hemorrhagic fever. *The American Journal of Tropical Medicine and Hygiene*, **75**:9–15.

Peterson, A.T., Papeş, M., Reynolds, M.G., Perry., N.D., Hanson, B., Regenery, R.L., et al. (2006b). Native-range ecology and invasive potential of *Cricetomys* in North America. *Journal of Mammalogy*, **87**:427–32.

Peterson, A.T., Sánchez-Cordero, V., Beard, C.B., and Ramsey, J.M. (2002). Ecologic niche modeling and potential reservoirs for Chagas disease, Mexico. *Emerging Infectious Diseases*, **8**:662–67.

Pikula, J., Treml, F., Beklová, M., Holesovska, Z., and Pikulova, J. (2003). Ecological conditions of natural foci of tularaemia in the Czech Republic. *European Journal of Epidemiology*, **18**:1091–95.

Poulin, R. (2006). Global warming and temperature-mediated increases in cercarial emergence in trematode parasites. *Parasitology*, **132**:143–51.

Pullan, R.L., Gething, P.W., Smith, J.L., Mwandawiro, C.S., Sturrock, H.J.W., Gitogna, C.W., et al. (2011). Spatial modeling of soil-transmitted helminth infections in Kenya: A disease control planning tool. *PLoS: Neglected Tropical Diseases*, **5**:e958.

Shuman, E.K. (2010). Global climate change and infectious diseases. *New England Journal of Medicine*, **362**:1061–63.

Smith, D.L., Lucey, B., Waller, L.A., Childs, J.E., and Real, L.A. (2002). Predicting the spatial dynamics of rabies epidemics on heterogeneous landscapes. *Proceedings of the National Academy of Sciences*, **99**:3668–72.

Soberón, J. and Peterson, A.T. (2005). Interpretation of models of fundamental ecological niches and species' distributional areas. *Biodiversity Informatics*, **2**:1–10.

Staubach, C., Thulke, H.H., Tackmann, K., Hugh-Jones, M., and Conraths, F.J. (2001). Geographic information system-aided analysis of factors associated with the spatial distribution of *Echinococcus multilocularis* infec-

tion of foxes. *The American Journal of Tropical Medicine and Hygiene*, **65**:943–48.

Swei, A., Meentemeyer, R., and Briggs, C.J. (2011). Influence of abiotic and environmental factors on the density and infection prevalence of *Ixodes pacificus* (Acari: Ixodidae) with *Borrelia burgdorferi*. *Journal of Medical Entomology*, **48**:20–28.

Triebenbach, A.M., Vogl, S.J., Lotspeich-Cole, L., Sikes, D.S., Happ, G.M., and Hueffer, K. (2010). Detection of *Francisella tularensis* in Alaskan mosquitos (Diptera: Culicidae) and assessment of a laboratory model for transmission. *Journal of Medical Entomology*, **47**:639–48.

Warner, R.E. (1968). The role of introduced diseases in the extinction of the endemic Hawaiian avifauna. *The Condor*, **70**:101–20.

Wobeser, G.A. (2007). *Disease in Wild Animals: Investigation and Management*, 2nd edn. Springer-Verlag, Berlin.

Yakob, L. and Mumby, P.J. (2011). Climate change induces demographic resistance to disease in novel coral assemblages. *Proceedings of the National Academy of Sciences*, **108**:1967–69.

Zinkl, J.G., Dey, N., Hyland, J.M., Hurt, J.J., and Heddleston, K.L. (1977). An epornitic of avian cholera in waterfowl and common crows in Phelps County, Nebraska, in the spring, 1975. *Journal of Wildlife Diseases*, **13**:194–98.

CHAPTER 8

Agent and Disease Detection—Laboratory Methods

All forms of analysis make assumptions which, if violated, can invalidate the conclusions. The difficulty is that you will still get a result even if it is incorrect.

(Clare C. Constantine 2003)

8.1 Introduction

There are many reasons why it may be important to know the status of wild animals with regard to the presence, prevalence, and distribution of a disease agent, or of the disease caused by that agent. Such reasons may range from basic science research into the epizootiology of a certain wildlife parasite, to urgent conservation investigations into the decline of an endangered species. Ultimately, understanding the epidemiology of wildlife diseases depends ineluctably on determining accurately the infection status of individual animals—obtaining this type of information requires testing animals. Here, "test" refers to any method of gathering data on an animal's status with regard to presence or absence of infection or disease. There are two general types of testing (Smith, 1991). *Diagnostic testing* "begins with diseased individuals," such as individuals found sick or dead, with no cause being evident. The need, in this situation, is to identify the nature and cause of the animal's condition, that is, to make a diagnosis. The person(s) who will carry out the sample analysis, usually in a diagnostic laboratory, should be contacted for advice on specimen submission. Ideally, the laboratory will be one that specializes in diseases of wild animals, but, because these are few in number, the analysis will more likely be done in a veterinary diagnostic laboratory, operated by the government or a university. Pathologists there will likely have experience with diverse species (mostly domestic), as well as the ancillary analytic services, such as bacteriology, virology, parasitology, and toxicology, needed to identify many disease agents, but may lack experience with wild animals. In general, freshly dead, unopened carcasses, kept chilled until submitted, are optimal for examination. If it is not practical to submit fresh carcasses, laboratory staff can recommend the best alternative, which might be to freeze intact specimens until they can be submitted, or to perform a field necropsy, with preservation of selected tissues in a specific manner.

Safety of the individual(s) handling the sick or dead animal is a major concern, as many diseases of wild animals are transmissible to humans. Suitable personal protective equipment, including rubber or plastic gloves, eye protection, a protective mask, and disposable or sterilizable outer garments need to be available and used to prevent contact with animal fluids or tissues; and workers need to be trained in their use (Bosch et al. 2013). Roffe and Work (2005)

Infectious Disease Ecology and Conservation. Johannes Foufopoulos, Gary A. Wobeser and Hamish McCallum, Oxford University Press.
© Johannes Foufopoulos, Gary A. Wobeser and Hamish McCallum (2022). DOI: 10.1093/oso/9780199583508.003.0008

provide an overview of techniques for many aspects of investigation of wildlife mortality.

Screening testing begins with taking samples from presumably healthy individuals and is used when high-quality information from live animals is required to support management choices, such as the risk in translocating animals to a new area, or the feasibility of eliminating a disease. The approach, type and level of sampling, and which tests are used, are defined by the reason for testing, features of the specific disease(s) of concern, and the availability of testing resources. To choose appropriate tests, it is necessary to "know what you are looking for," as there is no general test that will detect all agents or diseases, or be suitable for all purposes. Given the broad range of possible tests and the details involved in many of them, this chapter focuses primarily on common features and considerations of frequently used testing programs.

Successful diagnostic testing is a process with distinct preanalytical, analytical, and postanalytical components. All aspects of this total testing process (TTP) (Plebani 2010) should be considered *before* any sampling is done. Because of the specialized skills needed for most diagnostic tests, it is likely that most specimens collected during a disease investigation will be analyzed by someone other than the collector, so planning the TTP requires collaboration (see Figure 8.1). Thus, choosing the most suitable test(s), after assessing the strengths and weaknesses of those available, is probably the most critical aspect of planning.

8.2 Sampling—Preanalysis

Preanalytical errors account for 60–70% of problems in human laboratory diagnostics, most of which are attributable to "mishandling procedures during collection, handling, preparing or storing the specimens" (Lippi et al. 2011). Preanalytical errors are likely to be even more common when testing wild animals, because samples are collected under less controlled conditions than those from humans. To reduce errors, those doing the sampling and the laboratory analyses should agree, *before* sampling begins, on where, which, and how many animals to sample, what to sample, and how samples will be collected, preserved, and submitted

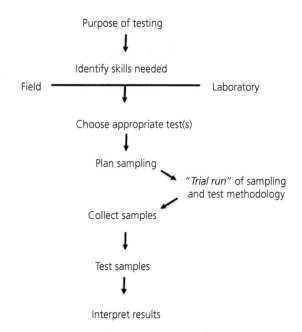

Figure 8.1 Steps in developing a testing program for a disease or agent. Once the purpose has been defined, the process should be collaborative and involve those who are knowledgeable about the species, and those who will conduct the test(s).

to the laboratory. The sampling design described by Springer et al. (2016) exemplifies the detail that may be necessary in planning disease surveillance. Both sampling and analytic methodology should be ideally tested by a "trial run" prior to the actual sampling.

The appropriate number of animals to sample depends upon the question to be answered. A common situation involves the need to know, with reasonable certainty, if an agent is present in a population or area. This often arises if animals are to be translocated, or when a wild species is suspected in playing a role in a disease of domestic animals or humans. Detecting an agent that is present at low prevalence requires examination of a much larger sample of animals than is generally appreciated and nondetection can be a serious problem. Detecting agents that have an aggregated distribution (see Chapter 2, Section 2.1, Figure 2.2), typical of many macroparasites, is also problematic. Methods are available for determining the minimum sample size required to be reasonably certain of detecting at least one diseased individual in a population of given size, among which the disease occurs at a specified prevalence (Hanley and Lippmann-Hand 1983; DiGiacomo and Koepsell 1986; Gu and Novak 2004; Thrushfield 2005). Chapter 4 discusses in more detail many of the considerations that go into determining sample size.

The search for avian influenza A viruses in wild birds provides a striking example of the sample size that may be required in some situations. After testing 22,892 samples from wild aquatic birds, Krauss et al. (2016) were careful to conclude that, although they had "95% confidence (assuming a 90% test sensitivity) that they would have detected" a highly pathogenic virus "at a prevalence of <0. 05%," they could not conclude that infection was *not* present in wild birds (Figure 8.2). To understand genetic diversity within the agent population may also require large samples. For instance, at least 112 subtypes of influenza A virus have been identified in wild birds. The sample size required to detect 75% of

subtypes in circulation in wild birds in the northern hemisphere was estimated to be 10,000–50,000 birds (Olson et al. 2014).

Requiring such a large sample may discourage any testing. The "rule of three" (Hanley and Lippmann-Hind 1983) is useful for considering what can be learned from a more modest sample. This rule is based on the observation that "if none of n patients shows the event about which we are concerned, we can be 95% confident that the chance of the event is at most 3 in n (i.e., $3/n$)" (see also Chapter 6, Section 6.3, Figure 6.1). The rule can be used in two ways: first, if a sample of 50 animals has been examined without finding evidence of disease, one can be 95% confident that the prevalence of disease in the population from which the sample was collected was not greater than 3/50, or 6% (assuming that the test used had a sensitivity of 100%, and that the sample was representative of the population). Alternatively, the rule can also be

Figure 8.2 A United States Geological Service employee collects an oral swab sample from a northern Pintail (*Anas acuta*) during a survey for avian influenza. Such testing is necessary to establish the identity and distribution of the different influenza strains circulating in waterfowl populations.

used to determine an appropriate sample size given a specified level of confidence, and a suspected prevalence of disease. If one wishes to be 95% confident of detecting at least one infected individual in a population in which the disease prevalence is 1%, a sample of 300 animals (3/0.01) would be required. Results of this type of test sample calculation may be useful in the planning stages of a more extensive program, as well avoiding the problem of inadequate sample size. For instance, examining a sample of 10 animals without finding infection would allow one to conclude that there is a 95% probability that no more than 30% of the population is infected!

In most instances, the samples should be representative of the studied host population (in terms of sex, age, and geographic distribution), which requires some knowledge of the total population, the dynamics of transmission of the infection, and the likelihood of bias (Courchamp et al. 2000; Connor et al. 2000). However, it may be advantageous to select a biased sample for some purposes. For instance, Thulke et al. (2009) suggested a "situation-based" approach to detecting the status of an area for infections that cause readily recognizable morbidity and mortality. In this approach, two types of host animal are differentiated: *indicator animals* (IA) (individuals suspected of having the disease, because of clinical signs or "suspicious" behavior, including being found dead, or even because they belong to a demographic group that is disproportionately likely to be infected); and *hunted animals* (HA) (individuals whose infection status is unknown, and assumed to be representative of the total host population). For example, if an infection is known to be present in a population, a systematic sampling of HA is appropriate to estimate objective values such as prevalence or parasite burden (Thulke et al. 2009).

Alternatively, when the presence of a disease is unknown, (either because it has never been detected or it was present and appears to have disappeared), the likelihood of detecting an infected individual is much greater if IA, rather than HA, are examined. This approach increases the probability of detecting the presence of a pathogen with reduced effort, but at the expense of being able to determine an unbiased estimate of prevalence in the host population.

Related to this is the approach of focusing sampling on the oldest (and in species with indeterminate growth, largest) individuals in a population: for life-long infections of low virulence, such as sexually transmitted pathogens and certain apicomplexan blood parasites, they will have the highest probability of testing positive for the parasite. Similarly, if sarcoptic mange has never been detected in a population of coyotes (*Canis latrans*), focused sampling of coyotes with visible skin lesions, rather than a random sample, will likely increase the probability of detecting infected animals. The same concept can be applied to invertebrate hosts: using traps that selectively target the relatively small subgroup of gravid, blood-fed—and therefore exposed—mosquitoes, substantially increases the probability of detecting vector-transmitted pathogens in an area (Springer et al. 2016) (see also Section 13.2 on disease surveillance).

8.3 Sampling—Analysis

Analysis of disease status may include considerations of history, physical examination, laboratory tests, function tests, or combinations of these. A test may be *pathognomonic* (an absolute predictor of the presence of the disease or disease agent) or *surrogate* (based on secondary changes, which it is hoped will predict the presence or absence of disease or disease agent) (Martin et al. 1987). For example, presence of a high concentration of lead in blood of a duck is pathognomonic for lead poisoning, while lesions detected at necropsy including impaction of the proventriculus, emaciation, and pallor of the viscera are presumptive or surrogate evidence of lead poisoning, and require toxicological confirmation. *Diagnostic accuracy* (ability of a test to discriminate between subjects with and without the condition) is the most important among many factors to consider in choosing a test. Accuracy is usually measured by the degree of agreement between the test under evaluation (the *index test*) and a *reference standard* which is "the best available method for establishing the presence or absence of the condition of interest" (Bossuyt et al. 2003), "by testing a standard set of well-documented samples using the techniques in parallel" (Conraths and Schares 2006). The index test should be based on a method different than that used to define the true status of the animals, because if both make the same misclassification errors, the apparent accuracy of the index test will be inflated (Vacek 1985). For example, Dalley et al. (2008) assessed a new test for detecting bovine tuberculosis (bTB) in European badgers by comparing results of the test to those obtained by culture of *Mycobacterium bovis* from postmortem tissue samples (the reference standard). Accuracy can be expressed in a number of ways: the two measures most often used are *sensitivity* and *specificity*, which are calculated from the proportion of infected and uninfected animals, respectively, that are diagnosed correctly (see also Chapter 6, Section 6.2, Table 6.1). Diagnostic accuracy is a measure of how well a test performs under a particular set of conditions. Because it cannot necessarily be assumed that a test developed for use in one species will be equally accurate in another species, or that two different laboratories will produce the same result, it is important to validate a test ahead of a study (LeSellier et al. 2008). For instance, Lyashchenko et al. (2008) found that the level of agreement between a serological test for bTB and culture of *M. bovis* from tissues varied among host species (74% for brushtail possums, 81% for European badgers, 90% for wild boar, and 97% for white-tailed deer). Probabilistic approaches for estimating sensitivity and specificity of a test, when no reference standard is available, have been proposed (Enoe et al. 2000); however, such estimates will be biased if certain assumptions are not met (Toft et al. 2005).

Which type of test is most suitable for a particular investigation depends on the specific aims of a project. Conraths and Schares (2006) outline the process of *test validation*, that is, the evaluation of

the fitness of a test for a given use. For example, the sensitivity and specificity of various tests, relative to a reference standard (necropsy), for detection of the causative agent of Johne's disease (caused by *Mycobacterium avium ss. paratuberculosis*) in cattle are shown in Table 8.1 (Collins et al. 2006). These data were used to choose the most appropriate test method for various investigational aims, such as determination of presence/absence of infection in a group, estimation of within-group prevalence, and disease control. Choosing the most appropriate test for a specific purpose might begin with a review of test methods that have been used by others in similar situations, followed by a detailed discussion with the laboratory scientists who will potentially examine the specimens. The aim is to establish which tests are available, and which might be most suitable. Accurate reference data (often called the "reference interval" or "reference range" against which test results can be compared, are essential for interpretation of the data. These background reference data should be species-specific, and preferably obtained from natural populations. Physiological reference intervals are available for many species in captivity (Teare 2013), but there may be important differences between captive and free-ranging animals of the same species (Beecher et al. 2009). However, sometimes when no reference data exist, they need to be generated for a specific situation (Friedrichs et al. 2012).

The challenges of using an appropriate test are illustrated in the case of *M. bovis*, the causative agent of bTB. Accurate diagnosis of this infection is "notoriously problematic" (Drewe et al. 2009). This bacterium infects and produces disease in most mammalian orders (Hickling 1995).

Table 8.1 Comparison of the accuracy of various tests, using specimens collected from live cattle, for detection of paratuberculosis (*Mycobacterium avium* subsp. *paratuberculosis* infection), using necropsy as the reference standard (adapted from Collins et al. 2006).

TEST	SENSITIVITY (%)	SPECIFICITY (%)
BACTERIAL CULTURE OF FECES	60 ± 5	99.9 ± 0.1
PCR OF FECES	30 ± 5	99.5 ± 0.5
ELISA ON SERUM OR MILK	30 ± 5	99.0 ± 1.0
MICROSCOPIC EVALUATION OF BIOPSY SPECIMEN	90 ± 5	100
NECROPSY	100[a]	100[a]

[a] The actual sensitivity and specificity of necropsy for detecting paratuberculosis was unknown.

Wild species are a reservoir of infection for cattle on several continents and the disease precipitates both species conservation and human health problems. For these reasons, sustained effort has been devoted to improve diagnostic methods for *M. bovis* infection. Attempts to develop tests for use in cattle began in the 1890s, but "no single ante- or (post) mortem test for TB can be expected, on its own, to detect every infected herd and every infected animal" (de la Rua-Domenech et al. 2006). Despite being less than perfectly accurate, no better test for screening cattle populations has been developed than the tuberculin skin test (TST), a test based on cell-mediated immunity. When TST has been used in a program of systematic testing of cattle herds at regular intervals, with compulsory slaughter of test "reactors," as well as movement restrictions on known infected herds, and slaughterhouse surveillance for diseased animals, *M. bovis* has been eliminated from cattle in many countries. This suggests that in the presence of a good regulatory framework, it is not necessary to have a perfect test to control or even eliminate an infection. However, elimination of tuberculosis from cattle in other countries has been hampered by the presence of *M. bovis* in wild animals. The rigorous testing program used in cattle is often impractical for wild species, because animals must be captured repeatedly for retesting; the TST must be measured 24–72 hours after tuberculin injection (requiring animals be held or recaptured for testing); furthermore, TST is unreliable in a number of wild species (Chambers 2009). Many alternative tests have been developed (e.g., Lyashchenko et al. 2008), but validation of any test for use in wild animals is hampered by difficulty in determining the true infection status of individual animals without postmortem examination, which "is neither desirable nor feasible in many cases" (Chambers 2009).

8.4 Sampling—Postanalysis

Interpretation of test results requires knowledge of the biology of the animal and the agent, the course of infection in the animals, and the accuracy and other features of the test, all under the particular circumstances. Ideally, all participants in the TTP should be involved in interpretation.

The disease tularemia (caused by the bacterium *Francisella tularensis*) provides an example of the need to understand the nature and course of disease in the host species when interpreting test results. Tularemia occurs most commonly in wild rodents and lagomorphs, and may infect other species, including humans. Serology has been used to quantify the prevalence of infection in wild species in areas of Europe where the disease is endemic. Kaysser et al. (2008) did not detect antibodies to *F. tularensis* in any of a sample of small rodents, although >4% of the animals were known to be infected. In contrast, 5% of European brown hares (*Lepus europaeus*) had antibodies, and all seropositive animals examined were infected (Gyuranecz et al. 2011). These studies are difficult to interpret, without knowing that tularemia is rapidly fatal in rodents, with few survivors; whereas hares develop chronic infection, and infected individuals have antibodies. Thus, serology is useful for detecting the presence of tularemia in hares, but is not a good indicator of infection in wild rodents.

Lastly, it is also important to understand what a test is exactly measuring in order to best interpret the results. For example, a common molecular diagnostic technique, polymerase chain reaction (PCR), may amplify DNA from dead as well as live agents. The presence of dead organisms may yield a positive test result that is diagnostically correct, but biologically is a false positive (Burkardt 2000), in terms of risk of ongoing transmission (Miller et al. 2015). Similarly, the presence of a positive antibody test (enzyme-linked immunosorbent assay (ELISA)), while pointing to a past infection, does not necessarily indicate that an active infection is present. As a result, PCR and ELISA tests, when administered to the same populations of hosts, may produce very different results. Without understanding the biases and constraints of each diagnostic test, interpretation of the analytical results will likely be flawed.

References

Beecher, B.R., Jolles, A.E., and Ezenwa, V.O. (2009). Evaluation of hematologic values in free-ranging African buffalo (*Syncerus caffer*). *Journal of Wildlife Diseases*, **45**:57–66.

Bosch, S.A., Musgrave, K., and Wong, D. (2013). Zoonotic disease risk and prevention practices among biologists and other wildlife workers—results from a national survey, U.S. National Park Service, 2009. *Journal of Wildlife Diseases*, **49**:475–85.

Bossyut, P.M, Reitsma, J.B., Bruns, D.E., Gatsonis, C.A., Glasizou, P.P., Irwing, L.M., et al. (2003). Towards complete and accurate reporting of studies of diagnostic accuracy: The STARD initiative. *American Journal of Roentgenology*, **181**:51–55.

Burkardt, H.-J. (2000). Standardization and quality control of PCR analyses. *Clinical Chemistry and Laboratory Medicine*, **38**:97–91.

Chambers, M.A. (2009). Review of the diagnosis and study of tuberculosis in non-bovine wildlife species using immunological methods. *Transboundary and Emerging Diseases*, **56**:215–27.

Collins, M.T., Gardner, I.A., Garry, F.B., Roussel, A.J., and Wells, S.J. (2006). Consensus recommendations on diagnostic testing for the detection of paratuberculosis in cattle in the United States. *Journal of the American Veterinary Medical Association*, **229**:1912–19.

Connor, M.M., McCarty, C.W., and Miller, M.W. (2000). Detection of bias in harvest-based estimates of chronic wasting disease prevalence in mule deer. *Journal of Wildlife Diseases*, **36**:691–99.

Conraths, F.J. and Schares, G. (2006). Validation of molecular-diagnostic techniques in the parasitological laboratory. *Veterinary Parasitology*, **136**:91–98.

Constantine, C.C. (2003). Importance and pitfalls of molecular analysis to parasite epidemiology. *Trends in Parasitology*, **19**:346–48.

Courchamp, F., Say, L., and Pontier, D. (2000). Detection, identification, and correction of a bias in an epidemiological study. *Journal of Wildlife Diseases*, **36**:71–78.

Dalley, D., Davé, D., Lesellier, S., Palmer, S., Crawshaw, T., Hewinson, R.G., et al. (2008). Development and evaluation of a gamma-interferon assay for tuberculosis in badgers (*Meles meles*). *Tuberculosis*, **88**:235–43.

de la Rua-Domenech, R., Goodchild, A.T., Vordermeir, H.M., Hewinson, R.G., Christiansen, K.H., and Clifton-Hadley, R.S. (2006). *Ante mortem* diagnosis of tuberculosis in cattle: A review of the tuberculin tests, γ-interferon assay and other ancillary diagnostic techniques. *Research in Veterinary Science*, **81**:190–210.

DiGiacomo, R.F. and Koepsell, T.D. (1986). Sampling for detection of infection or disease in animal populations. *Journal of the American Veterinary Medical Association*, **189**:22–23.

Drewe, J.A, Dean, G.S., Michel, A.L., and Pearce, G.P. (2009). Accuracy of three diagnostic tests for determining *Mycobacterium bovis* infection status in live-sampled wild meerkats (*Suricata suricatta*). *Journal of Veterinary Diagnostic Investigation*, **21**:31–39.

Enoe, C., Georgiadis, M.P., and Johnson, W.O. (2000). Estimation of sensitivity and specificity of diagnostic tests and disease prevalence when the true disease state is unknown. *Preventive Veterinary Medicine*, **45**:61–81.

Friedrichs, K.R., Harr, K.E., Freeman, K.P., Szladovits, B., Walton, R.M., Barnhart., K.F., et al. (2012). ASVP reference interval guidelines: Determination of de novo reference intervals in veterinary science and other related topics. *Veterinary Clinical Pathology*, **41**:441–45.

Gu, W. and Novak, R.J. (2004). Short report: Detection probability of arbovirus infection in mosquito populations. *American Journal of Tropical Medicine and Hygiene*, **71**:636–38.

Gyuranecz, M., Rigó, K., Dán, A., Földvári, G., Makrai, L., Denes, B., et al. (2011). Investigation of the ecology of *Francisella tularensis* during an inter-epizootic period. *Vector-Borne and Zoonotic Diseases*, **11**:1031–35.

Hanley, J.A. and Lippman-Hand, A. (1983). If nothing goes wrong, is everything all right? Interpreting zero numerators. *Journal of the American Medical Association*, **249**:1743–45.

Hickling, G. (1995). Wildlife reservoirs of bovine tuberculosis in New Zealand. In: Griffin, F. and De Lisle, G. (eds.), *Tuberculosis in Wildlife and Domestic Animals*, pp. 276–79 . University of Otago Press, Dunedin.

Kaysser, P., Seibold, E., Mätz-Rensing, K., Pfeffer, M., Essbauer, S., and Splettsoesser, W.D. (2008). Re-emergence of tularemia in Germany: Presence of *Francisella tularensis* in different rodent species in endemic areas. *BMC Infectious Diseases*, **8**:157.

Krauss, S., Stallknect, D.E, Slemons, R.D., Bowman, A.S., Poulson, R.L., Nolting, J.M., et al. (2016). The enigma of the apparent disappearance of Eurasian highly pathogenic H5 clade 2.3.4.4 influenza viruses in North American waterfowl. *Proceedings of the National Academy of Sciences*, **113**:9033–38.

LeSellier, S., Corner, L., Costello, E., Sleeman, P., Lyashchenko, K., Greenwald., et al. (2008). Antigen-specific immunological responses of badgers (*Meles meles*) experimentally infected with *Mycobacterium bovis*. *Veterinary Immunology and Immunopathology*, **122**:35–45.

Lippi, G., Chance, J.J., Church, S., Dazzi, P., Fontana, R., Giavarina, D., et al. (2011). Preanalytical quality improvement: From dream to reality. *Clinical Chemistry and Laboratory Medicine*, **49**:1113–26.

Lyashchenko, K.P., Greenwald, R., Esfandiari, J., Chambers, M.A., Vicente, J., Gortazar, C., et al. (2008). Animal-side serologic assay for rapid detection of *Mycobacterium bovis* infection in multiple species of free-ranging wildlife. *Veterinary Microbiology*, **132**:283–92.

Martin, S.W., Meek, A.H., and Willeberg, P. (1987). *Veterinary Epidemiology, Principles and Methods*. Iowa State University Press, Ames, IA.

Miller, M., Buss, P., Hofmeyer, J., Olea-Popelka, F., Parsons, S., and Van Helden, P. (2015). *Ante mortem* diagnosis of *Mycobacterium bovis* infection in free-ranging African lions (*Panthera leo*) and implications for transmission. *Journal of Wildlife Diseases*, **51**:493–97.

Olson, S.H., Parmley, J., Soos, C., Gilbert, M., Latore-Margalef, N., and Hall, J.S. (2014). Sampling strategies and biodiversity of influenza A subtypes in wild birds. *PLoS ONE*, **3**:e90826.

Plebani, M. (2010). The detection and prevention of errors in laboratory medicine. *Annals of Clinical Biochemistry*, **47**:101–10.

Roffe, T.J. and Work, T.M. (2005). Wildlife health and disease investigations. In: Braun, C.E. (ed.), *Techniques for Wildlife Investigations and Management*, 6th edn, pp. 197–212. The Wildlife Society, Bethesda, MD.

Smith, R.D. (1991). *Veterinary Clinical Epidemiology. A Problem-oriented Approach*. Butterworth-Heineman, Stoneham, MA.

Springer, Y.P., Hoekman, D., Johnson, P.T.J., Duffy, P.A., Hufft, R.A., Barnett, D.T., et al. (2016). Tick-mosquito, and rodent-borne parasite sampling designs for the National Ecological Observatory Network. *Ecosphere*, **75**:1–40.

Teare, J.A. (2013). *ISIS Physiological Reference Intervals for Captive Wildlife: A CD-ROM Resource*. International Species Information System, Bloomington, MN.

Thrushfield, M. (2005). *Veterinary Epidemiology*, 3rd edn. Blackwell Science, Oxford.

Thulke, H.-H., Eisinger, D., Freuling, C., Frölich, A., Globig, A., Grimm, V., et al. (2009). Situation-based surveillance: adapting investigations to actual epidemic situations. *Journal of Wildlife Diseases*, **45**:1089–103.

Toft, N., Jorgensen, E., and Hojsgaard, S. (2005). Diagnosing diagnostic tests: Evaluating the assumptions underlying the estimation of sensitivity and specificity in the absence of a gold standard. *Preventive Veterinary Medicine*, **68**:19–33.

Vacek, P.M. (1985). The effect of conditional dependence on the evaluation of diagnostic tests. *Biometrics*, **41**:959–68.

PART III

Modeling and Data Analysis

CHAPTER 9

Disease Modeling

All models are wrong, but some are useful...
George Box 1979, p202

9.1 Why use a model?

One of the fundamental ways in which disease epidemics can be understood and managed involves the use of mathematical models. There is an enormous scientific literature on modeling the dynamics of host–pathogen interactions both in human populations and in natural systems. For a broad background in the area, readers may consult Keeling and Rohani (2008). Not surprisingly, most of this literature is based around infectious diseases of humans and of livestock. Many of the same general approaches can be applied to modeling diseases in wildlife populations. There is, however, one essential difference in the dynamics of diseases in wildlife versus those in human or livestock populations. This is that infectious disease is rarely a primary driver of changes in population size in either humans or livestock. Population size can usually be considered to be constant throughout a disease epidemic in human populations and either constant or a variable to be manipulated (e.g., by culling) in livestock populations. In wildlife populations in contrast, infectious disease can be very important in driving population dynamics and, as a result, our key questions of interest are frequently concerned with the impact that the disease may have on population size. This means that the models need to be rather different in structure.

In any modeling exercise, it is essential to be clear about the purpose of the model. There is no such thing as a "correct" or "best" model for any population or host–parasite system. As the statistician George Box observed "all models are wrong, but some are useful" (Box 1979, p202). Models are a simplified representation of reality. This means that all models leave some details of reality out. In that sense, they are "wrong." And, of course, it is impossible to work out whether a model is "useful" until one has determined what its purpose is.

So, what can we use models of infectious diseases for in conservation biology? First and foremost, modeling needs to be regarded as a process, rather than simply a product (Restif et al. 2012). A successful research project on infectious disease will employ a series of models in an iterative approach over the course of the program. Initially, simple mathematical models can serve as a valuable tool for conceptualizing one's understanding of how an ecological system works. Even without doing any analysis of a model, the process of constructing it forces the investigator to formalize their assumptions about the major components of the system and how they interact to drive dynamics, identifying any gaps or uncertainties. Preliminary analysis of such simple models will identify the components of the host–parasite interaction that are

Infectious Disease Ecology and Conservation. Johannes Foufopoulos, Gary A. Wobeser and Hamish McCallum, Oxford University Press.
© Johannes Foufopoulos, Gary A. Wobeser and Hamish McCallum (2022). DOI: 10.1093/oso/9780199583508.003.0009

most important in determining the outcomes of the interaction. Such sensitivity analysis will then guide data collection and experimental design.

Once data have been collected from both the field and laboratory following priorities identified by simple models, more sophisticated models can be developed. In turn, analysis of these models will identify priorities for future data collection. This iterative approach has been usefully summarized as a "triangulation" approach by Plowright et al. (2008) (see Figure 9.1). Modeling is the "third leg of the stool" of disease ecology, complementing laboratory investigations and field studies, drawing data from each, and feeding back information on parameters that need to be estimated and hypotheses that require investigation.

Finally, management requires modeling. Managing any system means deciding between alternative management actions (keeping in mind that "do nothing" is a possible management action). To decide between management actions, it is necessary to have some way of predicting how the system might respond to each alternative management action. For example, if an infectious disease has emerged in a wildlife population, one potential action might be to cull the population. A model of some sort (even if not a formal mathematical model) would be needed to decide whether any culling would be necessary, and if so, what percentage of the population and on what spatial scale.

9.2 Types of models

9.2.1 Strategic versus tactical models

A distinction is often made between "strategic" and "tactical" models (see Figure 9.2). "Strategic" models are usually very simple and are intended to help in the understanding of broad concepts in population dynamics or epidemiology. This means that they can be applied quite broadly to a variety of different situations and are most useful in the early stages of the iterative approach to modeling when used to inform field and laboratory investigations. However, their relatively simple structure means that they are typically poor at making precise quantitative predictions about the behavior of any particular system. They have relatively few parameters and therefore do not require large amounts of data to be usable. In fact, many strategic models are so general that they do not use data from any particular real host–pathogen system. In contrast, "tactical" models are intended to provide quantitative predictions about the future behavior of a specific host–pathogen system in response to particular management actions. They are generally more complex in structure than strategic models and require substantial amounts of data from the study system in order to estimate the epidemiological and demographic parameters in the model. Tactical models are more appropriate in the later stages of the iterative modeling approach. Successful use of tactical models therefore relies very strongly on both the availability of good-quality data and the use of appropriate statistical tools to derive the necessary parameters from the available data. Chapter 10 deals with these statistical tools.

9.2.2 Deterministic versus stochastic models

Deterministic models do not include any random components, whereas stochastic models include random processes. Processes in the real world are driven by randomness, such as which individuals happen to interact with which other individuals, processes of birth and death, and environmental variation, such as the weather. Nevertheless, deterministic models are perfectly adequate for many applications as they are inherently simpler and the data generated are often sufficient to tell us about the mean or average properties of the system under investigation. For other purposes, such as determining the probability of extinction in a host population, it is essential to include stochastic factors. In fact, it is often the case that a particular problem needs to be attacked by a combination of deterministic and stochastic approaches. It is frequently quite difficult to understand why a stochastic model behaves in a particular way without first understanding the behavior of a deterministic version of the same model. This means that deterministic models are used more often in the earlier stages of an iterative modeling approach. Depending on the objectives of a research program, models used later in the iterative approach might be either stochastic or deterministic.

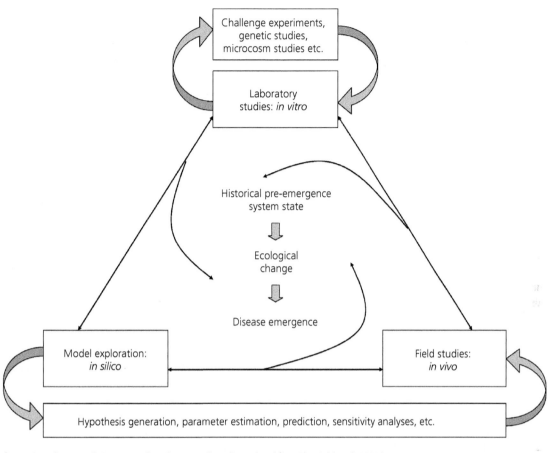

Figure 9.1 The triangulation approach to disease ecology. (Reproduced from Plowright et al. 2008.)

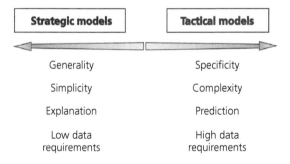

Figure 9.2 Strategic versus tactical models in the study of infectious disease epidemics.

A major practical issue in using stochastic models is that one run of a stochastic model is entirely uninformative. You need to run the model a large number of times (typically at least 1,000) and then examine the frequency distribution of the results. For complex models, computation time can therefore become a substantial problem.

9.2.3 Microparasite versus macroparasite models

Since the pioneering work of Anderson and May (Anderson and May 1978, 1979; May and Anderson 1978, 1979), it has been usual to divide epidemiological models into microparasite and macroparasite models. As is discussed in Chapter 2, and as their names suggest, microparasites are typically small: viruses, bacteria, protozoa, and fungi, whereas macroparasites are typically larger multicellular organisms such as worms. The simplest microparasite models divide the host population into compartments of "susceptible," "infectious," and "recovered" individuals, and track the rate at which individuals move between each of these compartments. Such models are often called SIR models. The key variable that needs to be modeled and measured is the prevalence of the disease (i.e., the proportion of individuals in the population affected by the disease).

In contrast, as discussed in Chapter 2, for macroparasites it is important to keep track of the number of hosts in the population, the number of parasites affecting the population (or equivalently, the mean parasite burden), and also the distribution of parasites among the hosts. In almost all cases of macroparasite infections, parasites are highly aggregated in the host population, with the majority of parasites being found in a relatively small number of heavily infected hosts (Anderson and May 1978). It is these heavily infected hosts (the "wormy" individuals) that are most affected by the parasite and, therefore, in order to understand the effect that the parasite is having on the population as a whole, estimating and modeling the extent of this aggregation is essential.

9.3 Microparasite models

The simplest microparasite models (Anderson and May 1979; May and Anderson 1979) have the general structure shown in Figure 9.3. This structure can be represented in equations as follows:

$$\frac{dS}{dt} = a\left(S + I + R\right) + vR - bS - \beta\left(S, I\right) \quad (9.1)$$

$$\frac{dI}{dt} = \beta\left(S, I\right) - I\left(b + \alpha\right) - vI \quad (9.2)$$

$$\frac{dR}{dt} = vI - R\left(b + v\right) \quad (9.3)$$

Here, S, I, and R are the numbers of susceptible, infectious, and recovered hosts. Susceptible hosts are born at a rate a per individual host and all classes of host suffer a death rate b independent of disease. In infectious hosts, the background death rate is increased by α because of the presence of the infection. The most important part of any epidemiological model is modeling the rate

of infection. Here, it is assumed to be a function β, which depends on the numbers of susceptible and infectious hosts. The way in which this dependence may operate depends on the nature of the microparasite.

One very common assumption is that infectious and susceptible hosts encounter each other through a process of random collision, so that the number of contacts is proportional to the density of both susceptible and infectious hosts. Thus, the transmission function is simply βSI. This is sometimes described as "mass action" transmission in the literature, but we will describe it as "density-dependent transmission" (McCallum et al. 2001). It is often an appropriate model for directly transmitted pathogens such as those spread by aerosol. Because microparasite infections are often eventually cleared by the immune system, hosts in the infected category will—if they don't die first—eventually transition into the recovered category at a rate v. Many microparasite infections also result in their host eventually developing an immunity of variable duration. Eventually, many recovered hosts may become susceptible again, thus transitioning into the susceptible category at a rate v.

Although Equations (9.1)–(9.3) are relatively simple in structure, it is not possible to write down an algebraic solution for the values of S, I, and R as functions of time. For particular values of the parameters and for particular starting values, it is possible to use numerical methods to solve the equations.

For some other types of microparasite, the number of encounters hosts have with others depends weakly, if at all, on population density. For example, some microparasites are sexually transmitted. The number of sexual encounters an individual animal has is largely determined by the mating system of the species in question and does not depend strongly on population density. This means that the number of contacts a susceptible host makes with other hosts is approximately constant and it is the proportion of those contacts that are with infectious hosts that determines the rate of the pathogen transmission. In this case, the transmission function is therefore $\beta S\, I/N$, where I/N is the proportion of infectious hosts among the total population N. This form of transmission is

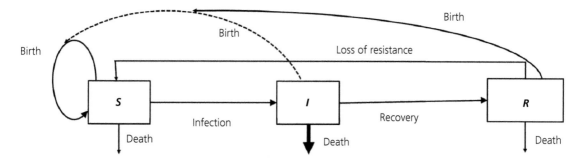

Figure 9.3 Basic structure of a compartmental model for microparasite infection in wildlife. The principal difference between this structure and most models of human infectious diseases is that the death rate of infected individuals is substantially increased beyond that of susceptible and recovered individuals, shown by the bold arrow. In human infections, the timescale is usually sufficiently short that reproduction can be discounted, whereas for many wildlife infections it needs to be taken into account. Reproduction from the infected class is shown with a dashed arrow because for some diseases, infected animals have reduced or no reproduction.

often described as frequency-dependent transmission. Vector-borne infections are, similar to sexually transmitted diseases, likely to have frequency-dependent transmission. Most vectors are capable of feeding on multiple species of host. As a result, vector populations do not depend strongly on host population size, meaning that the rate at which vectors bite a particular host species does not depend strongly on host population size. In other words, the number of blood meals that a vector takes is largely independent of the host population size, and the probability that this vector will pick up the parasite depends primarily on the proportion of those hosts that carry infection. Empirical data suggest that the transmission function for vector-borne pathogens may often be intermediate between "density-dependent" and "frequency-dependent" transmission (McCallum et al. 2001).

A flexible form of a transmission function that can encompass both density dependence and frequency dependence is:

$$F(I, S, N) = \beta S \frac{I}{N^\gamma} \tag{9.4}$$

Here, $F(I, S, N)$ is the transmission function, dependent on I, S, and N; where β is the transmission parameter and γ is a constant between zero and one, with zero corresponding to density-dependent transmission and one to frequency-dependent transmission.

Disease ecologists can add realism to these deterministic models of wildlife disease and account for environmental variation by utilizing a stochastic model approach. It is possible to develop stochastic versions of simple SIR models such as those in Equations (9.1)–(9.3). Each term in the equations represents a possible transition in the state of the system. For example, the first term in Equation (9.1) represents a transition that will increase the number of susceptible individuals by one, and occurs at a rate $a(S + I + R)$ per unit of time, whereas the infection term, the last term in Equation (9.1), simultaneously decreases the number of susceptible individuals by one and increases the number of infected individuals by one. To solve this stochastic version numerically, it is possible to use the Gillespie algorithm (Gillespie 1977). The idea here is very simple. If we represent all the n possible transitions by a_i, then the waiting time τ until the first transition happens is an exponentially distributed random variable with a mean of $\tau = \exp\left(-\sum_i a_i\right)$. Which transition happens is determined by the relative values of the rates at that particular time. For example, the probability that the first transition happens is just $a_i / \sum a_i$.

Simulating the system is then a matter of repeatedly sampling from an exponential distribution to see when an event happens, randomly choosing which event happens using the relative values of the rates, and updating the state variables S, I, and R accordingly. The only difficulty is that if population sizes are large and there are many transitions, then events may happen very rapidly, meaning that generating the solution is a slow process. Bear in mind that a single solution of a stochastic model is useless, and it will be necessary to run the entire model some hundreds or thousands of times to generate frequency distribution divisions of the possible outcomes.

9.3.1 Incorporating age structure and time delays

One of the major limitations in applying Equations (9.1)–(9.3) in practice is that they do not include age structure or time delays: individuals are assumed to be immediately able to reproduce or become infected as soon as they are born, and infected animals are assumed to be immediately able to transmit infection. These are quite limiting assumptions. Almost all organisms have a significant maturation period before they are able to reproduce and, unless this is included in a model, it is difficult to sensibly quantify the reproductive rate. Frequently, newborns are not susceptible to infection, either because they do not make appropriate contacts with infectious individuals or because maternal antibodies protect them from infection early in life. Many pathogens have a substantial latent or incubation period between infection and becoming infectious. It is relatively easy to add an "exposed" class E (infected, but not yet infectious) to the basic model, producing an SEIR model, but if this is represented simply with a constant rate of transition from the exposed class to the infectious class, it assumes that the latent period has an exponential distribution, which is unlikely to be appropriate in practice.

There are numerous ways in which time delays can be added to the basic framework. Wearing et al. (2005) describe a relatively simple approach that is particularly useful because it combines the algebraic tractability of differential equation models with realistic time delays. To model a time delay, they suggested dividing the relevant compartment into m different substages, each of which has a constant rate of recruitment into it and of maturation out of it. The resulting overall stage then has a delay that follows a gamma distribution. To illustrate this in more detail, adding a single exposed category to

Equations (9.1)–(9.3) entails replacing Equation (9.2) with two equations:

$$\frac{dE}{dt} = \beta(S, I) - E(b + \sigma) \tag{9.5}$$

$$\frac{dI}{dt} = E\sigma - I(b + \alpha) - vI \tag{9.6}$$

Here, E is the exposed class and α is the rate of maturation out of the exposed class, so that the mean incubation period is $1/\sigma$. The incubation period will have an exponential distribution, which means that exposed animals have the same probability of becoming infectious in the next short time interval whether they have just been infected or whether they were infected some time ago. This does not correspond with our usual understanding of an incubation period. To make this time delay more realistic, we divide E into m substages E_i. The first will be modeled by:

$$\frac{dE_1}{dt} = \beta(S, I) - E_1(b + m\sigma) \tag{9.7}$$

With the remainder modeled by:

$$\frac{dE_i}{dt} = m\sigma E_{i-1} - E_i(b + m\sigma) \tag{9.8}$$

This ensures that the overall incubation period maintains a mean of $1/\sigma$, but the variance in the incubation period is $1/m$, meaning that for large values of m the time delay approaches a fixed constant value $1/\sigma$.

An example applying this approach to conservation biology problem is provided by Beeton and McCallum (2011) in a model investigating the feasibility of culling to control Tasmanian devil facial tumor disease. In this model, both age structure and a latent period were modeled using a gamma distribution.

Many microparasites of interest to wildlife biologists are vector-transmitted, in which case it may be necessary to add some equations to represent the vector transmission. Rather than assuming exponential growth in the absence of infection, it is often more plausible to assume that other factors will limit host population growth. There is little point here in producing a catalogue of such models: the important point is that they can be produced.

9.4 Macroparasite models

As mentioned earlier, models for most macroparasite infections differ from microparasite models in that they keep track of how heavily infected the infectious hosts are. Anderson and May (1978) developed a basic framework for macroparasite infections, which models the size of the host population H, the size of the parasite population P, and the distribution of parasites among hosts. In its simplest version, the framework is as follows:

$$\frac{dH}{dt} = (a - b) H - \alpha P \qquad (9.9)$$

$$\frac{dP}{dt} = \lambda \frac{PH}{H + H_0} - (b + \mu) P - \alpha HE\left(i^2\right) \qquad (9.10)$$

Here, as with the corresponding microparasite model, the hosts are assumed to have a density-independent birth rate a and a death rate b, and each parasite is assumed to increment the death rate of its host by a further factor α. Each parasite is assumed to produce free-living infective stages at a rate λ. Transmission is subsumed in the term $H/(H + H_0)$, which is the proportion of infective stages that encounter another host before they die. This model assumes that parasites die when their host dies. The total number of parasite losses will therefore depend on the rate of parasite deaths when an infected host dies b, plus the natural death rate μ of parasites while on a living host, all multiplied by the number of parasites P. The final term in Equation (9.10) is perhaps the one requiring most explanation. Because the death rate of infected hosts is assumed to be linearly dependent on the parasite burden, the number of parasites that are removed by parasite-induced host death depends on the *square* of the number of parasites, because the death rate from parasitism of a host with i parasites is αi and this results in the death of i parasites. The final term is therefore the expected value $E(.)$ of the squared number of parasites per host. One of Anderson and May's major insights was to recognize that parasite distributions among hosts are usually aggregated and are frequently well described by a probability distribution known as the negative binomial, which has a parameter k that

inversely describes the extent of aggregation. If the assumption is made that k is constant, then Equation (9.10) can be written as:

$$\frac{dP}{dt} = \lambda \frac{PH}{H + H_0} - (b + \mu) P - \frac{\alpha P(k + 1)}{kH} \qquad (9.11)$$

While derivations of Equations (9.1)–(9.3) continue to form the basis of most microparasite models in the literature, Equations (9.9)–(9.11) are much less widely used. One reason for this is simply that microparasitic infections usually have more effect on their hosts than do macroparasitic infections and are therefore more frequently the focus of epidemiological investigations. However, another reason is that there are several simplifying assumptions made in Equations (9.9)–(9.11) that do not approximate real macroparasite infections. For example, the model assumes that there is no constraint on parasite numbers within hosts other than parasite-induced mortality. While one of the features of macroparasites is that complete immunity to infection is unusual, most vertebrate hosts are capable of mounting at least partial immunity that impacts parasite reproduction. In addition, there is evidence that many parasites compete with each other for scarce resources at least in heavily parasitized hosts. As a result, fecundity of parasites within heavily infected hosts is likely to be reduced. Anderson and May went some way to account for these factors in modifications of this basic model, but adequate modeling of the immune response to macroparasitic infections is still a topic of active research (Fenton and Perkins 2010; Cressler et al. 2014). A further limitation of this model is its assumption that the parameter k that describes the parasite distribution among hosts is a constant. In fact, the distribution of parasites among hosts is a dynamic variable, with parasite-induced death removing the upper tail of the parasite distribution, thereby reducing the degree of aggregation. Unfortunately, this cannot easily be included in the structure of the standard model. Finally, many macroparasites have complex life histories, with one or more intermediate hosts and often involve trophic transmission, in which an intermediate host has to be consumed by the next host for the life cycle to be completed. Variants of Equations (9.9)–(9.10) can be constructed to include intermediate hosts, but a separate pair of

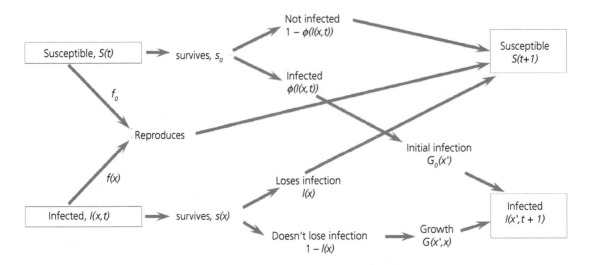

Figure 9.4 A flowchart of an integral projection model (IPM) for a host–parasite system, showing how a susceptible host at time t can transition at time $t + 1$ to be either still susceptible or become infected with a parasite burden of x'. The flowchart also shows how an infected host with parasite burden x at time t can transition either to have a parasite burden of x', or to lose infection. The function $\phi(I(x, t))$ describes the force of infection as a function of the parasite burden in the entire population and the function $G(x', x)$ describes the frequency distribution of parasite burdens x' at time $t + 1$ as a function of parasite burden x at time t. Survival to $t + 1$ is represented by s_0 for uninfected hosts and $s(x)$ for hosts infected with a parasite burden of x. Similarly, f_0 and $f(x)$ represent fecundity for susceptible and infected hosts, respectively. Loss of infection of hosts with parasite burden x is described by $l(x)$. Lastly, $G_0(x')$describes the initial frequency distribution of parasite burden x' for hosts first infected in the time interval $t \rightarrow t + 1$. (Adapted from Wilber et al., 2016.)

equations for each type of host and its correspond-ing parasite life history stage needs to be construct-ed. The primary value of these simple equations is that they identify the processes that must be quantified in order to understand the dynamics of host–macroparasite interactions.

9.5 Integral projection models

Integral projection models (IPMs) are a relative-ly recent development in modeling that offer the potential to bridge the gap between clas-sic microparasite and macroparasite models. They assume that demographic parameters of individu-als are affected by one or more continuous variables describing characteristics of those individuals and model dynamics in discrete time. In many IPMs, the continuous variable describing individuals is size (e.g., Rees and Ellner 2009; Coulson 2012). How-ever, parasite burden can also be considered as a continuous variable influencing host demography (Metcalf et al. 2016). Formulating a host–parasite model in this way combines some of the features

of microparasite and macroparasite models, in that all infected hosts need not to be considered identi-cal in their demographic properties (as is the case of microparasite models) and changes in parasite bur-den within hosts can be modeled without assuming that they are necessarily due to additional expo-sure to transmission stages (as is the case with most macroparasite models). In addition to modeling transmission, births, and deaths as functions of par-asite burden, IPMs also need to model changes in parasite burden on individual hosts from one time step to the next. While they have broad applicabili-ty to many wildlife host–parasite interactions, they are particularly useful for modeling the dynamics of fungal infections such as the amphibian chytrid fun-gus and white nose syndrome in bats (Wilber et al. 2016).

The basic structure of an IPM for a host pathogen interaction is shown in Figure 9.4. Three functions of parasite burden x need to be defined for infect-ed hosts over each time step $t \rightarrow t + 1$: the fecundity of infected hosts $f(x)$; the survival of infected hosts $s(x)$; and the loss of infection $l(x)$. In addition, the growth function $G(x',x)$ is an essential feature of an IPM. This is the probability distribution of parasite burden x' on an individual host at time $t + 1$ as a function of the parasite burden at the beginning of the time interval x. While this is described as a growth function, the parasite burden may decline, as well as grow. As with any host-parasite model, infection is a critical process and the proportion of uninfected hosts that become infected at the end of the time period and their resulting parasite burden x need to be described. Both of these will be functions of the overall disease burden in the population. Finally, the basic birth and death parameters for susceptible hosts need to be defined.

It should be evident that using an IPM for a host–parasite system is flexible but requires sufficient data to describe and estimate parameters for this substantial number of functions. Nevertheless, the approach has been used with some success for amphibian chytrid fungus (Wilber et al. 2016, 2017) and has potential applications to a number of other systems.

9.6 The basic reproductive number R_0

As discussed in Chapter 2, the most important quantity in epidemiology is the basic reproductive number R_0. To reiterate, for a microparasite, this is simply the number of secondary infections generated by each primary infection, when the pathogen is first introduced into an immunologically naive population. For a macroparasite, R_0 is the number of second-generation parasites per first-generation parasite when the parasite is rare.

For a simple, directly transmitted microparasitic infection, as can be represented by Equations (9.1)–(9.3), R_0 can simply be obtained by determining the conditions for dI/dt to be positive for small values of I. This ultimately gives us for density-dependent transmission dynamics,

$$R_0 = \frac{\beta N}{\alpha + b + v} \qquad (9.12)$$

Whereas for frequency-dependent transmission dynamics,

$$R_0 = \frac{\beta}{b + \alpha + v} \qquad (9.13)$$

It is important to note that R_0 is a dimensionless number, as it is the ratio of secondary cases to each primary case. A useful general way of representing R_0 for microparasites is:

$$R_0 = (\text{infection rate per unit time})$$
$$\times (\text{time infection persists}) \qquad (9.14)$$

9.6.1 R_0 for complex life cycles and multiple hosts

Most parasites and infectious diseases that are important in conservation biology are not directly transmitted with only a single host species. In fact, theory shows that a pathogen with only a single host and density-dependent transmission cannot cause host extinction on its own (McCallum and Dobson 1995; de Castro and Bolker 2005): once the host become sufficiently rare, R_0 drops below one and the parasite can no longer persist in the host population. To manage most parasites and pathogens capable of causing host extinction, it is therefore important to be able to define and estimate R_0 for systems with multiple host species and for life cycles more complex than simple direct transmission.

Elimination of a pathogen from a host community requires driving R_0 for the multi-species system to below one (see Chapter 13). For practical management purposes, writing down an expression for R_0 is extremely valuable, even if it is not possible to produce numerical estimates of each of its components. Once an expression is written down in algebraic or matrix form, a range of standard tools for sensitivity and elasticity analysis (Caswell 2001) can be used to identify where control strategies can most usefully be applied (see, e.g., Matser et al. 2009).

The next-generation matrix provides a general approach for calculating R_0 for multihost pathogens and pathogens with complex life histories. In its full generality, this approach is quite complex, as it needs to be able to handle all the diversity of parasite transmission biology (see Diekmann et al. 2010). In Section 9.6.1.1, we describe how to define and derive the next-generation matrix. This recipe largely follows Hartemink et al. (2008) and Dobson and Foufopoulos (2001), which are written in a style accessible to ecologists. For more formal descriptions, see Diekmann et al. (2010) and Roberts and Heesterbeek (2013).

9.6.1.1 Constructing a next-generation matrix and deriving R_0

First, identify the different "types" of the hosts that may be able to become infected. These may be different host species, including vectors, they may be different life history stages of a single host species, or they may be different sexes or genotypes of the one host species. In Hartemink et al. (2008) these are described as "types-at-birth," but this does not mean "birth" in the usual demographic sense; rather it is the type at the point that infection is first acquired. The next-generation matrix **K** will have as many rows and columns as there are different types.

Define each of the elements k_{ij} in the next-generation matrix **K**. The element k_{ij} is the expected number of newly infected hosts of type i, produced by each individual that was first infected when it was of type j, during the entire infectious period of the host in question. This is the critical and most tricky point in constructing the matrix—this term will include infections that originate from the given host first infected when it was of type j, even though it is no longer of that type. For example, if the host in question was first infected at the egg stage, k_{ij} will include new infections of hosts of type i that may have occurred when the source host has matured to become a larva, nymph, or adult. Each element k_{ij} may therefore itself be quite complex, including terms from multiple routes of transmission from the source host j as it matures and survives through a number of life history stages. Each of these terms may also depend on the density of hosts of each type.

Derive the dominant eigenvalue of the matrix **K**. This will be R_0. Eigenvalues can be calculated numerically using most standard mathematical computer packages, such as Matlab or R. For matrices that are not too large or have a substantial number of zero entries, eigenvalues can also be derived algebraically. R_0 for the multitype system has the same interpretation as it does for a single-host system: if $R_0 > 1$, the epidemic will proliferate, whereas if $R_0 < 1$, it will decay.

There are some close parallels, but also some subtle differences between the next-generation matrix in infectious disease ecology and standard population projection matrices in population ecology (Matser et al. 2009). Both share a common threshold property, in that if the dominant eigenvalue exceeds one, then the population will increase, or the epidemic will proliferate. One key difference is that population projection matrices operate on a fixed time step, meaning that it is possible to derive a growth rate through time from the dominant eigenvalue. In the case of a next-generation matrix, there is no fixed time step, as what is being calculated is the number of infections in the next generation of infection (the time between infection generations is often called the serial interval in the epidemiological literature). Where there are multiple routes of transmission, there is no simple definition of the length of an infection generation—the next-generation matrix therefore does not provide any information about the speed at which an epidemic may propagate. A full dynamic model is required to do this.

A further limitation of this approach is that it is implicitly concerned with what will happen in the long run when all of the elements of the matrix **K** remain constant. If transmission fluctuates seasonally, or if host population numbers are variable, then the method is not applicable. Again, a full dynamic model would be required to determine the circumstances under which an epidemic might occur.

9.6.1.2 Examples

As these ideas may a little difficult to follow in the abstract, we provide here three examples: a pathogen infecting several host species, a pathogen transmitted solely through heterosexual transmission, and a vector-transmitted pathogen.

9.6.1.3 Pathogens affecting several host species

This example is modified from Dobson (2004). It is based on an extension of the microparasite model presented in Equations (9.1)–(9.3), with the parameters as defined in those equations. We have presented it for only two host species, but generalization to any number of host species is obvious. Each element k_{ij} in the next-generation matrix consists of two components: (i) the rate per unit time at which species j transmits infection to species i, multiplied by (ii) the time for which species j remains infectious. In this simple model, infected hosts are removed from the infectious pool either by death ($\alpha_i + b_i$: disease-induced mortality and natural mortality, respectively) or recovery (v_i). Transmission may be dependent on the density of infected hosts i and j, or on the proportion of infected hosts of types i and j, following the lines of Equation (9.4). The rate per unit time at which hosts of species j infect species i is $\beta_{ij}p_{ij}$. The next-generation matrix is then:

$$\mathbf{K} = \begin{pmatrix} \dfrac{\beta_{11}p_{11}}{(\alpha_1+v_1+b_1)} & \dfrac{\beta_{12}p_{12}}{(\alpha_2+v_2+b_2)} \\ \dfrac{\beta_{21}p_{21}}{(\alpha_1+v_1+b_1)} & \dfrac{\beta_{22}p_{22}}{(\alpha_2+v_2+b_2)} \end{pmatrix} \tag{9.15}$$

The terms in the leading diagonal of this matrix are essentially R_0 for each species individually (see Equations (9.12) and (9.13)), but there is one important difference, "hiding" in the terms p_{11} and p_{22}: transmission within species may sometimes be reduced by "wasted contacts" with another species that is a less competent host. This is known as the dilution effect (see Figure 9.5, and Keesing et al. 2006; Johnson et al. 2013).

9.6.1.4 A pathogen transmitted solely through heterosexual contact

This example is from Diekmann et al. (2010). It is an SEI model, including hosts that are susceptible, exposed (infected but not yet infectious), and infectious. Thus, it models a chronic infection in which there is no recovery with subsequent host immunity. The total population size is N, which is assumed to be constant for each sex. Assuming frequency-dependent contact, with a transmission rate β_F from males to females and β_M from females to males, a mortality rate in both sexes of μ, sex-specific incubation rates of v_M and v_F, and sex-specific recovery rates of γ_M and γ_F, the system can be written as:

$$\frac{dE_F}{dt} = \beta_F S_F \frac{I_M}{N_M} - (v_F + \mu) E_F$$

$$\frac{dI_F}{dt} = v_F E_F - (\gamma_F + \mu) I_F \tag{9.16}$$

$$\frac{dE_M}{dt} = \beta_M S_M \frac{I_F}{N_F} - (v_M + \mu) E_M$$

$$\frac{dI_M}{dt} = v_M E_M - (\gamma_M + \mu) I_M$$

Each newly infected female E_F has a probability $v_F/(v_F + \mu)$ of entering the infectious class, and then will infect males at a rate $\beta_M N_M/N_F$ for a period $1/(\gamma_F + \mu)$ and vice versa for newly infected males. Thus,

$$\mathbf{K} = \begin{pmatrix} 0 & \dfrac{v_M \beta_F N_F}{(v_M+\mu)(\gamma_M+\mu)N_M} \\ \dfrac{v_F \beta_M N_M}{(v_F+\mu)(\gamma_F+\mu)N_F} & 0 \end{pmatrix} \tag{9.17}$$

Given the assumption of purely heterosexual transmission, neither males nor females can infect members of the same sex, although the generalization to include homosexual transmission should be obvious. In this particular case, the two zeros in the matrix mean that it is very straightforward to derive an algebraic expression for the largest eigenvalue and thus:

$$R_0 = \sqrt{\frac{v_F v_M \beta_F \beta_M}{(v_F + \mu)(v_M + \mu)(\gamma_F + \mu)(\gamma_M + \mu)}} \tag{9.18}$$

9.6.1.5 A vector-transmitted pathogen: *Borrelia burgdorferi*

Hartemink et al. (2008) derive R_0 for the bacterium *B. burgdorferi*, the causative agent of Lyme disease, which is a significant public health problem. This pathogen is transmitted by ticks of different stages (i.e., larvae, nymphs, or adults) normally between different species of wildlife, but now increasingly also to nearby human populations (see Figure 9.6). They include five "types": ticks first infected (via transovarial transmission) as eggs (1), or as larvae (2), nymphs (3) or adults (4), and vertebrate hosts (5). An important feature of the biology to keep in mind in the following discussion is that ticks feed on a vertebrate host only once per life history stage. The next-generation matrix has the following form:

$$\mathbf{K} = \begin{bmatrix} k_{11} & k_{12} & k_{13} & k_{14} & 0 \\ k_{21} & k_{22} & k_{23} & 0 & k_{25} \\ k_{31} & k_{32} & k_{33} & 0 & k_{35} \\ k_{41} & k_{42} & k_{43} & 0 & k_{45} \\ k_{51} & k_{52} & k_{53} & 0 & 0 \end{bmatrix} \tag{9.19}$$

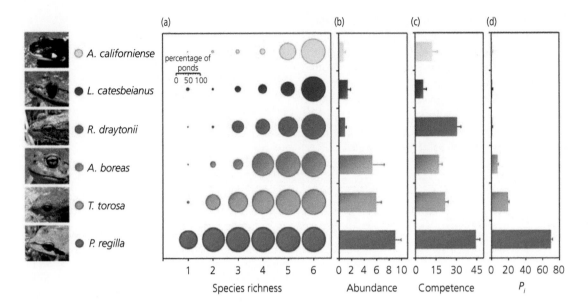

Figure 9.5 Among amphibian communities infected with parasites, increased levels of species diversity protect against disease (dilution effect). (a) Percentage of ponds within each richness category (on the x axis) that support each amphibian species. Percentage is represented by the size of the circle. The amphibian species involved, *Ambystoma californiense*, *Lithobates (Rana) catesbeianus*, *Rana draytonii*, *Anaxyrus boreas*, *Taricha torosa*, and *Pseudacris regilla*, are listed in increasing order of abundance from top (rare even in large, species-rich ponds), to bottom (common even in small ponds). (b) Mean abundance for each host species. (c) Each species' laboratory-measured competence for supporting the parasite *Ribeiroia ondatrae*. (d) An index P_i of each host species' contribution to overall community competence (P_i combines the fraction of wetlands occupied by a host, its relative abundance when present, and host competence—scaled between 0 and 100%). (Source: Johnson et al. 2013.)

Some of the elements of the matrix are necessarily zero. For example, $k_{15} = 0$ because infected vertebrate hosts (type 5) cannot infect tick eggs (type 1); nor can they infect other vertebrate hosts, so $k_{55} = 0$. However, if susceptible ticks of any other life history stage feed on an infected vertebrate they can acquire infection, so k_{25}, k_{35}, and k_{45} all may be nonzero. In the fourth column, newly infected (female) adult ticks can give rise to infected eggs, but because they feed only once in the adult life history stage, if they have acquired infection in that stage, they cannot further transmit to any other type, so the remaining terms are all zero. Describing the derivation of a few of the remaining nonzero terms is sufficient to explain how the process works—Hartemink et al. (2008) provide a more complete exposition. Consider k_{51}, the number of the vertebrate hosts (5) infected by a tick infected as an egg (1). Vertebrate hosts cannot acquire infection directly from the egg, but they can from larvae, nymphs, or adults that were first infected as eggs, so the term k_{51} is nonzero. Defining "transmission

efficiency" q_j as the number of competent vertebrate hosts infected by each infected tick in life history stage j, denoting survival from the previous life history stage through to life history stage j as s_j, and the fraction of bites on hosts that are competent as h_C,

$$k_{51} = \left(s_2 q_2 + s_2 s_3 q_3 + s_2 s_3 s_4 q_4\right) h_C \qquad (9.20)$$

It is important to note that the survival terms s_j and transmission efficiency terms q_j may well be functions of the population size or density of both ticks and vertebrate hosts and that h_C will also depend on the composition of the vertebrate community (including noncompetent hosts).

So, for a tick first infected as an egg to infect a vertebrate host as an adult, it must have survived through all the previous life history stages. There is an additional subtle complication here in that only the adult female ticks take a blood meal, so that s_4 needs to incorporate the probability that a tick is female. The remainder of the last row of the matrix is derived similarly.

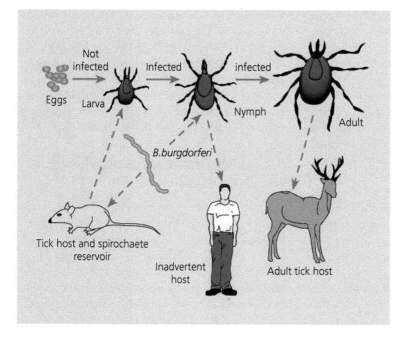

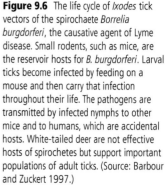

Figure 9.6 The life cycle of *Ixodes* tick vectors of the spirochaete *Borrelia burgdorferi*, the causative agent of Lyme disease. Small rodents, such as mice, are the reservoir hosts for *B. burgdorferi*. Larval ticks become infected by feeding on a mouse and then carry that infection throughout their life. The pathogens are transmitted by infected nymphs to other mice and to humans, which are accidental hosts. White-tailed deer are not effective hosts of spirochetes but support important populations of adult ticks. (Source: Barbour and Zuckert 1997.)

The number of infected eggs produced per infected egg k_{11} requires an infected egg to survive through to become an adult female, which then produces an average number of eggs E, a fraction r of which are infected. Thus:

$$k_{11} = s_2 s_3 s_4 E r \qquad (9.21)$$

And the remainder of the first row is derived similarly.

Ticks in life history stages other than eggs can be infected by feeding on infected vertebrate hosts. For host to tick transmission, k_{i5}, relevant variables are the average number of ticks of stage i per host N_i and the transmission efficiency from a vertebrate host to a given tick in stage i, p_i. Thus,

$$k_{i5} = p_i N_i \qquad (9.22)$$

Finally, a susceptible tick feeding on the same host as an infected tick may acquire infection in a process known as co-feeding. Here, it is necessary to describe C_{ij}, the number of ticks in life history stage i co-feeding in proximity to each infected tick in life history stage j, and the probability that transmission will occur as a result θ_{ij}. In addition, the infected tick will need to have survived from first infection through to the stage at which

the co-feeding transmission occurs. So, transmission from a tick first infected as an egg (1) to a susceptible nymph (3) will be:

$$k_{31} = \left(s_2 \theta_{23} C_{32} + s_2 s_3 \theta_{33} C_{33} + s_2 s_3 s_4 \theta_{43} C_{43} \right) h_C \qquad (9.23)$$

This is a complex example, but illustrates how the approach can be used when there are multiple modes of transmission, different life history stages, and more than one host species. For an example of how the approach can be used to determine the elasticity of R_0 depending on the magnitude of particular routes of transmission, see Matser et al. (2009).

9.6.2 R_0 for network models

Models that divide the host population into compartments, such as SIR models, usually make the assumption that the population mixes randomly, so that each infectious individual has an equal probability of coming into contact with any susceptible individual. Clearly, this is an assumption that frequently cannot be justified. One way of getting around this constraint is to assume that the population is divided into a number of different subcategories, which may have different rates of mixing between them. Provided that all individuals of one subcategory can be assumed to have the same contact rate with all individuals of another subcategory, then it is possible to use the approaches described in the previous section to calculate R_0, with each subcategory treated as a different "type."

Another approach is to use information from the structure of a contact network between individuals to calculate R_0. In recent years, there have been rapid advances in network theory, which is increasingly being applied to infectious disease epidemiology. A network is simply a number of nodes connected by links. In the context of infectious disease, nodes usually represent individuals. These can have a range of properties such as infection status, age, species, etc. Edges represent contacts between individuals, which can be bidirectional or unidirectional and can be of varying strength. In network analysis theory, there is a rather bewildering variety of statistics that can be used to describe properties of networks as a whole such as the degree distribution (the frequency distribution of the number of edges coming from each node) or connectedness (the proportion of all possible links between the ages that actually exist). Other statistics describe properties of individual nodes within the network, such as "closeness," defined as the shortest path between that node and all others in the network, and "betweenness," defined as the number of shortest paths between all other individuals that pass through the node in question (see, e.g., Wasserman and Faust 1994). That some of these properties will be important in disease transmission is obvious. For example, an individual with very high betweenness will be able to transmit infection to a large number of other individuals, potentially being a "superspreader" (Lloyd-Smith et al. 2005). If such individuals can be identified, then interventions targeted against them will be particularly effective in controlling disease spread.

Properties of the network as a whole also have implications for calculating R_0. If there is a heterogeneous degree distribution, meaning that some individuals have much higher numbers of contacts than others, then R_0 will be greater than if the degree distribution is random (as would be the case in a randomly mixing population). May (2006) shows that:

$$R_0 = \rho_0 \left(1 + CV^2\right) \qquad (9.24)$$

Where ρ_0 is a constant and CV is the coefficient of variation (i.e., the variance divided by the mean) of the degree distribution. If you have information on the degree distribution, you can therefore improve estimates of R_0, which will be important for disease management. In practice, obtaining information on contact network structure is not easy. Some possible approaches are described in Chapter 10.

9.7 Adding pathogens to population viability analysis models

Population viability analysis (PVA) has become a widely used modeling tool to guide conservation management. In very broad terms, the approach involves using a stochastic modeling approach to evaluate the likelihood of a population falling below specified thresholds in a given time frame. There are two general approaches to PVA modeling (Morris and Doak 2002). One uses time series analysis to estimate the parameters associated with a single-species model of population size of the species in question over historical time, and then projects the model over the time frame of the PVA. This approach has the advantage of being statistically rigorous but because it does not include a mechanistic description of population dynamics it is not well suited to explore the consequences on extinction probability of alternative management strategies. The second approach builds a stochastic age- or stage-structured model, estimating the relevant stage-specific fecundity and survival parameters from data other than the simple trajectory of the population through time. This approach is more complex but is better suited to evaluating alternative management actions. Reviews of PVA can be found in Beissinger and Westphal (1998); Ludwig (1999).

One approach to including disease threats in PVA is simply to treat the arrival of disease as a "catastrophe." The manual for Vortex (Lacy and Pollock (2021), probably the most widely used software package for PVA, gives "disease" as one of the forms of catastrophe that may affect a population of an endangered species. While this approach might be helpful in some cases, catastrophes are modeled in Vortex as occurring with some specified probability and then influencing survival and fecundity by reducing them to some predetermined fraction of their previous level. This means that a key feature of infectious disease, which is that its impact may depend on population density, cannot be modeled and that there is no modeling of epidemic dynamics. Furthermore, the most appropriate use of PVA is to compare the probability of population extinction in response to alternative management actions, rather than simply to calculate the probability of extinction of the population as it stands. Without explicitly including disease dynamics, it is very difficult to appropriately model management actions targeted at disease. In the light of these limitations, "Outbreak," a package specifically designed to model disease dynamics in wildlife, is downloadable from https://scti.tools/outbreak/. An example of its application can be found in Bradshaw et al (2012).

One of the principal difficulties in incorporating epidemic dynamics into PVA is the very different timescale on which typical microparasitic infections operate, compared with the timescale of most vertebrate populations. Most PVA models for vertebrate populations, especially seasonal breeders, use a 1-year time step. For most microparasitic infections, this would encompass many generations of the parasite. This means that it is difficult to simultaneously model the dynamics of a microparasite and its host using the discrete-time approach implicit in standard matrix type models. As May (1986) observed, highly nonlinear processes such as infectious diseases can lead to very complicated dynamics in discrete time. A second approach, which might prove valid for many infectious diseases, is to assume that the dynamics of infection are so fast that any epidemic essentially runs to completion within a single year or time step of the host population (Gerber et al. 2005). If this is the case, then a simple SIR model with no recruitment but potentially removals can be imposed within each time step of an age-structured model. This approach would be applicable to pathogens such as the distemper virus that caused massive mortalities in North Sea seal populations (Swinton et al. 1998), which spread over 1,000 km within a single season, causing epidemics that ran to completion within individuals seal colonies in 2–3 months.

A difficulty arises with those microparasitic infections that operate on a sufficiently slow timescale that epidemics are spread over several years. Unfortunately, several of the most significant wildlife diseases fall into this category, including Tasmanian devil facial tumor disease (McCallum et al. 2009), chronic wasting disease in deer (Miller et al. 2006), and herpesvirus infections in turtles (Aguirre and Lutz 2004). Similar considerations are likely to apply to macroparasites because of their typically longer life cycles than microparasites. In their modeling study of bovine tuberculosis in water buffalo in the Northern Territory of Australia, Bradshaw et al. (2012) show how "Outbreak," an epidemiological model operating on a short time frame, can be linked to the PVA program "Vortex," which operates on a longer timeframe.

9.8 Individual-based models

The most flexible way to construct an ecological model is by using an individual-based or agent-based approach, in which every individual in the population is explicitly represented, each described by a number of state variables (Grimm et al. 2006), with explicit rules describing how those state variables may change through time. State variables may include fixed properties, such as species or sex; properties that may change predictably through time, such as age; and those that may change according to random factors and interactions with other agents, such as infection status, reproductive status, location in space, and survival.

There have been relatively few such models developed to describe the dynamics of disease in wildlife populations. A notable example is a model of classical swine fever (CSF) developed to investigate alternative hypotheses for the persistence of the disease through time in wild boar populations (Kramer-Schadt et al. 2009). The model was spatially explicit and allowed spatial distribution, infection status, and survival of individual boar to change on a weekly time frame, with recruitment occurring on a seasonal basis. The results suggested that the most plausible explanation for disease persistence was the existence of heterogeneity in infectiousness between individuals. Inevitably, this would be a difficult process to model using mean field assumptions. More recently, individual-based approaches were used by Lambert et al. (2020) to model brucellosis in alpine ibex; also Wells et al. (2019) modeled Tasmanian devil facial tumor disease with a stochastic individual-based model. The principal advantage of using individual-based models in both of these systems is that they allow incorporation of heterogeneity between individuals in susceptibility and transmission, spatial structure, and stochasticity.

Such models have substantial demands for both computing power and storage capacity, but these are no longer major constraints except for simulations based on very large populations. They also require some programming ability, but there are now computer packages available to handle many of the routine programming tasks (e.g., Simecol in R; Petzoldt and Rinke 2007). They are in some ways easier to construct than compartmental models because it is easier to describe what happens to individuals in ecological or epidemiological interactions than to describe implications for entire categories of individuals, particularly when heterogeneity and stochasticity are involved.

The disadvantages of such models are those of any complex model. First, they are demanding in terms of the number of parameters that need to be estimated in order to use the models. We will return to this issue in Chapter 10. Second, sensitivity analysis is essential for models that cannot be written down algebraically and becomes both increasingly more important and more difficult the more complex the model and the more parameters there are that need to be investigated. Finally, one of the problems with any complex model is that understanding why it behaves as it does can be very difficult. While there is no doubt that individual-based and agent-based models will become increasingly more used in wildlife epidemiology, it is best to use them in association with simpler models. The simple models will provide insights into why the individual-based models behave the way they do, and individual-based models are helpful in ensuring that the behavior of the simple models is not an artifact of the approximations made in developing the simple models.

9.9 Models for spatial spread

Beyond being interested in knowing how disease prevalence might change with time in a particular population or at a particular place, conservation biologists may also want to model how a disease epidemic spreads across space. This is particularly important for emerging disease threats, where a key management action might be to implement strategies to slow or prevent disease progression into unaffected areas. Models are important in predicting the extent to which alternative strategies might slow or limit spread.

The simplest and most abstract models of spatial spread are so-called diffusion models. They are derived, as the name suggests, from analogous models of diffusion used in the physical sciences. An exponentially increasing and uniformly spreading pathogen population can be modeled in two dimensions using the following partial differential equation (van den Bosch et al. 1992):

$$\frac{\partial n}{\partial t} = rn + \frac{1}{2}s \left\{ \frac{\partial^2 n}{\partial x_1^2} + \frac{\partial^2 n}{\partial x_2^2} \right\} \qquad (9.25)$$

Here, n is the population size at a given point in time and space, r is the intrinsic rate of growth per unit of time, s is the diffusion coefficient, and x_1 and x_2 are the two spatial coordinates. As with most highly abstract models, is difficult to apply Equation (9.25) directly to any particular system. However, its general properties are important. The principal result from this simple model is that the square root of area occupied is predicted to increase as a linear function of time, with a rate coefficient:

$$C = \sqrt{2rs} \qquad (9.26)$$

As a corollary, the rate of linear spread in any particular direction is a simple linear function of time. Equation (9.26) shows that this rate of spread depends on the intrinsic rate of increase at r at any particular point in space. This means that any management strategy that reduces R_0 will also reduce the rate of spatial spread. The diffusion coefficient for infectious diseases is dependent on the rate of movement and mixing of susceptible and infectious hosts. However, it is difficult to derive directly from movement data and is usually estimated from time series data on the spatial spread of the pathogen in question (see, e.g., Childs et al. 2000). Management strategies that decrease the rate of movement of hosts by putting barriers in place or by otherwise decreasing the permeability of environment will decrease the diffusion coefficient.

Moving beyond these highly abstract diffusion models, spatial dynamics and spatial spread can be modeled with a range of approaches, including network models, metapopulation models, cellular automata, and individual-based models. White et al. (2018) conducted a systematic review of the wildlife disease spatial modeling literature, finding that individual-based models have been the most widely used approach, followed by metapopulation models. Theoretical studies were most common, followed by rabies as the specific pathogen most investigated.

References

Aguirre, A.A. and Lutz, P.L. (2004). Marine turtles as sentinels of ecosystem health: Is fibropapillomatosis an indicator? *EcoHealth*, **1**:275–83.

Anderson, R.M. and May, R.M. (1978). Regulation and stability of host-parasite interactions. I. Regulatory processes. *Journal of Animal Ecology*, **47**:219–47.

Anderson, R.M. and May, R.M. (1979). Population biology of infectious diseases. Part I. *Nature*, **280**:361–67.

Barbour, A.G. and Zückert, W.R. (1997). Genome sequencing: New tricks of tick-borne pathogen. *Nature*, **390**:553.

Beeton, N. and McCallum, H. (2011). Models predict that culling is not a feasible strategy to prevent extinction of Tasmanian devils from facial tumour disease. *Journal of Applied Ecology*, **48**:1315–23.

Beissinger, S.R. and Westphal, M.I. (1998). On the use of demographic models of population viability in endangered species management. *Journal of Wildlife Management*, **62**:821–41.

Box, G.E.P. (1979). Robustness in the strategy of scientific model building. In: Launer, R.L. and Wilkinson, G.N. (eds.), *Robustness in Statistics*, pp. 201–36. Academic Press, Madison, WI.

Bradshaw, C.J.A., McMahon, C.R., Miller, P.S., Lacy, R.C., Watts, M.J., Verant, M.L., et al. (2012). Novel coupling of individual-based epidemiological and demographic models predicts realistic dynamics of tuberculosis in alien buffalo. *Journal of Applied Ecology*, **49**:268–77.

Caswell, H. (2001). *Matrix Population Models: Construction, Analysis and Interpretation*, 2nd edn. Sinauer, Sunderland, MA.

Childs, J.E., Curns, A.T., Dey, M.E., Real, L.A., Feinstein, L., Bjornstad, O.N., et al. (2000). Predicting the local dynamics of epizootic rabies among raccoons in the United States. *Proceedings of the National Academy of Sciences*, **97**:13666–71.

Coulson, T. (2012). Integral projections models, their construction and use in posing hypotheses in ecology. *Oikos*, **121**:1337–50.

Cressler, C.E., Nelson, W.A., Day, T., and McCauley, E. (2014). Disentangling the interaction among host resources, the immune system and pathogens. *Ecology Letters*, **17**:284–93.

De Castro, F. and Bolker, B. (2005). Mechanisms of disease-induced extinction. *Ecology Letters*, **8**:117–26.

Diekmann, O., Heesterbeek, J.A.P., and Roberts, M.G. (2010). The construction of next-generation matrices for compartmental epidemic models. *Journal of the Royal Society Interface*, **7**:873–85.

Dobson, A. (2004). Population dynamics of pathogens with multiple host species. *American Naturalist*, **164**:S64–78.

Dobson, A. and Foufopoulos, J. (2001). Emerging infectious pathogens of wildlife. *Philosophical Transactions of the Royal Society of London B: Biological Sciences*, **356**:1001–12.

Fenton, A. and Perkins, S.E. (2010). Applying predator-prey theory to modelling immune-mediated, within-host interspecific parasite interactions. *Parasitology*, **137**:1027–38.

Gerber, L., McCallum, H., Lafferty, K., Sabo, J., and Dobson, A. (2005). Exposing extinction risk analysis to pathogens: Is disease just another form of density dependence? *Ecological Applications*, **15**:1402–14.

Gillespie, D.T. (1977). Exact stochastic simulation of coupled chemical-reactions. *Journal of Physical Chemistry*, **81**:2340–61.

Grimm, V., Berger, U., Bastiansen, F., Eliassen, S., Ginot, V., Giske, J., et al. (2006). A standard protocol for describing individual-based and agent-based models. *Ecological Modelling*, **198**:115–26.

Hartemink, N.A., Randolph, S.E., Davis, S.A., and Heesterbeek, J.A.P. (2008). The Basic reproduction number for complex disease systems: Defining R_0 for tick-borne infections. *The American Naturalist*, **171**:743–54.

Johnson, P.T.J., Preston, D.L., Hoverman, J.T., and Richgels, K.L.D. (2013). Biodiversity decreases disease through predictable changes in host community competence. *Nature*, **494**:230–33.

Keeling, M.J. and Rohani, P. (2008). *Modeling Infectious Diseases in Humans and Animals*. Princeton University Press, Princeton, NJ.

Keesing, F., Holt, R.D., and Ostfeld, R.S. (2006). Effects of species diversity on disease risk. *Ecology Letters*, **9**:485–98.

Kramer-Schadt, S., Fernandez, N., Eisinger, D., Grimm, V., and Thulke, H.H. (2009). Individual variations in infectiousness explain long-term disease persistence in wildlife populations. *Oikos*, **118**:199–208.

Lacy, R.C., and J.P. Pollak. 2021. Vortex: A stochastic simulation of the extinction process. Version 10.5.5. Chicago Zoological Society, Brookfield, Illinois, USA.

Lambert, S., Gilot-Fromont, E., Toigo, C., Marchand, P., Petit, E., Garin-Bastuji, et al. (2020). An individual-based model to assess the spatial and individual heterogeneity of Brucella melitensis transmission in alpine ibex. *Ecological Modelling*, **425**:14.

Lloyd-Smith, J.O., Schreiber, S.J., Kopp, P.E., and Getz, W.M. (2005). Superspreading and the effect of individual variation on disease emergence. *Nature*, **438**:355–59.

Ludwig, D. (1999). Is it meaningful to estimate a probability of extinction? *Ecology*, **80**:298–310.

Matser, A., Hartemink, N., Heesterbeek, H., Galvani, A., and Davis, S. (2009). Elasticity analysis in epidemiology: An application to tick-borne infections. *Ecology Letters*, **12**:1298–305.

May, R.M. (1986). When two and two do not make four: Nonlinear phenomena in ecology. *Proceedings of the Royal Society of London B: Biological Sciences*, **228**:241–66.

May, R.M. (2006). Network structure and the biology of populations. *Trends in Ecology & Evolution*, **21**:394–99.

May, R.M. and Anderson, R.M. (1978). Regulation and stability of host–parasite interactions. II. Destabilizing processes. *Journal of Animal Ecology*, **47**:249–67.

May, R.M. and Anderson, R.M. (1979). Population biology of infectious diseases. Part II. *Nature*, **280**:455–61.

McCallum, H.I. and Dobson, A.P. (1995). Detecting disease and parasite threats to endangered species and ecosystems. *Trends in Ecology & Evolution*, **10**:190–94.

McCallum, H., Barlow, N.D., and Hone, J. (2001). How should transmission be modelled? *Trends in Ecology & Evolution*, **16**:295–300.

McCallum, H., Jones, M., Hawkins, C., Hamede, R., Lachish, S., Sinn, D.L., et al. (2009). Transmission dynamics of Tasmanian devil facial tumor disease may lead to disease-induced extinction. *Ecology*, **90**:3379–92.

Metcalf, C.J.E., Graham, A.L., Martinez-Bakker, M., and Childs, D.Z. (2016). Opportunities and challenges of integral projection models for modelling host–parasite dynamics. *Journal of Animal Ecology*, **85**:343–55.

Miller, M.W., Hobbs, N.T., and Tavener, S.J. (2006). Dynamics of prion disease transmission in mule deer. *Ecological Applications*, **16**:2208–14.

Morris, W.F. and Doak, D.F. (2002). *Quantitative Conservation Biology: Theory and Practice of Population Viability Analysis*. Sinauer, Sunderland, MA.

Petzoldt, T. and Rinke, K. (2007). Simecol: An object-oriented framework for ecological modeling in R. *Journal of Statistical Software*, **22**:1–31.

Plowright, R.K., Sokolow, S.H., Gorman, M.E., Daszak, P., and Foley, J.E. (2008). Causal inference in disease ecology: Investigating ecological drivers of disease emergence. *Frontiers in Ecology and the Environment*, **6**:420–29.

Rees, M. and Ellner, S.P. (2009). Integral projection models for populations in temporally varying environments. *Ecological Monographs*, **79**:575–94.

Restif, O., Hayman, D.T.S., Pulliam, J.R.C., Plowright, R.K., George, D.B., Luis, A.D., et al. (2012). Model-guided fieldwork: Practical guidelines for multidisciplinary research on wildlife ecological and epidemiological dynamics. *Ecology Letters*, **15**:1083–94.

Roberts, M.G. and Heesterbeek, J.A.P. (2013). Characterizing the next-generation matrix and basic reproduction number in ecological epidemiology. *Journal of Mathematical Biology*, **66**:1045–64.

Swinton, J., Harwood, J., Grenfell, B.T., and Gilligan, C.A. (1998). Persistence thresholds for phocine distemper virus infection in harbour seal *Phoca vitulina* metapopulations. *Journal of Animal Ecology*, **67**:54–68.

Van Den Bosch, F., Hengeveld, F.R., and Metz, J.A.J. (1992). Analysing the velocity of animal range expansion. *Journal of Biogeography*, **19**:135–50.

Wasserman, S. and Faust, K. (1994). *Social Network Analysis: Methods and Applications*. Cambridge University Press, Cambridge.

Wearing, H.J., Rohani, P., and Keeling, M.J. (2005). Appropriate models for the management of infectious diseases. *PLoS Medicine*, **2**:621–27.

Wells, K., Hamede, R.K., Jones, M.E., Hohenlohe, P.A., Storfer, A., and McCallum, H.I. (2019). Individual and temporal variation in pathogen load predicts long-term impacts of an emerging infectious disease. *Ecology*, e02613.

White, L.A., Forester, J.D., and Craft, M.E. (2018). Dynamic, spatial models of parasite transmission in wildlife: Their structure, applications and remaining challenges. *Journal of Animal Ecology*, **87**:559–80.

Wilber, M.Q., Knapp, R.A., Toothman, M., and Briggs, C.J. (2017). Resistance, tolerance and environmental transmission dynamics determine host extinction risk in a load-dependent amphibian disease. *Ecology Letters*, **20**:1169–81.

Wilber, M.Q., Langwig, K.E., Kilpatrick, A.M., McCallum, H.I., and Briggs, C.J. (2016). Integral projection models for host-parasite systems with an application to amphibian chytrid fungus. *Methods in Ecology & Evolution*, **7**:1182–94.

CHAPTER 10

Estimating Basic Epidemiological Parameters

> With four parameters I can fit an elephant, and with five I can make him wiggle his trunk.
> **Attributed to John von Neumann by Enrico Fermi**

10.1 Introduction

To apply an epidemiological model to any real host–parasite systems requires first describing the current state of the system. At a minimum, this means estimating the size of the host population; for microparasites the prevalence of infection; and for macroparasites both the mean parasite burden and distribution of parasites within the host population. Understanding the dynamics of the system further requires estimation of the parameters that lead to changes in both populations, which include birth and death rates of the host population, the transmission dynamics of the pathogen, and the recovery rate from infection. In many systems, there may be multiple host populations, including intermediate hosts and vectors, and there may be multiple pathogen species.

In order to provide a path through this tangle of interacting factors, we have structured this chapter by first describing approaches to estimate the current state of a host–parasite interaction, which includes the size and structure of the host population and the extent of parasite infection within that host population. Next, we describe methods for estimating birth, death, and population growth of a host population, with the complication that

these are likely to be affected by parasite infection. Finally, we deal with approaches to estimate the dynamics of parasite infection within the host population, including the rate at which hosts acquire infection and the rate at which they may recover from infection.

10.2 Estimating host population size and infection status

Two standard approaches are usually applied to estimate the size of a free-living vertebrate population: transect-based counts and mark–recapture methods. These are described at length in most standard field ecology texts. An excellent guide to the range of techniques available to use, and with a particular emphasis on suitability for various taxa, is provided by Sutherland (2006). For the purposes of this chapter, it is sufficient to note that methods based solely on sighting animals will usually not be able to determine the infection status at the same time, which means that independent samples will be needed to determine the level of parasite infection within that host population. Most modern uses of transect-based data recognize that animals toward the edge of a sampling strip are more likely

Infectious Disease Ecology and Conservation. Johannes Foufopoulos, Gary A. Wobeser and Hamish McCallum, Oxford University Press.
© Johannes Foufopoulos, Gary A. Wobeser and Hamish McCallum (2022). DOI: 10.1093/oso/9780199583508.003.0010

to be missed than those in the middle of the strip and that therefore a sightability curve should be employed to estimate population size. The standard software package used for this purpose is DISTANCE and a detailed description of its application can be found in Thomas et al. (2010). Spatially explicit capture–recapture methods are a relatively recent development combining some of the features of distance sampling and mark–recapture approaches, and are particularly worth considering if data consist of detections or captures on a spatial array (Efford et al. 2009; Gopalaswamy et al. 2012). N-mixture models offer the prospect of estimating population size based on counts without being able to identify individuals (Royle, 2004; Zipkin et al. 2014) and are being increasingly widely used. However, they rely on strong assumptions about the nature of the underlying probability distributions and should be used with caution (Knape et al. 2018; Link et al. 2018).

Mark–recapture methods can be applied to a variety of data types. All but the simplest analysis requires that animals should be marked so that they are individually recognizable. Most of these data sources do involve physically catching the animal at some point to individually mark it, at which time it is usually possible to make some assessment of infection status, often using a blood sample (see Chapter 6). As a more recent development in the past few years, relatively cheap "trail cameras" have become available. These incorporate digital cameras with some form of motion detection device. Particularly when they are used in video mode, it is often possible to identify individual animals on the basis of their markings, permitting the use of mark–recapture analysis (Kelly 2008; O'Brien and Kinnaird 2008; Naing et al. 2019). Mark–recapture methodology has also been applied to identification information based on individual DNA profiles from hairs or scats (Mills et al. 2000; Brøseth et al. 2010; Marucco et al. 2011; Carroll et al. 2018). It is also possible to collect infection information from scats (whether worm counts or immunological information from scats) in conjunction with the collection of DNA profiles for individual recognition, as was done by Liccioli et al. (2015) in a study of *Echinococcus granulosus* in urban coyotes.

Chapter 6 describes some of the methods available to detect the presence or level of infection of a pathogen in an individual. It is common to use the prevalence or intensity of infection in a sample of animals that has been captured as an estimate of the prevalence or intensity of infection in the population as a whole. This implicitly assumes that infection does not influence catchability. If it does, then any estimate of prevalence in a population based on a sample will be biased (Jennelle et al. 2007). Depending on the pathogen, the capture methodology, and the biology of the host, infection may either make animals more catchable, if it makes them less able to evade capture, or alternatively less trappable, if it impairs their ability to find traps, or their desire to enter traps. As described later in this chapter, modern mark–recapture methods allow testing of whether infection affects catchability and, if so, can be used to correct for this bias.

10.3 Mortality and fecundity: Basic demographic parameters

Any epidemiological model for a free-ranging population will require estimates of basic demographic parameters such as mortality and fecundity rates. There are discussions of how to estimate these parameters in the absence of disease in a number of texts (e.g., McCallum 2000). In this section, we concentrate on estimating the impact of infection on these basic demographic parameters, although most of the approaches we discuss can be used to estimate mortality and fecundity in healthy animals.

10.3.1 Estimating pathogen- or parasite-induced mortality

Figure 10.1 provides a simple decision tree to aid in selecting an appropriate analytical method to determine the effect of parasite infection on host survival. The first decision point is whether one intends to, or is able to manipulate and control infection: the most satisfactory way of estimating the effect of parasites or pathogens on the death rate of free-ranging animals involves following the fate of individuals *after* infection status or parasite burden has been manipulated. Simply estimating survival in animals with

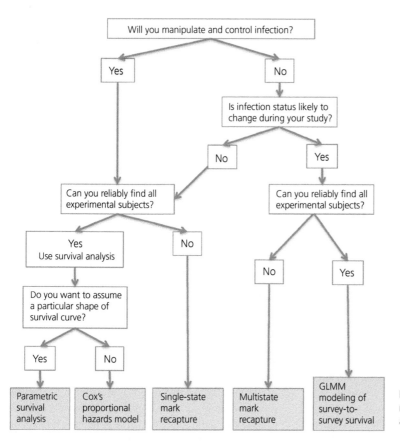

Figure 10.1 A flowchart for selecting methods to estimate the effect of parasites and pathogens on the survival of wildlife.

differing observed infection status is less informative, because there is the possibility that some other factor (e.g., nutritional status) may affect both the survival and parasite burden. Because it is often ethically unacceptable to experimentally infect wild animals, the most common way to manipulate parasite burden is to randomly select some individuals to be treated with an antiparasitic agent, while others are given placebos (Foufopoulos 1998). Not all treatments to remove or prevent infection are completely effective, but following the "yes" pathway at this point assumes that the infection status of the animals in a study will be controlled by the investigator. If one is unable to manipulate infection, then the next decision point depends on whether or not infection status is likely to change in many individuals during the course of the study. This will depend on the force of infection relative to the time frame of the study. Following the "no" pathway at this point assumes that the infection status of

the individuals in your study is primarily a fixed property of the individuals throughout the course of the study. On the other hand, following the "yes" pathway assumes that infection status may change and that therefore estimating the rate at which individuals acquire infection or lose it will likely be an objective of the study.

The next decision point applies irrespective of the previous decisions, although the pathway to be followed consequent on the decision does depend on the previous decisions. This decision is whether all (or almost all) individuals in the study can be detected each time the investigator wants to check on the status (alive or dead) of the animals. This may be possible if the animals in the study require little effort to observe or capture, or the total population is sufficiently small. For example, in the classic studies of Soay sheep on St. Kilda (an island off the coast of Scotland), almost the entire population was observed in each survey period (Gulland 1992). Alternatively, radio telemetry can be used to follow study animals. In this case, there will often be a very high probability of detection of an animal with a functioning radio transmitter. Even if the animal cannot be located or observed, one can use mortality-sensing radio collars that will change the emitted pulse interval if the collar has been immobile for a set period of time. Similarly, Global Positioning System (GPS) (e.g., Sommer et al. 2016) or satellite (e.g., Welbergen et al. 2020) tracking devices are often capable of downloading information to the mobile phone network or base stations, allowing the status of the animal to be determined without the necessity for recapture.

An inevitable issue with of telemetry studies is the budgetary constraints that limit the number of animals that can be monitored. The sample size necessary to detect a difference in survival between infected and uninfected animals will depend on the details of the method of analysis (see Section 10.3.2). Recall the useful rule of thumb from Chapter 3 where a 95% confidence interval for a proportion of around 0.5 is $\pm 1/\sqrt{n}$ (where n is the sample size). To estimate a proportion (and survival is essentially a proportion) to a precision of $\pm 10\%$ will require a sample size of at least 100. This would be considered a very large telemetry study. If one can reliably find all experimental subjects, and infection status does not change during the study, either because natural infection occurs at a low rate, or because one can control infection experimentally, then the appropriate method is survival analysis, described in Section 10.4.2.

In most studies that involve capturing animals and then releasing them back into the wild, not all individuals in a study group can be located each time one looks for them, even though they may still be present in the population. This restricts the use of standard survival analysis and instead requires estimating the probability of an animal present in the population being caught, in addition to estimating the probability it is still present in the population. If infection status remains constant during the study, then the appropriate method is single-state mark–recapture analysis with infection status as a "covariate" (a fixed property of the individual). This is described in Section 10.4.3.

If infection status may change during the study and one cannot reliably locate all of the study individuals, then one will also need to simultaneously estimate capture probability, survival of infected and uninfected individuals, rates of acquisition of infection, and rates of recovery. This is inevitably a complex problem and requires multistate mark–recapture methods. These are described in Section 10.5.4.

Finally, it is possible that infection status may change during the study, while one is able to reliably locate individuals. In this case, the objective will be to estimate the probability of survival from one survey to the next depending on infection status and also, given the individual has survived, the probabilities of it becoming infected or recovering if infected. This is a problem of estimating proportions, dependent on a range of predictor variables, and is suited to a logistic model, as described in Section 10.2.1. An additional complication is that the data are derived from repeated observations of the same individual animals. This requires a "repeated measures" design, because there will be additional random variation in survival, infection, and recovery between individual animals. A generalized linear mixed model (GLMM) is the appropriate way to handle this problem. An excellent and comprehensive guide to applying models of this type in R can be found in Zuur et al. (2009).

10.3.2 Survival analysis

If it is possible to determine whether most animals in a study are alive or dead at any given time, then a wide range of survival analysis methods are available. These can be used to investigate whether various factors or covariates, such as parasite burden, age, or sex affect survival, and also to estimate mortality rates depending on parasite burden. A common feature of survival data is "censoring." Some animals may be still alive at the conclusion of the study, in which case it is known that they survived for at least a certain period (until the end of the study), but it is not known when they actually died. Such observations are known as right-censored. It is also often the case that animals are added to the study at different times (e.g., they may be radio-collared on capture over the course of the study). This is known as staggered entry. Most methods of survival analysis can handle both censoring and staggered entry.

Parametric survival analysis produces an estimate of survival as a continuous function of age or time of entry of the subject into the study. The first step in the process is to select an explicit functional form for the hazard function, which is the death rate as a function of age (or time elapsed since entry into the study). Once this is done, one will need to estimate the parameters of the relevant hazard function, which may include effects of infection and other covariates. The simplest hazard function is that the risk of death ρ remains constant through time for a given individual:

$$h(t) = \rho \qquad (10.1)$$

This leads to exponential survivorship:

$$S(t) = e^{-\rho t} \qquad (10.2)$$

In many situations, particularly for large vertebrates, the risk of death changes with age, perhaps as the animal senesces. A flexible way of modeling this sort of survival is a with a Weibull hazard function:

$$h(t) = \kappa\rho(\rho t)^{\kappa-1} \qquad (10.3)$$

Here, there is a death rate parameter ρ, which has units of t^{-1} and κ is a dimensionless "shape" parameter. If $\kappa > 1$, the risk of death increases with age (corresponding to the Type I survival curve described in basic ecology texts (e.g., Begon et al. 2006), whereas if $\kappa < 1$, the risk of death decreases with age (corresponding to a Type III survival curve). If $\kappa = 1$, Equation (10.3) reduces to Equation (10.1), producing a constant hazard function and exponential survival and a Type II survival curve. The survivorship curve for a Weibull hazard function can be written as:

$$S(t) = \exp\left(-(\rho t)^{\kappa}\right) \qquad (10.4)$$

The effect of parasite infection or any other covariates is usually modeled by making the rate parameter an exponential function of the covariates. The exponential ensures that the hazard remains positive. For example, the effect of a parasite burden X on survivorship would be modeled as:

$$\rho(X) = \exp(\alpha + \beta X) \qquad (10.5)$$

Where α and β are two parameters to be estimated.

An alternative approach is to use Cox's proportional hazards model (Cox and Oakes 1984), which is a flexible way to assess the impact that infection or other covariates such as sex or age have on host survival, without assuming any particular form for the survival function itself. As the name suggests, the only assumption that needs to be made is that, at each time since the commencement of the study, the relative risk of dying depending on the differing covariates remains constant. For example, if the only covariate in the model were infection status, the model would assume that the amount by which infection would increase the death rate of infected animals relative to uninfected animals remains the same through time. A substantial limitation of this approach, particularly if the objective is to use the results of the analysis in a model, is that it does not produce an actual estimate of survival: merely (for example) the survival of infected animal relative to an uninfected one.

An R package "survival" is available to perform both parametric survival analysis and Cox proportional hazards analysis. Chapter 25 of Crawley (2007) provides a description of the package in a form accessible to ecologists.

10.4 Mark–recapture methods

In many studies of wild animals, such as those involving trapping, only a fraction of the animals in the study population can be detected each time the investigator searches for them or runs a trapping survey. Most ecologists and wildlife managers are very familiar with the Lincoln–Peterson method that is used to estimating population size from such data (Krebs 1999).

The real power of mark–recapture methods, however, is to estimate survival rates rather than just population size. A key challenge is to distinguish between an animal not observed in a particular survey session that has died, versus one that was alive and present but was not observed. Fortunately, there is a broad range of sophisticated mark–recapture methods that, given certain assumptions, allow the probability that an animal survives from one trapping session to the next to be estimated separately from the probability that an animal is detected in any given trapping session (Lebreton et al. 2009).

To be able to separate detection probability from survival probability in a mark–recapture experiment, there are two key requirements: (i) animals must be uniquely marked or otherwise individually recognizable; and (ii) a bare minimum of three discrete sampling periods is necessary. Whatever their sophistication, mark–recapture methods separate mortality from nondetection using the number of animals that are missed in a given sampling period that are subsequently rediscovered. This means that to obtain one estimate of survival, three sampling periods are needed, and that survival cannot be estimated between the penultimate and last sampling periods. The sample size necessary for a given level of precision in survival estimates depends on details of the experimental design and the method of analysis, but a useful rule of thumb is that estimates are likely to be very imprecise unless at least 30% of the population under study can be marked.

The raw data for estimating survival from a mark–recapture study consist of capture histories for each individual. A capture history is simply a series of zeros and ones denoting whether or not a particular individual was detected or not on each sampling occasion. For example, suppose there are five sampling occasions. The capture history:

$$00101$$

simply states that the animal was first detected on the third sampling occasion, not seen on the fourth occasion, but detected again on the fifth sampling occasion. The task is to model the number of individuals with particular capture histories, depending on survival and recapture probabilities.

Typically capture–recapture studies consist of multiple field trips spread across time. Within each field trip, capture events may be closely spaced, often occurring daily. This is usually done for logistical convenience, but this design has also a number of theoretical advantages and is known as the "robust design" (Kendall and Pollock 1992; Kendall et al. 1995). Within field trips, it is often possible to assume that the population is closed, with no

immigration, emigration, births, or deaths, which means that a variety of "closed population" methods can be used to estimate population size. To analyze survival between field trips, the data can be aggregated so that "1" identifies that an animal was caught at least once and "0" indicates that the animal was not caught at any time during a field trip.

A range of software packages are available to fit mark–recapture models. One of the most widely used is the program MARK (http://www.phidot.org/software/mark). Cooch and White (2018) provide a "gentle introduction." In general, single-state mark–recapture methods can estimate two broad classes of parameters: survival rates from one capture interval to the next (usually represented by the Greek symbol Φ); and capture probabilities (usually represented by p), which is the probability of an animal being captured in a particular trapping interval, given that it is present in the population. These can either be constant through time, or different for each time interval, and may depend upon covariates specific to an individual (such as its sex, size, or infection status), or may be influenced by factors specific to a particular sampling time such as the weather or season. Consequently, there is an enormous range of possible models, many of which may have a large number of parameters.

Mark–recapture models compare the observed number of individuals with a specific capture history to the numbers predicted by a model derived from survivorship and capture probabilities. MARK provides several tools for comparing these models based on the Akaike information criterion (AIC). It is often the case that no single model will unequivocally be identified as the best for a particular dataset. In these cases, there are methods to estimate parameters averaged across the range of plausible models that reasonably fit the data. Bayesian approaches provide an alternative way of analyzing mark–recapture models, which may be more appropriate for the small sample sizes typical of many mark–recapture studies, as well as the ability to include information based on prior knowledge of parameters. A good introduction to these methods is provided by Kéry and Schaub (2011). Software for Bayesian analysis of mark–recapture data is also available (Colchero et al. 2012).

Using infection status or parasite burden as a covariate, mark–recapture methods can be used to compare survival between infected and uninfected animals. A limitation of this "single-state" approach is that it is necessary to assume that infection status is a fixed property of the individuals in the population—once an animal is caught for the first time and its infection status is determined, it neither becomes infected nor recovers. Dependent on the pathogen, this clearly can be quite a limiting assumption.

Multistate mark–recapture approaches allow individuals to change infection status and also can estimate rates of transition. We describe these in Section 10.5. Regardless of whether a multistate or single-state mark–recapture method is used, it is important to be aware that a difference in survival in a natural environment between infected and uninfected individuals does not necessarily imply that the infection itself is responsible for the difference. It may be the case that some other factor (e.g., stress or nutritional status) may be responsible for both infection and reduced survival. Because they also estimate the probability of an animal being captured on a given trapping occasion, mark–recapture methods can also be used to determine whether infection status (or any other covariate) has an influence on trappability.

10.5 Quantifying transmission

The key process in any host–parasite interaction is transmission of infection from an infected individual or infectious stage to a susceptible host. Although transmission is one of the most difficult processes to quantify in a study of wildlife disease, understanding it is critical for wildlife management. For conservation purposes, it is important to know the relationship between transmission and host population density (see Chapter 9).

10.5.1 What does one need to estimate and why?

The first question to address is which aspect of transmission is one aiming to quantify. One of the more straightforward concepts in transmission is the *force of infection*. This is the rate per unit time at which susceptible individuals in a particular population at a particular time acquire infection. This quantity can be directly estimated and does not rely on any particular model of transmission. However, in order to use information about transmission for managing an infectious disease and population level, it is usually necessary to know how the force of infection depends on the population size or density. This means that you will need to describe the nature of the *transmission function* (see Equation (9.4)).

It is rarely appropriate to apply estimates of transmission derived from captive or enclosed populations to free-ranging populations. For a directly transmitted pathogen, transmission is dependent on contacts between susceptible and infected hosts and therefore is dependent on the behavior of the hosts. Behavior of captive animals in enclosed areas is unlikely to be representative of the behavior of their free-ranging conspecifics.

Transmission from an infected host to a susceptible host involves at least three distinct steps (McCallum et al. 2017): (i) an infected host must contact a susceptible host in such a way that the pathogen could potentially be transferred; (ii) the pathogen must be successfully transferred between infected and susceptible host; and (iii) following transfer of infectious stage, it must develop to become a new infection in the susceptible host. Most methods of quantifying transmission either estimate only one or two of these processes or confound the estimation of all three.

10.5.2 Estimating contact rates

Proximity-sensing radio collars are a relatively recent technological development that enables estimation of the first of these three components of the transmission rate (see, e.g., Hamede et al. 2009). These radio collars work by logging the identity of an individual with a radio collar that comes within a predetermined distance (a distance appropriate to the disease of interest) of another radio-collared individual. Because mere proximity cannot determine whether or not the pathogen is successfully transferred, nor whether it has successfully developed within the host, this approach cannot provide complete estimates of the transmission rate. However, proximity-sensing radio collars can provide very useful information on heterogeneity in contact rates between individuals and how contact rates vary seasonally or with population density. Contact can sometimes also be inferred if animals have been otherwise recorded in close proximity within a short interval of time. For example, Perkins et al. (2009) assumed that contact relevant to disease transmission had occurred in yellow-necked mice (*Apodemus flavicollis*) if they had been caught in the same or an adjacent trap within the same trapping session.

10.5.3 Estimating the force of infection

A common approach to quantifying transmission in livestock is through the use of "sentinels," which are uninfected tracer animals introduced into the population in which transmission is to be estimated. This approach is problematical in wild populations as it is unlikely that introduced sentinels will interact normally with the wild population. A similar approach might be to capture animals from the wild population, identify those that are free of infection, and then release these individuals back into the population to determine the probability that they pick up the infection. While various variants on this approach have been tried, if susceptibility to infection is heterogeneous in a wild population (as is usually the case) individuals that are disease-free at the time of first capture may be less susceptible to infection than an average individual, and the force of infection will therefore probably be underestimated.

An alternative approach to using sentinels is to use the relationship between host age and the prevalence of infection (in microparasites) or parasite burden (in macroparasites) to estimate the force of infection. Based on the idea that newborns can be considered as uninfected tracers, this idea has been widely used in the study of human infections (Anderson and May 1985; Grenfell and Anderson 1985). In the simplest case, where the force of infection can be considered to be constant with age, the proportion of individuals F in a cohort that has experienced infection by age a is given by:

$$\frac{dF}{da} = \Lambda\left(1 - F\right) \qquad (10.6)$$

Where Λ is the force of infection. Solving this equation shows that Λ is simply the reciprocal of the average age at which individuals acquire infection.

Multistate mark–recapture methods, described in Section 10.5.4, are a powerful way of estimating the force of infection from field data.

10.5.4 Multistate mark–recapture methods

Infection is usually not a fixed state: animals become infected throughout the course of the study and may also recover. Multistate mark–recapture methods (Lebreton et al. 2009) enable us to estimate the rate of transition from "healthy" to "infected" status and from "infected" to "recovered," from one capture occasion to the next. They allow the state of an animal to change during a mark–recapture study and therefore permit us to estimate both the rate of transition between stages, as well as the possibility of survival and the variation of catchability dependent on state. The rate of transition from "healthy" to "infected" status is essentially the force of infection. The one qualification is that this transition rate is estimated conditional on the animal still being alive at the time of the next capture occasion. This means that the force of infection may be underestimated for diseases that are so highly pathogenic that there is a high probability that an animal may both become infected and die between two capture occasions.

Multistate mark–recapture methods allow us to overcome the limitations of simply treating infection status as a covariate (a property of the individual animal) as is the case in a "single-state" mark–recapture analysis. These models have been applied in a number of studies of wildlife disease (Faustino et al. 2004; Murray et al. 2009; Ozgul et al. 2009; Lachish et al. 2010, 2011). Below, we describe applying these to a specific case study: Tasmanian devils (Figure 10.2a) and Tasmanian devil facial tumor disease. Unfortunately, multistate mark–recapture methods can be unreliable if the intervals between capture occasions are unequal. The problem is essentially that individuals can undertake more than one transition per time interval and, if time intervals are quite different, the probability of doing so will be very different between capture occasions (see Cooch and White 2018, Section 10.6).

(a)

(b)

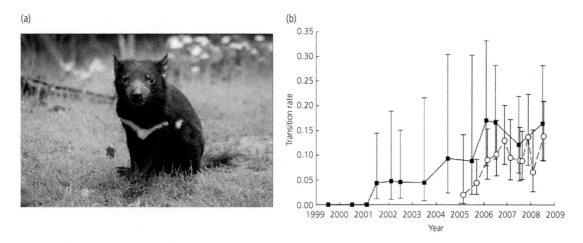

Figure 10.2 (a) The Tasmanian devil (*Sarcophilus harrissii*) is the largest species of surviving marsupial carnivore. It is acutely endangered by a lethal infectious cancer. (Photo by Mathias Appel.) (b) Model averaged estimates from multistate mark–recapture analysis showing transition rates from healthy adults to diseased adult at the Freycinet site (where no removals took place; filled squares, solid line) and the Forestier Peninsula (where all diseased animals were removed; open circles, dashed line) (Source: Lachish et al. 2010.)

Quantifying transmission from field data is a difficult problem and is the subject of much current research. DiRenzo et al. (2019) describe the application of N-mixture models to estimate transitions to and from diseased status (in addition to survival and recruitment) without the necessity of marking individuals using a "robust" design in which counts are made repeatedly at sites within seasons and between seasons. This approach is likely to be sensitive to the assumptions made about the probability distributions of detections, as are the N-mixture model approaches used to estimate abundance described earlier in the chapter. Gamble et al. (2020) describe an approach that combines information available from longitudinal data (such as mark–recapture) with information from cross-sectional data, for which larger sample sizes are easier to obtain, to improve the precision of estimators.

10.5.5 Applying multistate mark–recapture to Tasmanian devils and facial tumor disease

The Tasmanian devil (*Sarcophilus harrissii*), the largest surviving species of marsupial carnivore (Fig 10.2a), is threatened with extinction by a unique infectious cancer. The cancer is directly transmitted, presumably through biting during encounters for food and mating. Tumors are externally visible and appear to be universally lethal within 6 months.

This case study (Lachish et al. 2010) occurred on the Freycinet Peninsula in eastern Tasmania, where the Tasmanian devil population has been monitored from 1999 onwards (with the disease first detected in 2002), and compared transition rates with the nearby Forestier Peninsula, where an attempt was made to control the disease starting in 2004, by removing all diseased individuals on capture. The critical issue for management was whether these removals succeeded in reducing the transition rates at Forestier Peninsula relative to those observed at Freycinet.

The devil population was classified into three states: subadults (in which disease is rare); healthy adults; and diseased adults. Possible transitions from one survey to the next (aside from mortality) are from subadult to healthy adult, subadult to diseased adult, and healthy adult to diseased adult.

Analysis of the dataset was undertaken in several stages.

1. First, a model was employed allowing recapture, survival, and transition rates to vary with site, state, and time including all two-way interactions. The goodness of fit of this model was tested using the median ĉ approach (see Cooch and White 2018, chapter 5). This step is necessary because mark–recapture data are frequently "overdispersed" meaning that there is

excess variation in the numbers of individuals with particular capture histories beyond that assumed by the model. Before ranking models using the AIC, this needs to be corrected for, producing a small-sample AIC corrected for overdispersion (QAICc).

2. The most strongly supported model for recapture rates was selected, while allowing for all possible effects of site and time on survival and transition.

3. Using this recapture model, levels of support for various models of survival, including effects of site, infection status, age, season, and time were examined (note as survival of diseased adults is very low, this was not explicitly modeled).

4. Using this best-supported model for survival and recapture, transition rates as functions of time, site, and overall disease prevalence were modeled.

This stepwise procedure is necessary because of the extremely large number of potential models. It is not guaranteed to find an "optimal" model, and indeed the entire philosophy of this approach is that more than one model may be an adequate description of the results.

Two models gave plausible explanations (with a QAIC value within two of the best model) of the transition from healthy adult to diseased adult. In one model, the transition varied between the time intervals, and in the other the transition depended on the prevalence of disease P, but in a somewhat different way at each site. When there is more than one plausible model, the appropriate course of action is to use model averaging, in which the parameter estimates are averaged across these plausible models, weighted using the model weights.

The results are shown in Figure 10.2b. Comparison of these transition rates is hampered by the large error bars and the fact that the disease arrived at a different time at each site. Nevertheless, the conclusion is that there is no clear evidence of a difference in the rate at which healthy animals become diseased between the two sites, showing that removal of diseased animals was not a successful management strategy.

This example illustrates both the potential power and the challenges associated with multistate mark–recapture modeling. Effort and sample size were quite large: 448 devils captured at Forestier, of which 145 were diseased; and 633 captured at Freycinet, of which 115 were diseased. Despite these large numbers and the high probability of capturing a particular devil given it was present in the population (between 0.54 and 0.94, depending on site and time), the error bars for transition rates remained frustratingly wide and two rather different models plausibly represented the data.

10.6 Estimating R_0

As already explained in Chapter 2, understanding R_0, the number of secondary cases per primary case when disease is first introduced into a naive population, is critical for management. As will be discussed further in Chapter 13, disease elimination can be achieved if R_0 can be driven below one. It is important to note first that R_0 is a property of the pathogen in the specific population, environment, and ecological community under investigation. Therefore, it is not solely an intrinsic characteristic of the pathogen itself. One approach to estimating R_0 is to estimate all the parameters that contribute to R_0 in a model appropriate to the specific host–pathogen interaction under consideration. Disadvantages of this approach are not only that many separate parameters must be estimated, but also that, because each will be estimated with uncertainty, quantifying the resulting overall uncertainty in the final estimate of R_0 is difficult.

An alternative approach is to try to estimate R_0 directly. Two very different avenues are possible here, one that is appropriate if the infection has only just appeared (i.e., it is an emerging disease in the population under investigation—see Section 10.6.1), and the second that is applicable if it can be assumed that the host–parasite system is approximately at equilibrium (see Section 10.6.1).

10.6.1 Direct estimation of R_0 for emerging diseases

As the definition of R_0 already suggests, it should be possible to estimate R_0 directly from the rate at which new cases occur in a population following the first introduction of disease. Indeed, this can be accomplished, though it is not entirely straightforward.

The quantity that can most easily be directly estimated from epidemiological data is the initial rate of increase in prevalence with time, which we will call r_0. Provided that this underlying rate of increasing prevalence with time is exponential, it can be estimated by simply regressing the natural log of prevalence versus time. The resulting estimate has units of time^{-1}, whereas R_0 is a dimensionless quantity. To derive an estimate of R_0 from r_0 it is thus necessary to multiply it by a quantity that has units of time— for example, an estimate of the serial interval, which is the mean generation time of the infection (i.e., the average time from an individual acquiring infection to it passing the infection on to a second individual). Unfortunately, this is not a quantity that can be directly observed so it is necessary to make some assumptions about the model describing the transmission process. The simplest relationship, which is frequently used in practice, is:

$$R_0 \approx \exp\left(r_0 T_G\right) \qquad (10.7)$$

Where T_G is the generation time of the infection. Roberts and Heesterbeek (2007) denote the expression in Equation (10.7) as R_0^+. They also note that a simpler expression:

$$R_0^- = 1 + rT_G \qquad (10.8)$$

is also sometimes used in cases when rT_G is small. Note that R_0^- is simply the first term in a Taylor series expansion of Equation (10.7), which means that R_0^- is always smaller than R_0^+. Roberts and Heesterbeek (2007) show that Equation (10.7) is accurate if the generation time for all infections is exactly the same. In reality, of course, this will never be the case because some infected animals will pass on infection earlier than others. They show that a more general expression is:

$$R_0 \int_0^\infty \exp\left(-r_0 t\right) f(t)\,dt = 1 \qquad (10.9)$$

Where $f(t)$ is the probability distribution for the generation interval and t is time. Equation (10.9) is just a version of the Euler equation for population growth and similarly does not have a simple algebraic solution (except in some special cases). Instead, it needs

to be solved by a process of successive approximation. In practice, the probability distribution $f(t)$ is not directly observable in an epidemic. Instead, it needs to be derived from on an epidemiological model. Roberts and Heesterbeek (2007) provide a detailed discussion of this problem.

A second approach to estimating R_0 from the initial phases of an epidemic is appropriate if regularly collected data on the number of new cases occurring are available (Ferrari et al. 2005). Suppose that for a series of regular time intervals t_0, t_1 ... t_n from the onset of the epidemic at t_0, data on the number of new infectious individuals I_0, I_1 ... I_n are available (further assuming that $I_0 = 1$). Following the logic of removal methods in ecology, it is possible to then estimate R_0 by regressing the ratio of infectious cases in successive time intervals I_{t+1}/I_t against the cumulative number of new infectious cases, using the following equation:

$$\frac{I_{t+1}}{I_t} = R_0 - \beta \sum_{i=1}^{t} I_i \qquad (10.10)$$

Thus, R_0 is the intercept on the y axis from the regression and β is the slope of the regression relationship. There are some minor complications, however. First, the resulting estimate (denoted by $\hat{R}_0$, using the statistical convention that an estimate of a parameter is represented by using a circumflex above the symbol for the parameter) is biased because of the way in which a continuous epidemic is approximated by data from discrete intervals. This can be corrected for approximately as follows:

$$\hat{R}_0^{corrected} \approx \hat{R}_0 2\left(1 - e^{-1}\right) - e^{-1} \qquad (10.11)$$

Second, a weighted regression should be used, with weights proportional to the number of cases in each observation interval. Third, it is rarely the case that an epidemic is completely observed. Provided a constant fraction of the new infectious cases are reported (as would be the case, for example, if the available data were mortalities, and the proportion of infectious individuals that died remained constant through time), Ferrari et al. show that this method of analysis remains valid. For details of the theoretical justification for this approach, see Ferrari et al. (2005).

Alternatively, an option is to sidestep the generation time issue by using the reproductive rate per unit of time (denoted by r_0) in an appropriate model to investigate potential control options, rather than attempting to estimate the reproductive number R_0. This was the approach taken in Beeton and McCallum (2011), in which an age-structured model with a gamma-distributed latent period was used to investigate whether culling could be an appropriate strategy to control Tasmanian devil facial tumor disease. In this model, an estimate of r_0 derived directly from the rate of increase in prevalence through time when disease first entered the population was used to quantify transmission.

10.6.2 Estimation of R_0 for infections at equilibrium

By definition, when a host–pathogen system is in equilibrium, the mean number of secondary cases per primary case R is exactly one. Otherwise, the prevalence would be either increasing or decreasing. In a microparasitic epidemic occurring in a homogeneously mixing population, the number of secondary cases produced by an infectious individual will be directly proportional to the probability that any contact is with a susceptible individual (Anderson and May 1991). Hence,

$$R = R_0 x \qquad (10.12)$$

Where x is the proportion of susceptible individuals in the population. Rearranging Equation (10.12) when $R = 1$,

$$R_0 = 1/x^* \qquad (10.13)$$

Where x^* is the proportion of susceptible individuals in the population at equilibrium. The principal limitation with this approach is that it assumes that the system is at equilibrium. Whether this is the case will be difficult to determine in a wildlife population.

An approach similar in principle, but more complex in practice, can be taken for macroparasites. At equilibrium, each parasite must replace itself with exactly one other parasite in the next generation. Hence,

$$R_0 f(M^*) = 1 \qquad (10.14)$$

Where $f(M)$ is a function describing the joint effect of all density-dependent factors influencing the parasite population at a mean parasite burden of M, and M^* is the mean parasite burden at equilibrium. Density dependence in macroparasites can come from a wide variety of sources but one of the most common is density dependence in egg production. In that specific case, Anderson and May (1991) show that:

$$f(M) = \left\{ 1 + \frac{M(1 - \exp[-\gamma])}{k} \right\}^{-(k+1)} \qquad (10.15)$$

Where k is the parameter of a negative binomial distribution inversely describing the degree of aggregation, and γ is the exponent in the following relationship between per capita mean parasite fecundity $\lambda(i)$ and parasite burden i:

$$\lambda(i) = \lambda_0 \exp(-\gamma[i - 1]) \qquad (10.16)$$

If you can obtain data on egg count as a function of parasite burden, then γ can be estimated from a regression of log egg count on parasite burden. Given estimates of M^* and k, both of which will be obtainable from a frequency distribution of parasite burden in the host population, R_0 can then be estimated as:

$$R_0 = \frac{1}{f(M^*)} \qquad (10.17)$$

10.6.3 Estimating R_0 in multiple host systems

The approaches described above can be used to estimate within- and between-host components of the overall R_0 for a pathogen affecting several host species using the next-generation matrix approach outlined in Chapter 9. The value of this to managers of wildlife disease is that it can be used to identify what interventions on which species would be most effective to reduce overall disease prevalence or impact on a focal species. For example, Rudge et al. (2013) estimated the components of R_0 for *Schistosomiasis japonica*, an indirectly transmitted zoonotic helminth, in two regions of China. The life cycle involves definitive mammalian hosts belonging to a number of species (in these regions of China, humans, cattle, water buffalo, humans, dogs, cats, and rodents) and an intermediate snail host. Given a model of the system, information on population size, and prevalence of infection in each host, and assuming that the system was close to equilibrium, they were able to use a modified version of Equation (10.17) to estimate transmission rates between snails and each definitive host. This enabled them to calculate R_0 for each definitive host (the terms on the leading diagonal of the next-generation matrix) and the overall R_0 (the dominant eigenvalue of the next-generation matrix). They were then able to explore the consequences for the overall R_0 of removing transmission from each of the definitive hosts and were also able to explore predicted impacts of management strategies such as vaccination targeted on key definitive hosts, such as cattle and water buffalo.

A broadly similar approach conceptually was taken by Wilber et al. (2020) to estimate R_0 for the amphibian chytrid fungus at species, patch, and landscape level across 77 amphibian communities in California. By assuming that the fungus was essentially at equilibrium in these populations, they were able to derive estimates of R_0 based on estimates of host density, pathogen prevalence, and pathogen load, using a multispecies version of Equation (10.13). The principal conclusion of the study, which has direct relevance for managing the pathogen in California, is that one species, the Pacific tree frog (*Pseudacris regilla*) is the main maintenance host for the fungal pathogen.

10.7 Estimating the parameters for spatial spread

It will often be important to predict the rate at which an emerging wildlife disease is likely to spread through a susceptible population. As discussed in Chapter 9, the simplest model of spatial spread is a reaction–diffusion model, which predicts that the rate of spatial spread should be a simple linear function of time, so the radial distance d has spread by time t from the point of the first introduction can be represented as:

$$d = t\,2\sqrt{2r_0 s} \qquad (10.18)$$

Where r_0 is the intrinsic rate of increase with time in prevalence when disease is first introduced and s is

the diffusion coefficient. This means that if the point and time of first introduction are known, together with times and places of subsequent appearance of the infection, and the environment and host densities are reasonably constant, a simple regression constrained to pass through the origin will enable the prediction of future spread. A variation on the problem above is attempting to identify the likely point of disease introduction, given data on disease arrival at various points. An example of this approach is given in Lips et al. (2008).

If high-quality data are available on the previous rate of spatial spread through the environment, then it may be possible to use approaches more sophisticated than simply applying Equation (10.18). For example, Smith et al. (2002) were able to use data on the first time of arrival of raccoon rabies in the 167 townships in Connecticut to build a stochastic model of rabies transmission that allowed the rate of local spread to be positively affected by human population (which is a good proxy for raccoon density) and negatively affected by the presence of rivers (which form barriers to spread). This type of information can be used to inform management strategies to prevent or slow rabies spread.

In practice, however, managers will often want to know how rapidly an invasive pathogen might spread into an entirely new environment, rather than simply trying to project the rate of future expansion of a pathogen that is already present. For example, an important issue in Australia is to predict the likely rate of spread of foot and mouth disease (which is not present in Australia) through feral pig populations (e.g., Doran and Laffan 2005). In this case, Equation (10.18) is of limited value. While estimates of r_0 and the diffusion coefficient s may be available from other places, they will almost certainly be very different in the environment of Australia.

10.8 Bayesian approaches

Linking epidemiological and ecological models to data continues to be an area of active research. A full exploration of the most recent approaches is beyond the scope of this book. Cooch et al. (2010) provide

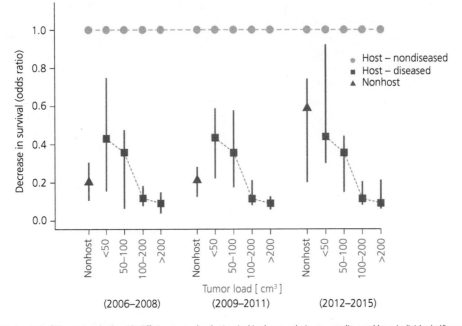

Figure 10.3 Survival of Tasmanian devils with differing tumor loads. Survival is shown relative to nondiseased host individuals ("nondiseased hosts" are prospective hosts prior to the onset of tumor growth, whereas "nonhost" individuals are those that never become infected). Triangles and squares are posterior modes for the odds ratio of survival compared with survival of nondiseased host individuals. Vertical bars represent 95% credible intervals. (Source: Wells et al. 2017.)

a review of some more sophisticated mathematical approaches to estimating parameters for wildlife disease models.

One area that is rapidly gaining in importance is the application of Bayesian approaches (e.g., McCarthy 2007). The key idea in Bayesian statistics is that one can combine prior knowledge—that is, what is already known about the plausible range for a particular parameter—with further information obtained by observations or experiments to produce a posterior distribution, which is an updated idea of the likely range of the parameter in the light of further information. This seems a logical idea, but has received opposition on the basis that it is "subjective" in the sense that results are influenced by prior beliefs (Dennis 1996). Until recently, Bayesian statistics were also difficult to apply computationally, in comparison with the relatively simple application of "canned" statistical packages to standard (so-called frequentist) statistical approaches. In the past few years, the development of statistical packages for Bayesian analysis such as BUGS (http://www.mrc-bsu.cam.ac.uk/software/bugs),

together with some excellent introductory texts (McCarthy 2007; Kéry 2011; Kéry and Schaub 2011) has made these methods much more accessible to general ecologists. Bayesian methods also allow for complex hierarchical models with multiple sources of uncertainty.

For example, Wells et al. (2017) used a hierarchical Bayesian approach to estimate the effect of the tumor mass of Tasmanian devil facial tumor disease on the survival of Tasmanian devils, based on mark–recapture information collected over 10 years. The survival probability of a Tasmanian devil from one time period until the next cannot be directly observed, as not all animals present are captured at any one time period, and given that they may not necessarily be captured, their tumor size at that time can also not be directly observed. However, using a hierarchical Bayesian approach it is possible to use Monte Carlo Markov chain methods (Lunn et al. 2009) to derive estimates of these parameters and the capture probability. As this is a Bayesian approach, prior distributions for each parameter need to be specified although they can

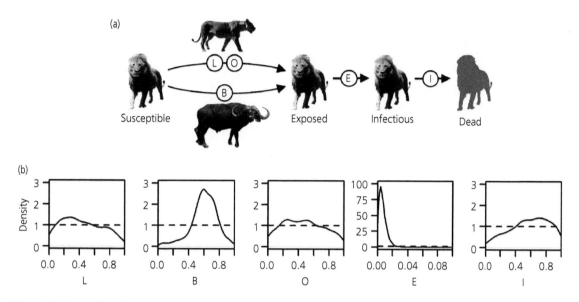

Figure 10.4 Approximate Bayesian computation to estimate parameters for a stochastic model of bovine tuberculosis in lions in Kruger National Park. (Source: Kosmala et al. 2016.) (a) Schematic representation of model showing the parameters to be estimated. L is the probability of lion-to-lion transmission within a pride, O is the probability of transmission between prides, B is the probability of transmission from an infected buffalo, E is the rate of transition from exposed to infectious state, and I is the mortality rate. (b) Posterior distributions of the five disease parameters (solid lines) and their uninformative prior distributions (dashed lines).

be "uninformative" in the sense that all biologically plausible values can be assigned equal prior probability. The method returns distributions of posterior estimates for the parameters from which "credible" intervals can be derived. In theory, credible intervals are different from the confidence intervals derived from classical statistics, but in practice they can be treated similarly. Figure 10.3 shows the survival of Tasmanian devils with tumors of different sizes derived from this analysis, relative to nondiseased animals that eventually become infected. Strikingly, devils with small tumors have higher survival rates than those that never become infected, indicating that devils that eventually become infected with the tumor are in fact fitter than those that never become infected. This is likely to be because socially dominant animals are more likely to bite others and acquire infection (Hamede et al. 2013). Further, the effect of tumors of given size on survival has remained the same over the 10 years of the study, indicating that there is no evidence for an increase in tolerance of infection. This example shows the power of these approaches to extract

ecologically important information from complex datasets.

A recent approach that shows enormous promise for estimating parameters for complex epidemiological models of wildlife is Approximate Bayesian Computation (ABC) (Beaumont 2010; Sun et al. 2015; Kosmala et al. 2016). This approach can be used to estimate parameters for models that are too complex to allow a formal likelihood to be written down for the parameters in the model and can be used for stochastic or individual-based models. In particular, it offers the promise of being able to estimate the transmission rate of pathogens in a field situation, which, as we have seen above, is otherwise an almost intractable problem.

ABC involves choosing sets of parameters randomly selected from the plausible range of the parameters (these constitute a prior distribution), and then simulating the relevant model. The model outputs are then compared with the observed population trajectories. This process is repeated a very large number of times for different parameter sets, and a small subset for which the simulated results are closest to the observed results is used to generate posterior estimates for the relevant parameters. Beaumont (2019) provides an overview of the approach.

Kosmala et al. (2016) used this approach to estimate parameters for a stochastic model of an outbreak of bovine tuberculosis in a lion population in Kruger national park. The model is shown schematically in Figure 10.4a, which shows the five parameters to be estimated. The available data were bovine tuberculosis prevalence in lions over 2 years in three regions. Figure 10.4b shows the posterior distributions of these parameters obtained from the approximate Bayesian approach. More recently, Wells et al. (2017) used a similar approach to parameterize an individually based model of the dynamics of Tasmanian devil facial tumor disease.

References

Anderson, R.M. and May, R.M. (1985). Helminth infections of humans: Mathematical models, population dynamics, and control. *Advances in Parasitology*, **24**: 1–99.

Anderson, R.M. and May, R.M. (1991). *Infectious Diseases of Humans*. Oxford University Press, Oxford.

Beaumont, M.A. (2010). Approximate Bayesian Computation in evolution and ecology. *Annual Review of Ecology, Evolution, and Systematics*, 41:379–406.

Beaumont, M. A. 2019. Approximate Bayesian Computation. Annual Review of Statistics and Its Application 6:379–403.

Beeton, N. and McCallum, H. (2011). Models predict that culling is not a feasible strategy to prevent extinction of Tasmanian devils from facial tumour disease. *Journal of Applied Ecology*, **48**:1315–23.

Begon, M., Townsend, C.R., and Harper, J.L. (2006). *Ecology: From Individuals to Ecosystems*, 4th edn. Blackwell Science, Oxford.

Brøseth, H., Flagstad, Ø., Wärdig, C., Johansson, M., and Ellegren, H. (2010). Large-scale noninvasive genetic monitoring of wolverines using scats reveals density dependent adult survival. *Biological Conservation*, **143**:113–20.

Carroll, E.L., Bruford, M.W., DeWoody, J.A., Leroy, G., Strand, A., Waits, L., and Wang, J., (2018). Genetic and genomic monitoring with minimally invasive sampling methods. *Evolutionary Applications*, **11**: 1094–119.

Colchero, F., Jones, O.R., and Rebke, M. (2012). BaSTA: An R package for Bayesian estimation of age-specific survival from incomplete mark–recapture/recovery data with covariates. *Methods in Ecology and Evolution*, 3 (3):466–70.

Cooch, E. and White, G. (2018). *Using MARK—A Gentle Introduction*, 17th edn. Accessed at: http://www.phidot.org/software/mark/docs/book (accessed August 1, 2018).

Cooch, E.G., Conn, P.B., Ellner, S.P., Dobson, A.P., and Pollock, K.H. (2010). Disease dynamics in wild populations: Modeling and estimation: A review. *Journal of Ornithology*, **152**:S485–509.

Cox, D.R. and Oakes, D. (1984). *Analysis of Survival Data*. Chapman and Hall, London.

Crawley, M.J. (2007). *The R Book*. John Wiley & Sons, Chichester.

Dennis, B. (1996). Discussion: should ecologists become Bayesians? *Ecological Applications*, **6**:1095–103.

DiRenzo, G.V., Che-Castaldo, C., Saunders, S.P., Grant, E.H.C., and Zipkin, E.F. (2019). Disease-structured N-mixture models: A practical guide to model disease dynamics using count data. *Ecology and Evolution*, **9**: 899–909.

Doran, R.J. and Laffan, S.W. (2005). Simulating the spatial dynamics of foot and mouth disease outbreaks in feral pigs and livestock in Queensland, Australia, using a susceptible-infected-recovered cellular automata model. *Preventive Veterinary Medicine*, **70**:133–52.

Efford, M.G., Dawson, D.K., and Borchers, D.L. (2009). Population density estimated from locations of individuals on a passive detector array. *Ecology*, **90** (10): 2676–82.

Faustino, C.R., Jennelle, C.S., Connolly, V., Davis, A.K., Swarthout, E.C., Dhondt, A.A., et al. (2004). *Mycoplasma gallisepticum* infection dynamics in a house finch population: Seasonal variation in survival, encounter and transmission rate. *Journal of Animal Ecology*, **73**: 651–69.

Ferrari, M.J., Bjornstad, O.N., and Dobson, A.P. (2005). Estimation and inference of R_0 of an infectious pathogen by a removal method. *Mathematical Biosciences*, **198**: 14–26.

Foufopoulos, J. (1998). Host–parasite interactions in the mountain spiny lizard *Sceloporus jarrovi*. Unpublished PhD thesis, University of Wisconsin, Madison, WI.

Gamble, A., Garnier, R., Chambert, T., Gimenez, O., and Boulinier, T., (2020). Next-generation serology: Integrating cross-sectional and capture–recapture approaches to infer disease dynamics. *Ecology*, **101**:p.e02923.

Gopalaswamy, A.M., Royle, J.A., Hines, J.E., Singh, P., Jathanna, D., Kumar, N.S., and Karanth, K.U. (2012).

Program SPACECAP: Software for estimating animal density using spatially explicit capture–recapture models. *Methods in Ecology and Evolution*, 3:1067–72.

Grenfell, B.T. and Anderson, R.M. (1985). The estimation of age-related rates of infection from case notifications and serological data. *Journal of Hygiene*, **95**: 419–36.

Gulland, F.M. (1992). The role of nematode parasites in Soay sheep (*Ovis aries* L.) mortality during a population crash. *Parasitology*, **105**:493–503.

Hamede, R.K., Bashford, J., McCallum, H., and Jones, M. (2009). Contact networks in a wild Tasmanian devil (*Sarcophilus harrisii*) population: Using Social Network Analysis to reveal seasonal variability in social behaviour and its implications for transmission of devil facial tumour disease. *Ecology Letters*, **12**:1147–57.

Hamede, R.K., McCallum, H., and Jones, M. (2013). Biting injuries and transmission of Tasmanian devil facial tumour disease. *Journal of Animal Ecology*, **82**: 182–90.

Jennelle, C.S., Cooch, E.G., Conroy, M.J., and Senar, J.C. (2007). State-specific detection probabilities and disease prevalence. *Ecological Applications*, **17**:154–67.

Kelly, M.J. (2008). Design, evaluate, refine: Camera trap studies for elusive species. *Animal Conservation*, **11**: 182–84.

Kendall, W.L. and Pollock, K.H. (1992). The robust design in capture-recapture studies: A review and evaluation by Monte Carlo simulation. In: McCullough, D.R. and Barrett, R.H. (eds), *Wildlife 2001: Populations*, pp. 31–43. Elsevier Applied Science, London.

Kendall, W.L., Pollock, K.H., and Brownie, C. (1995). A likelihood-based approach to capture-recapture estimation of demographic parameters under the robust design. *Biometrics*, **51**:293–308.

Kéry, M. (2011). *Bayesian Population Analysis Using WinBugs: A Hierarchical Perspective*. Elsevier Academic Press, San Diego, CA.

Kéry, M. and Schaub, M. (2011). *Bayesian Population Analysis Using WinBUGS: A Hierarchical Perspective*. Academic Press, Waltham, MA.

Knape, J., Arlt, D., Barraquand, F., Berg, Å., Chevalier, M., Pärt, T., et al. (2018). Sensitivity of binomial N-mixture models to overdispersion: The importance of assessing model fit. *Methods in Ecology and Evolution*, **9**: 2102–14.

Kosmala, M., Miller, P., Ferreira, S., Funston, P., Keet, D., and Packer, C. (2016). Estimating wildlife disease dynamics in complex systems using an Approximate Bayesian Computation framework. *Ecological Applications*, **26**:295–308.

Krebs, C.J. (1999). *Ecological Methodology*, 2nd edn. Benjamin Cummings, Menlo Park, CA.

Lachish, S., Knowles, S.C.L., Alves, R., Wood, M.J., and Sheldon, B.C. (2011). Infection dynamics of endemic malaria in a wild bird population: Parasite species-dependent drivers of spatial and temporal variation in transmission rates. *Journal of Animal Ecology*, **80**: 1207–16.

Lachish, S., McCallum, H., Mann, D., Pukk, C., and Jones, M. (2010). Evaluation of selective culling of infected individuals to control Tasmanian devil facial tumor disease. *Conservation Biology*, **24**:841–51.

Lebreton, J.D., Nichols, J.D., Barker, R.J., Pradel, R., and Spendelow, J.A. (2009). Modeling individual animal histories with multistate capture-recapture models. *Advances in Ecological Research*, **41**: 87–173.

Liccioli, S., Rogers, S., Greco, C., Kutz, S.J., Chan, F., Ruckstuhl, K.E., et al. (2015). Assessing individual patterns of *Echinococcus multilocularis* infection in urban coyotes: Non-invasive genetic sampling as an epidemiological tool. *Journal of Applied Ecology*, **52**: 434–42.

Link, W.A., Schofield, M.R., Barker, R.J., and Sauer, J.R. (2018). On the robustness of N-mixture models. *Ecology*, **99**:1547–51.

Lips, K.R., Diffendorfer, J., Mendelson, J.R., and Sears, M.W. (2008). Riding the wave: Reconciling the roles of disease and climate change in amphibian declines. *PLoS Biology*, 6:e72.

Lunn, D., Spiegelhalter, D., Thomas, A., and Best, N. (2009). The BUGS project: Evolution, critique and future directions. *Statistics in Medicine*, **28**:3049–67.

Marucco, F., Boitani, L., Pletscher, D.H., and Schwartz, M.K. (2011). Bridging the gaps between non-invasive genetic sampling and population parameter estimation. *European Journal of Wildlife Research*, **57**:1–13.

McCallum, H. (2000). *Population Parameters: Estimation for Ecological Models*. Blackwell Science, Oxford.

McCallum, H., Fenton, A., Hudson, P.J., Lee, B., Levick, B., Norman, R., et al. (2017). Breaking beta: Deconstructing the parasite transmission function. *Philosophical Transactions of the Royal Society of London B: Biological Sciences*, **372**:20160084.

McCarthy, M.A. (2007). *Bayesian Methods for Ecology*. Cambridge University Press, Cambridge.

Mills, L.S., Citta, J.J., Lair, K.P., Schwartz, M.K., and Tallmon, D.A. (2000). Estimating animal abundance using noninvasive DNA sampling: Promise and pitfalls. *Ecological Applications*, **10**:283–94.

Murray, K.A., Skerratt, L.F., Speare, R., and McCallum, H. (2009). The impact and dynamics of disease in species threatened by the amphibian chytrid fungus, *Batrachochytrium dendrobatidis*. *Conservation Biology*, **23**:1242–52.

Naing, H., Ross, J., Burnham, D., Htun, S., and Macdonald, D. (2019). Population density estimates and conservation concern for clouded leopards *Neofelis nebulosa*, marbled cats *Pardofelis marmorata* and tigers *Panthera tigris* in Htamanthi Wildlife Sanctuary, Sagaing, Myanmar. *Oryx*, **53**:654–62. doi:10.1017/S0030605317001260

O'Brien, T.G. and Kinnaird, M.F. (2008). A picture is worth a thousand words: The application of camera trapping to the study of birds. *Bird Conservation International*, **18**:S144–62.

Ozgul, A., Oli, M.K., Bolker, B.M., and Perez-Heydrich, C. (2009). Upper respiratory tract disease, force of infection, and effects on survival of gopher tortoises. *Ecological Applications*, **19**:786–98.

Perkins, S.E., Cagnacci, F., Stradiotto, A., Arnoldi, D., and Hudson, P.J. (2009). Comparison of social networks derived from ecological data: Implications for inferring infectious disease dynamics. *Journal of Animal Ecology*, **78**:1015–22.

Roberts, M.G. and Heesterbeek, J.A.P. (2007). Model-consistent estimation of the basic reproduction number from the incidence of an emerging infection. *Journal of Mathematical Biology*, **55**:803–16.

Royle, J.A. (2004). N-mixture models for estimating population size from spatially replicated counts. *Biometrics*, **60**:108–15.

Rudge, J.W., Webster, J.P., Lu, D.B., Wang, T.P., Fang, G.R., and Basanez, M.G. (2013). Identifying host species driving transmission of *Schistosomiasis japonica*, a multihost parasite system, in China. *Proceedings of the National Academy of Sciences*, **110**:11457–62.

Smith, D.L., Lucey, B., Waller, L.A., Childs, J.E., and Real, L.A. (2002). Predicting the spatial dynamics of rabies epidemics on heterogeneous landscapes. *Proceedings of the National Academy of Sciences*, **99**:3668–72.

Sommer, P., Liu, J., Zhao, K., Kusy, B., Jurdak, R., McKeown, A., and Westcott, D. (2016). Information bang for the energy buck: Towards energy- and mobility-aware tracking. *EWSN*, 193–204.

Sun, L.B., Lee, C., and Hoeting, J.A. (2015). Parameter inference and model selection in deterministic and stochastic dynamical models via approximate Bayesian computation: Modeling a wildlife epidemic. *Environmetrics*, **26**:451–62.

Sutherland, W.J. (2006). *Ecological Census Techniques: A Handbook*, 2nd edn. Cambridge University Press, Cambridge.

Thomas, L., Buckland, S.T., Rexstad, E.A., Laake, J.L., Strindberg, S., Hedley, S.L., et al. (2010). Distance software: Design and analysis of distance sampling surveys for estimating population size. *Journal of Applied Ecology*, **47**:5–14.

Welbergen, J.A., Meade, J., Field, H.E., Edson, D., McMichael, L., Shoo, L.P., et al. (2020). Extreme mobility of the world's largest flying mammals creates key challenges for management and conservation. *BMC Biology*, **18**:1–13.

Wells, K., Hamede, R.K., Kerlin, D.H., Storfer, A., Hohenlohe, P.A., Jones, M.E., et al. (2017). Infection of the fittest: Devil facial tumour disease has greatest effect on individuals with highest reproductive output. *Ecology Letters*, **20**:770–78.

Wilber, M.Q., Johnson, P.T.J., and Briggs, C.J. (2020). Disease hotspots or hot species? Infection dynamics in multi-host metacommunities controlled by species identity, not source location. *Ecology Letters*, **23**:1201–11. https://doi.org/10.1111/ele.13518

Zipkin, E.F., Thorson, J.T., See, K., Lynch, H.J., Grant, E.H.C., Kanno, Y., et al. (2014). Modeling structured population dynamics using data from unmarked individuals. *Ecology*, **95**:22–29.

Zuur, A.F., Ieno, E.N., Walker, N.J., and Smith, G.M. (2009). *Mixed Effects Models and Extensions in Ecology with R*. Springer, New York.

PART IV

Managing Wildlife Disease

CHAPTER 11

Disease Management: Introduction and Planning

'There are known knowns; there are things we know we know. We also know there are known unknowns; that is to say, we know there are some things we do not know. But there are also unknown unknowns – the ones we don't know we don't know.'

—Donald Rumsfeld, United States Secretary of Defense, February 2002
(Donald Rumsfeld was widely criticized for this statement, but it provides a useful framework for discussing management of wildlife diseases.)

11.1 Identifying the problem: Known knowns

There are some infectious agents that we know are likely to cause substantial conservation problems if they are present in a population or ecological community. These are "known knowns." If they are not already entrenched in a population or community, then the management strategy should be to prevent their arrival (see Chapter 12). If the infectious agent is established in a population, the management objective can be to eliminate it from the population. This is discussed in Chapter 13. If it is not feasible to completely eliminate the infectious agent, then the management strategy may have to allow the affected species to live with the infection. This issue of disease control is discussed in Chapter 14.

11.2 Known unknowns

In many cases, however, a parasitic organism is known to be present in an ecological community, but the extent to which it actually poses a real conservation threat is unknown. The vast majority of free-living species harbor at least one, and potentially many parasite or pathogen species, which means that simply identifying that an infectious agent is present does not imply that there is a management problem requiring a solution. Indeed, parasites that are endemically present in wild populations at high prevalence are probably not causing substantial conservation problems (McCallum and Dobson 1995).

For these "known unknowns," the first step in management is to determine whether, and in what circumstances, they may pose a conservation threat in the particular species or community. This first step will likely involve using some of the modeling approaches discussed in Chapters 9 and 10. Usually, the effect of infection on survival, growth, or fecundity of animals needs to be determined in the field, and then this effect needs to be included in some form of model to determine what the consequences are likely to be at a population level (Foufopoulos 1998). In some cases, experimentation may be able to directly investigate the effect of infection at the level of the population rather than the individual. If the parasite or pathogen and its impacts are indeed

Infectious Disease Ecology and Conservation. Johannes Foufopoulos, Gary A. Wobeser and Hamish McCallum, Oxford University Press.
© Johannes Foufopoulos, Gary A. Wobeser and Hamish McCallum (2022). DOI: 10.1093/oso/9780199583508.003.0011

identified as a conservation threat, then it becomes a "known known."

11.3 Unknown unknowns

Finally, there is the situation where pathogenic organisms previously unknown to the scientific community impact a species. Typically such pathogens come to the attention of investigators when there is a sudden unexplained population decline. A dramatic example of this was recently observed in the western Mediterranean where, starting in 2017, unexplained mass mortalities of the endemic giant fan mussel *Pinna nobilis* took place (Vázquez-Luis et al. 2017). These catastrophic mortality events, caused by an unknown agent, spread rapidly across the whole Mediterranean Basin imperiling the giant fan mussel as a species. The outbreaks were ultimately traced to a previously undocumented protozoan pathogen that has been now formally described under the name *Haplosporidium pinnae* (Darriba 2017; Catanese et al. 2018).

Such pathogens can be considered to be "unknown unknowns." We have already discussed the example of chytridiomycosis-driven declines of frogs earlier in this book (see Chapter 1, Section 1.4.2). Indeed, early disappearances of numerous frog species in apparently pristine tropical and subtropical rainforest environments in Australia and Central America from about 1980 onward presented an urgent conservation mystery for wildlife biologists. On the basis of morbid individuals, the rapidity of decline at individual sites, and the apparent wave-like spread of the declines, Laurance et al. (1996) proposed that an infectious disease was responsible for the extirpations, but did not identify a specific pathogen. It was not until 2 years later that Berger et al. (1998) identified the novel amphibian chytrid fungus *Batrachochytrium dendrobatidis* as the causative agent for these declines. Documenting such connections is neither obvious, nor easy. In some cases, the causative pathogen still remains unknown: infectious disease has been proposed as having caused both historical and current declines of several Western Australian mammal species (Abbott 2006; Thompson et al. 2010), but no specific agent has yet been identified.

An alternative situation of an "unknown unknown" can occur when increased mortality and morbidity are observed in a population, suggesting impacts of a disease, but where no infectious agent is implicated. A well-known example is the decline of vultures and other scavenging birds on the Indian subcontinent commencing in the 1990s. Populations of the Oriental white-backed vulture (*Gyps bengalensis*) collapsed by over 95%, causing the species to become critically endangered (Pain et al. 2003). Disease was indeed responsible—though it was not infectious disease but rather poisoning by the common anti-inflammatory veterinary drug diclofenac, which was ingested by vultures feeding on carcasses of livestock that had been treated with the drug (Oaks et al. 2004). Following banning of the drug in 2007, there is now evidence that these population declines have slowed or reversed (Chaudhry et al. 2012; Prakash et al. 2012; Bowden 2020; Galligan et al. 2020).

Problems of this type are particularly challenging given our broader lack of knowledge about host–parasite systems in nature. Addressing such issues, which will inevitably continue to arise, is best done by using the combination of approaches described earlier in this book.

11.4 Steps in managing threats to wildlife from infectious disease

What should be done if evidence arises that a wildlife health problem is occurring?

To answer this question, the OIE—the World Organisation for Animal Health—and the IUCN—the International Union for the Conservation of Nature, have created formal protocols to analyze the risk that a disease may pose to a wildlife species (Jakob-Hoff et al. 2014). The overall process of Disease Risk Analysis (DRA) involves several sequential stages, which are summarized in Figure 11.1. While all stages are important in this analysis, one particularly important aspect is the centrality of risk communication, which in practice means regularly engaging with relevant experts and stakeholders. Risk communication is important for two reasons. First, it is essential to draw together the diversity of expertise and interests necessary to identify and define the problem and its possible solutions;

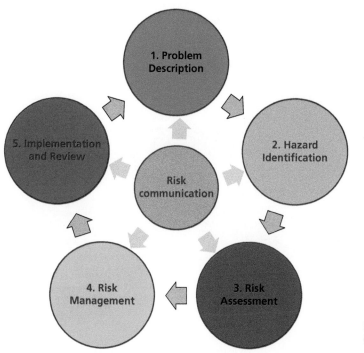

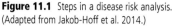

Figure 11.1 Steps in a disease risk analysis. (Adapted from Jakob-Hoff et al. 2014.)

and second, it is critical to ensure broad support and understanding of any management actions proposed by both experts and stakeholders in order for the actions to be implemented successfully (for details, see Chapter 16). An excellent way to accomplish this is by conducting a formal DRA workshop, to which a full range of experts and stakeholders are invited. Jakob-Hoff et al. (2014) provide advice and examples of how to conduct such workshops.

A more applied approach to manage an infectious wildlife disease outbreak is given by McCallum and Jones (2006) (see Figure 11.2). The figure shows a decision tree for managing an epizootic, from the first suspicion that disease might be responsible for a conservation problem, to dealing with a known threat from an established infectious agent. Suggested actions are determined by the responses to specific key questions. In this decision tree, proposed actions are shaded according to the associated economic and conservation costs (high–dark brown; medium–tan; low–yellow. The first step in the process is to determine whether the conservation threat is sufficiently serious to justify action. If not, then the appropriate strategy is to continue

monitoring at some level and to reevaluate the situation periodically to ensure that the threat has not become more serious. This decision box is coded as having a medium-level cost if the decision is incorrect, because if no action is taken when instead some action was justified, time will have been lost and the host population is likely to have declined. This first decision point essentially encompasses the first three DRA steps from Figure 11.1: problem description, hazard identification, and risk assessment. The remainder of the Figure 11.2 flowchart then corresponds to the remaining DRA steps—risk management, implementation, and review.

If, in this initial decision tree evaluation, the threat is deemed to be serious, then, without any further knowledge, an appropriate strategy is to establish insurance populations: such captive or translocated populations would be isolated from any possible disease threat and can ensure the survival of the focal species. For example, captive populations of Tasmanian devils (*Sarcophilus harrisii*) were established as an early part of the management response to Tasmanian devil facial

Decision tree for an emerging wildlife disease

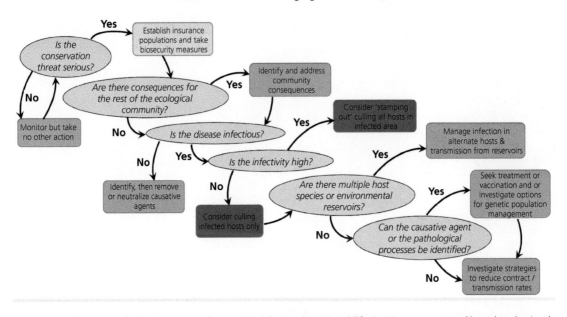

Figure 11.2 A decision tree for the management of an emerging infectious disease in wildlife. Decisions are represented by ovals and actions by rectangles. Shading of rectangles represents the cost (both financial and for conservation) associated with the specified action, if it proves to be the result of an incorrect decision. Light brown shading of a rectangle indicates low costs, medium shading indicates intermediate costs, and dark brown indicates high costs. (Modified from McCallum and Jones 2006.)

tumor disease (Jones et al. 2007). These were later supplemented with a free-ranging translocated population on Maria Island, off the east coast of Tasmania (Thalmann et al. 2016). It may also be sensible to put biosecurity precautions in place to prevent further spread of a possible infectious agent. If this decision is incorrect and the disease turns out not to constitute a serious conservation threat, then these management actions will have resulted in some financial cost but no additional threat to the diseased species itself. This action is therefore coded as low risk.

The next decision point, deliberately placed quite early in the hierarchy, is to determine whether there are consequences for the remainder of the ecological community. If there are, then these may be as serious, or more serious, than the impacts on the diseased species itself. It is therefore important to address any such community consequences as early as possible. If they are not properly identified, then the consequences could be severe, leading to this decision being flagged as medium risk.

Irrespective of whether there are any consequences for the remainder of the ecological community, the next step is to determine whether or not the agent under consideration is infectious. This will likely eventually require the taxonomic identification of the infectious agent, though this may not be immediately necessary. Several actions will be appropriate for any infectious agent, even if unidentified, whereas other approaches will be needed if the disease-causing agent is not infectious (such as the diclofenac example discussed earlier). If the disease-causing agent is not infectious, then the appropriate strategy is to identify and remove it, or put in place strategies to mitigate its effect. Strategies to manage infectious agents such as culling are likely to be counterproductive if deployed against a noninfectious disease-causing agent. Making the right decision at this step in the process is therefore essential.

The next decision point is one with particularly high potential costs. If infectivity is sufficiently high, then a broad "stamping out" approach may be the most sensible management strategy. This is the strategy employed against incursions of highly infectious pathogens in domestic stock and involves broad-scale culling of all individuals that are infected or might have been in contact with infected individuals (see Chapter 13, Section 13.4). The strategy was successfully employed against the incursion of foot and mouth disease into the UK around the turn of the millennium, but with very high financial and social costs (Ferguson et al. 2001; Woolhouse and Donaldson 2001; Haydon et al. 2004). Because of the heavy population losses potentially associated with this approach, it is best used when the disease incursion is recent and only a limited geographic area is affected. The conservation consequences of incorrectly employing the strategy against a disease incursion into endangered wildlife populations could be extremely severe; hence the risk of taking this action has been coded as high.

For infectious diseases that are less highly infectious or are already well established in a population, an alternative is to cull only infected animals. This is known as "test and cull" in the veterinary literature. The application of this strategy and its record of success in managing disease in wildlife populations are also discussed in Chapter 13, Section 13.4. Given the relatively high stakes of this approach, it has been coded as medium to high risk. The choice between the two aforementioned strategies can be difficult given the trade-offs. If the less heavy-handed "test and cull" action is applied when the more severe "stamping out" strategy would have been necessary, valuable time may be lost, allowing the disease to spread further into the population. At that point, the window of time in which "stamping out" would have been possible is likely to have closed, as it would necessitate the culling of an unacceptably high number of animals. Conversely, if the "stamping out" strategy is applied to a species of high conservation value when other, less severe strategies would have been equally successful, a substantial number of animals are likely to have been removed from the population, increasing extinction risk.

The next decision point requires determining whether or not there are multiple host species involved or whether there exist environmental reservoirs for the infectious disease. If there are,

then attempting to manage the disease in a single target species is likely to be unsuccessful. Instead, appropriate strategies will need to involve managing infection in alternative hosts, or in the environment reservoir, or managing transmission from these other sources into the species of conservation significance.

We have placed identification of the etiological agent late in this decision tree. Identifying the causative agent of an infectious disease is highly desirable and is often the best way of determining whether or not the disease is infectious, the likely infectivity, and whether or not there are multiple host species. Nevertheless, the point is that the management strategies earlier in the decision tree can be applied without necessarily knowing the precise identity of the etiological agent. In any emerging disease problem, time is critical—all management does not need to be postponed until the agent is identified. In addition to the strategies earlier in the decision tree, it may still be possible to reduce contact or transmission rates without knowing the identity of the agent. After the agent has been identified, then various additional management strategies open up, such as vaccination, genetic management (artificial selection for resistance or assisted spread of resistant genotypes through wild populations), and potentially treatment of wild individuals. These strategies are further discussed in Chapters 13 and 14.

11.5 Management plans

It is important that any disease management program is based on a written plan. Otherwise, management is likely to proceed on an ad hoc basis, resulting in both a potential waste of resources and a failure to meet goals. Most wildlife disease problems ultimately require long-term solutions, meaning that inevitable changes in personnel over time are likely to result in protocol drift and loss of direction in the absence of formal written goals. It is noteworthy that management of any wild population or ecosystem involves considerations other than simply scientific ones. There are economic, ethical, and social issues that need to be considered and incorporated into the final management program. Some of these are discussed more fully in Chapter 16.

11.5.1 Elements of a management plan

Clear and realistic objectives. What are the goals that the campaign is aiming to achieve? Elimination of infectious disease, prevention of spread to new populations, or persistence of the host population in the presence of disease are all very different objectives and require distinct approaches. The plan may have more than one goal and may involve fallback positions. For example, the initial goal may be to eliminate disease, but if this proves not to be possible, the goal may then be to allow the host population to persist with as wide a distribution as is possible in the face of disease. Without clear goals agreed to by stakeholders, a management plan is unlikely to succeed.

Evaluation of costs, benefits, and risks. Management actions need to be explicitly directed at one or more stated goals. It is likely several potential actions exist that may achieve a specific goal. For example, if the goal is elimination of an emerging disease from a population, then selective culling, nonselective culling, or vaccination are three potential actions that all may accomplish this particular goal. Actions need to be evaluated against clear criteria (Bottrill et al. 2008) including their benefit to biodiversity if successful, their risk to biodiversity if unsuccessful, and their probability of success. Criteria also need to include the cost of taking the action, a consideration that may include both financial and nonfinancial components.

Capacity for adaptation. The importance of adaptive management has been increasingly recognized since Holling first coined the term in the 1970s (Williams 2011a). In its simplest form, it means that there should be the capacity to adapt the management strategy in the light of new knowledge acquired during the management process. Management is thus iterative. Adaptation can happen at two levels: first, regular evaluation and adaptation of the actions implemented (e.g., field manipulations; termed technical learning); and, second, a more infrequent reconsideration of the overall project set-up elements (e.g., general objectives or stakeholders; termed institutional learning; see

Figure 11.3). A distinction is also sometimes made between passive and active adaptive management (Williams 2011b). Both modify management actions in the light of new information, but in active adaptive management, one of the criteria in designing management actions is learning about the behavior of the system. When a management intervention is deliberately designed as an experiment, this constitutes active adaptive management.

Clear criteria for success or failure. It is important to establish stopping rules as part of a management plan. For example, if the initial objective is elimination of a pathogen that has entered a population of conservation importance, the manager will at some stage need to decide either that the eradication program has been successful and therefore the management intervention can cease, or that eradication has failed. If victory is declared too early and management ceases, the disease is likely to re-emerge and the previous investment in eradication will be wasted. If managers concede that the program has failed, alternative conservation strategies that involve living with the pathogen need to be employed. Indefinite application of strategies intended to eradicate disease will not only waste resources but may also result in less favorable conservation outcomes than using the same resources with the objective of enabling the population to persist in the face of the disease threat.

11.5.2 Existing management plans for wildlife diseases: The example of Australia

Australia, as a relatively isolated continent with a unique biota, free of a number of pathogens that are elsewhere widespread, provides a good exemplar of the management planning process for animal diseases, both those already present and for those that might be introduced. Regarding the first category, under the Australian Federal Environment Protection and Biodiversity Conservation Act (Australian Government 1999), two wildlife diseases, psittacine beak and feather disease (PBFD) and amphibian chytrid fungus, have been listed

Adaptive management in a disease control campaign

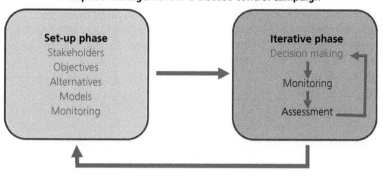

Figure 11.3 Overview of the adaptive management approach during a wildlife disease outbreak. Following planning of the project in the set-up phase (L) the actual implementation occurs in the iterative phase (R). Evaluation and adaptation may occur in two different stages. First, technical learning occurs in the iterative phase where insights stemming from regular monitoring and evaluation are incorporated into the decision-making process. Second, process and institutional learning also happen during periodic reconsiderations of the whole management project and its set-up elements (e.g., objectives, alternatives etc.; bottom arrow). (Adapted from Williams 2011a.)

as Key Threatening Processes (Australian Government Department of the Environment and Energy 2018).

PBFD is caused by a virus in the family Circoviridae. It is widespread in wild populations of Australian parrots but can in some circumstances cause significant mortality, potentially threatening several endangered parrot species (Raidal and Peters 2018). A formal threat abatement plan was prepared in 2005, but was withdrawn after a review in 2012 (Australian Government 2005; Australian Government Department of Sustainability, Environment, Water, Population and Communities 2012) and replaced by a nonstatutory Threat Abatement Advice. The original plan identified two goals:

1. To ensure that PBFD does not increase the likelihood of extinction or escalate the threatened species status of psittacine birds.
2. To minimize the likelihood of PBFD becoming a key threatening process for other psittacine species.

Although some of the objectives subsidiary to these goals were partially met, the conclusion of the review was that the threat abatement plan had not succeeded in abating the threat. While the review attributes this failure to a number of issues, including the intractability of the problem and lack of resources, the plan did not include explicit evaluation of costs, benefits, and risks, capacity to adaptation, or clear success and failure criteria.

The second pathogen listed is the previously discussed fungus *B. dendrobatidis*, which has already caused the extinction of many frog species, both in Australia and Central America, and that threatens many other frog taxa worldwide (Kilpatrick et al. 2010). The current Threat Abatement Plan (Australian Government 2016) identifies four objectives:

1. Improve understanding of the extent and impact of infection by amphibian chytrid fungus and reduce its spread to uninfected areas and populations.
2. Identify and prioritize key threatened amphibian species, populations, and geographical areas and improve their level of protection by implementing coordinated, cost-effective on-ground management strategies.
3. Facilitate collaborative applied research that can be used to inform and support improved management of amphibian chytrid fungus.
4. Build scientific capacity and promote communication among stakeholders.

The plan recognizes from the outset that eradication of the fungus is not a feasible goal. An extensive series of actions is associated with each of the objectives, together with projected costs and performance indicators. With clearly stated objectives, the plan thus addresses the first of the aforementioned essential elements of a management plan and goes some way to also meeting the other elements of a good plan. The emphasis on research implies a capacity for adaptive management and modification of plans. Ultimately, successful management plans need to be "living documents" that are regularly updated and revised as more knowledge becomes available.

Beyond these two wildlife pathogens, which are well entrenched on the continent, Australia has also a detailed framework for dealing with other nonnative, potentially invasive pathogenic organisms. Written management plans that deal with disease incursions into populations where they are not currently present are particularly important. When dealing with any invading organism, speed of response before the species becomes well established is essential. The presence of a management plan in place, with agreed actions, *before* an incursion actually occurs, maximizes the chance of successfully eradicating an invasive disease. Australia has a formal, well-developed scheme, AUSVETPLAN (Animal Health Australia 2018) for responding to disease incursions and emergency disease events. While the plan is directed primarily at diseases posing a threat to human health or livestock, the key components are relevant to handling an infectious disease incursion into any animal population, including wildlife. The default response to a disease incursion is stamping out, supplemented (or replaced when stamping out is not appropriate) by disease control measures such as vaccination or treatment. While response plans to individual diseases differ, the general structure of a typical disease-specific management strategy under AUSVETPLAN is summarized in Box 11.1. The documentation of the plan also includes a specific operational manual dedicated to managing disease incursions in wild animals (Animal Health Australia, 2011). Although this document is written from the perspective of managing a livestock disease or zoonotic diseases that happen to have a reservoir in wild animals, it nevertheless also contains information useful for managing incursions of pathogens that affect primarily wildlife.

Box 11.1 Structure of a typical Australian Veterinary Emergency Plan (AUSVETPLAN) for an emerging disease (Animal Health Australia 2018)

- The nature of the disease (etiology, host range, geographic distribution, incubation period, transmission, clinical signs, diagnosis, the development of immunity, and the availability of vaccination and treatment).
- Principles of control and eradication (critical factors for formulating a response strategy, and options for control or eradication based on these factors).
- Policy and rationale (the agreed Australian response policy and strategies for its implementation—such as stamping out, tracing and surveillance, vaccination, treatment, disposal, decontamination, wild animal and vector control, public awareness, and media).
- Use of declared areas and premises classifications.
- Recommended quarantine and movement controls.
- Procedures for surveillance and proof of freedom.

References

Abbott, I. (2006). Mammalian faunal collapse in Western Australia, 1875–1925: The hypothesised role of epizootic disease and a conceptual model of its origin, introduction, transmission, and spread. *Australian Zoologist*, **33**:530–61.

Animal Health Australia (2011). Wild Animal Response Strategy (Version 3.3). Australian Veterinary Emergency Plan (AUSVETPLAN), edition 3, Primary Industries Ministerial Council, Canberra, ACT. https://www.animalhealthaustralia.com.au/our-publications/ausvetplan-manuals-and-documents (accessed December 27, 2020).

Animal Health Australia (2018). Overview (Version 4.0). Australian Veterinary Emergency Plan (AUSVETPLAN), edition 4, National Biosecurity Committee, Canberra, ACT. https://www.animalhealthaustralia.com.au/our-publications/ausvetplan-manuals-and-documents (accessed December 27, 2020).

Australian Government (1999). Environment protection and biodiversity conservation act. Page 486. Commonwealth of Australia, Canberra, ACT. https://www.legislation.gov.au/Details/C2015C00422 (accessed December 27, 2020).

Australian Government (2005). Threat abatement plan for Psittacine beak and feather disease affecting endangered psittacine species. Department of Environment and Heritage, Commonwealth of Australia, Canberra, ACT. http://www.environment.gov.au/system/files/resources/5764cda0-5e94-48c7-8841-49b09ff7398c/files/beak-feather-tap.pdf (accessed December 7, 2018).

Australian Government (2016). Threat abatement plan for infection of amphibians with chytrid fungus resulting in chytridiomycosis, Commonwealth of Australia 2016. Commonwealth of Australia, Canberra, ACT. http://www.environment.gov.au/system/files/resources/d7506904-8528-411e-a3f4-19d4379935f9/files/tap-chytrid-fungus-2016.pdf (accessed December 27, 2020).

Australian Government Department of the Environment and Energy. (2018). Species Profile and Threats Database—Listed key threatening processes. Canberra, ACT. http://www.environment.gov.au/cgibin/sprat/public/publicgetkeythreats.pl. (accessed December 7, 2018).

Australian Government Department of Sustainability, Environment, Water, Population and Communities. (2012). Review of the Threat Abatement Plan for Psittacine Beak and Feather Disease Affecting Endangered Psittacine Species (2005). http://www.environment.gov.au/biodiversity/threatened/publications/tap/beak-and-feather-disease-affecting-endangered-psittacine-species#review (accessed August 26, 2020).

Berger, L., Speare, R., Daszak, P., Green, D.E., Cunningham, A.A., Goggin, C.L., et al. (1998). Chytridiomycosis causes amphibian mortality associated with population declines in the rain forests of Australia and Central America. *Proceedings of the National Academy of Sciences*, **95**:9031–36.

Bottrill, M.C., Joseph, L.N., Carwardine, J., Bode, M., Cook, C., Game, E.T., et al. (2008). Is conservation triage just smart decision making? *Trends in Ecology and Evolution*, **23**:649–54.

Bowden, C.G. (2020). Asian vulture crisis update: Populations respond to effective removal of diclofenac but the threat from non-steroidal anti-inflammatory drugs remains. *BirdingAsia*, **33**:46–50.

Catanese, G., Grau, A., Valencia, J.M., García-March, J.M., Álvarez, E., Deudero, S., et al., (2018). *Haplosporidium pinnae* sp. nov., a haplosporidan parasite associated with massive mortalities of the fan mussel, *Pinna nobilis*, in the Western Mediterranean Sea. *Journal of Invertebrate Pathology*, **157**:9–24.

Chaudhry, M.J., Ogada, D.L., Malik, R.N., Virani, M.Z., and Giovanni, M.D. (2012). First evidence that populations of the critically endangered long-billed

vulture *Gyps indicus* in Pakistan have increased following the ban of the toxic veterinary drug diclofenac in south Asia. *Bird Conservation International*, **22**:389–97.

Darriba, S. (2017). First haplosporidan parasite reported infecting a member of the Superfamily Pinnoidea (*Pinna nobilis*) during a mortality event in Alicante (Spain, Western Mediterranean). *Journal of Invertebrate Pathology*, **148**:14–19.

Ferguson, N.M., Donnelly, C.A., and Anderson, R.M. (2001). The foot-and-mouth epidemic in Great Britain: Pattern of spread and impact of interventions. *Science*, **292**:1155–60.

Foufopoulos, J. (1998). Host–parasite interactions in the Mountain Spiny lizard *Sceloporus jarrovii*. Unpublished PhD thesis, University of Wisconsin, Madison, WI.

Galligan, T.H., Bhusal, K.P., Paudel, K., Chapagain, D., Joshi, A.B., Chaudhary, I.P., et al. (2020). Partial recovery of critically endangered *Gyps* vulture populations in Nepal. *Bird Conservation International*, **30**:87–102.

Haydon, D.T., Kao, R.R., and Kitching, R.P. (2004). The UK foot-and-mouth disease outbreak—The aftermath. *Nature Reviews Microbiology*, **2**:675–80.

Jakob-Hoff, R.M., MacDiarmid, S.C., Lees, C., Miller, P.S., Travis, D., and Kock, R. (2014). Manual of Procedures for Wildlife Disease Risk Analysis. World Organisation for Animal Health, in association with the International Union for Conservation of Nature and the Species Survival Commission, Paris.

Jones, M., Jarman, P., Lees, C., Hesterman, H., Hamede, R., Mooney, N., et al. (2007). Conservation management of Tasmanian devils in the context of an emerging, extinction-threatening disease: Devil facial tumor disease. *EcoHealth*, **4**:326–37.

Kilpatrick, A.M., Briggs, C.J., and Daszak, P. (2010). The ecology and impact of chytridiomycosis: An emerging disease of amphibians. *Trends in Ecology and Evolution*, **25**:109–18.

Laurance, W.F., McDonald, K.R., and Speare, R. (1996). Epidemic disease and the catastrophic decline of Australian rain forest frogs. *Conservation Biology*, **10**:406–13.

McCallum, H. and Dobson, A. (1995). Detecting disease and parasite threats to endangered species and ecosystems. *Trends in Ecology and Evolution*, **10**:190–94.

McCallum, H. and Jones, M. (2006). To lose both would look like carelessness: Tasmanian devil facial tumour disease. *PLoS Biology*, **4**:1671–74.

Oaks, J.L., Gilbert, M., Virani, M.Z., Watson, R.T., Meteyer, C.U., Rideout, B.A., et al. (2004). Diclofenac residues as the cause of vulture population decline in Pakistan. *Nature*, **427**:630–33.

Pain, D.J., Cunningham, A.A., Donald, P.F., Duckworth, J.W., Houston, D.C., Katzner, T., et al. (2003). Causes and effects of temporospatial declines of *Gyps* vultures in Asia. *Conservation Biology*, **17**:661–71.

Prakash, V., Bishwakarma, M.C., Chaudhary, A., Cuthbert, R., Dave, R., Kulkarni, M., et al. (2012). The population decline of *Gyps* vultures in India and Nepal has slowed since veterinary use of diclofenac was banned. *PLoS ONE*, **7**:e49118.

Raidal, S.R. and Peters, A. (2018). Psittacine beak and feather disease: Ecology and implications for conservation. *Emu*, **118**:80–93.

Thalmann, S., Peck, S., Wise, P., Potts, J.M., Clarke, J., and Richley, J. (2016). Translocation of a top-order carnivore: Tracking the initial survival, spatial movement, home-range establishment and habitat use of Tasmanian devils on Maria Island. *Australian Mammalogy*, **38**:68–79.

Thompson, R.C., Lymbery, A.J., and Smith, A. (2010). Parasites, emerging disease and wildlife conservation. *International Journal for Parasitology*, **40**:1163–70.

Vázquez-Luis, M., Alvarez, E., Barrajon, A., García-March, J.R., Grau, A., Hendriks, I.E., et al. (2017). S.O.S. *Pinna nobilis*: A mass mortality event in western Mediterranean Sea. *Frontiers in Marine Science*, **4**:109.

Williams, B.K. (2011a). Adaptive management of natural resources—Framework and issues. *Journal of Environmental Management*, **92**:1346–53.

Williams, B.K. (2011b). Passive and active adaptive management: Approaches and an example. *Journal of Environmental Management*, **92**:1371–78.

Woolhouse, M.E. and Donaldson, A. (2001). Managing foot-and-mouth. *Nature*, **410**:515–16.

CHAPTER 12

Preventing New Disease Occurrences

Κάλλιον το προλαμβάνειν ή το θεραπεύειν.
[It is better to prevent than to cure.]

Hippocrates (460–370 BCE)

An Ounce of Prevention is worth a Pound of Cure.

Benjamin Franklin (*ca.* 1733)

Preventing the introduction of new health problems, as opposed to intervening once a situation has already reached the crisis point, can and should be a bigger part of what we do as conservation biologists.

Deem et al. (2001)

12.1 Background and definitions

Preventing the emergence of infectious diseases in new situations is pre-damage control, as distinct from disease reduction, elimination, and eradication, which apply after an infectious agent is already established. Although prevention is usually designed to protect a species of concern from a particular disease, successful preventative measures must often not only take into account multiple disease agents that might be involved, but also consider a community of species at risk, including wild and domestic animals, as well as humans. In this case, risk includes explicitly both the probability and consequences of a disease emergence event.

Most infectious diseases are clearly restricted in space, time, and host range, a concept termed the "Natural Nidality of Transmissible Disease" by Pavlovsky (1966). A disease may have evolved in its *nidus*, or arrived there at some time in the past (Audy 1958), but typically further spread is limited by some combination of: (i) geophysical barriers, for example, an ocean, desert, or mountain range;

(ii) ecological barriers, such as lack of suitable hosts, habitat, or climatic conditions, that limit contact, exposure, and transmission; and (iii) species barriers including the specificity of receptor molecules on cell membranes, innate immune defenses, as well as behavioral separation, all of which limit spread between species (Webby et al. 2004; Parrish et al. 2008). Expansion of infection into new areas and hosts arises when a barrier is circumvented by disease agents moving or being moved from a nidus to a location where there are susceptible hosts; or by susceptible hosts moving or being moved into a disease nidus; or when a barrier is rendered ineffective. Effectiveness of the first two barrier types has decreased as "growth of human populations and the development of rapid transportation systems have made the world's biota more connected than at any time in earth's history" (Kilpatrick, 2011). Indeed, the diversity of ways in which these barriers are breaking down today needs to be considered when attempting to prevent wildlife disease outbreaks (Daszak et al. 2000; Figure 12.1).

Infectious Disease Ecology and Conservation. Johannes Foufopoulos, Gary A. Wobeser and Hamish McCallum, Oxford University Press.
© Johannes Foufopoulos, Gary A. Wobeser and Hamish McCallum (2022). DOI: 10.1093/oso/9780199583508.003.0012

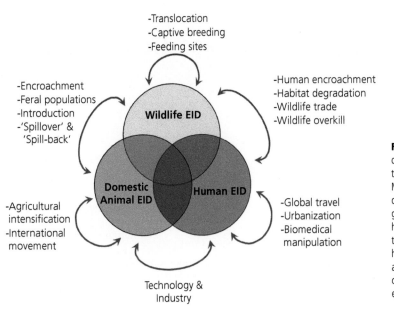

-Translocation
-Captive breeding
-Feeding sites

-Encroachment
-Feral populations
-Introduction
-'Spillover' &
 'Spill-back'

-Human encroachment
-Habitat degradation
-Wildlife trade
-Wildlife overkill

-Agricultural
 intensification
-International
 movement

-Global travel
-Urbanization
-Biomedical
 manipulation

Technology &
Industry

Figure 12.1 Patterns of host range overlap (shown as circles) for different types of emerging infectious diseases (EID). Most pathogens will infect a broad range of hosts categorized here as three broad groups (wildlife, domestic animal, and human hosts). Arrows indicated transmission pathways within and between host groups; accompanying lists provide anthropogenic factors that promote disease emergence. (Adapted from Daszak et al. 2000.)

Assessing disease importation risks is challenging. The World Organisation for Animal Health has developed a system (see Figure 11.2) for assessing disease risks in general. It can also be applied to assess dangers associated with the importation of animals or related materials (OIE 2013), based on a model by Covello and Merkhoher (1993). This system can be adapted for any situation in which a barrier might be circumvented or altered. Assessment has four components that will be discussed in more detail in the text that follows: hazard identification, risk assessment, risk management, and risk communication.

Hazard identification involves identifying the disease agent(s), which in the specific situation could potentially produce adverse consequences. Determining these potentially dangerous agents for wild animals is more difficult than for domestic species, because only a fraction of the agents present in most wild species have been described. For instance, it is estimated that less than half of all parasites (viruses, bacteria, protozoa, macroparasites) of wild carnivores ("one of the best-studied mammal groups") have been discovered to date (Huang et al. 2015). Agents that cause no obvious ill-health in their native host are especially likely to go undetected, and many important agents, including the Nipah, Hendra, HIV, and severe

acute respiratory syndrome (SARS) viruses, were unknown in their wild host until disease emerged in humans or domestic animals. Even a competent diagnostician may overlook infection by previously undescribed agents (Arthur 1995). Even infection caused by a known pathogen does not guarantee timely recognition of an outbreak. History is replete with examples of diseases that, despite being known, have been spread through human actions. When this occurs, "what is blameworthy is not that one was ignorant, but that one did not make an effort to reduce the ignorance" (UNESCO 2005).

Risk assessment includes estimation of the probability and consequences of an adverse event associated with each hazard identified. Qualitative assessment ("a reasoned, systematic and logical discussion of the relevant contributory factors and epidemiology of a hazard, in which the likelihood of its release and exposure, and the magnitude of its consequences are expressed using non-numerical terms such as high, medium, low, or negligible") should be done first (Peeler et al. 2007). If this is inadequate for decision making, it can be followed by a quantitative assessment, provided that resources and data are available. Risk assessment has four components:

Entry assessment—description of the biological pathway(s) necessary for introduction of the pathogen into a particular environment, and estimation of the probabilities of the complete introduction process occurring.

Exposure assessment—description of the biological pathway(s) necessary for exposure of animals and humans at the destination of the pathogen, and estimation of the probability of exposure.

Consequence assessment—description of potential consequences of a given exposure, and estimation of the probability of these occurring. There are no general rules to predict which potential hosts (across various taxonomic groups) might be susceptible to a specific agent, although agents with a broad host range, such as many single-stranded RNA viruses, are more prone to jump species barriers (Woolhouse et al. 2005). Species that are phylogenetically related, or occupy a similar ecologic niche, are also more likely to share diseases. Nevertheless, predicting host shifts with sufficient accuracy to be useful practically is currently not possible (Longdon et al. 2014). Moving individuals between subpopulations of the same species is especially risky, because there is no species barrier to protect individuals at the destination. For instance, the "two most extensive mass mortalities recorded for marine organisms" occurred when pilchards (*Sardinops sagax*) in a subpopulation near Australia became infected by a previously unknown herpesvirus introduced by pilchards from another subpopulation (Jones et al. 1997; Gaughan 2002). In general, the risk of translocating diseases is much greater when moving wild animals than domestic species, because potential disease hazards are less well known, systems for testing and certification—even for known agents—are less established, and tests with known reliability in detecting disease agents are less likely to be available.

Risk estimation—this involves integration of results from the entry assessment, exposure assessment, and consequence assessment into a holistic evaluation of expected risk.

Risk communication is the process by which information regarding hazards and risks is gathered and the results of risk assessment and risk management measures, together with the assumptions and uncertainty, are communicated to decision makers. This is central to the process for two reasons. First, it is essential to draw together the diversity of expertise and interests necessary to identify and define the problem and possible solutions; and, second, it is critical to ensure broad support and understanding of any management actions proposed by both experts and stakeholders in order for the actions to be implemented successfully. One way in which this can be accomplished is by conducting a formal disease risk assessment workshop, to which a full range of experts and stakeholders are invited. Jakob-Hoff et al. (2014) provide advice and examples of how to conduct these workshops. Further details on the best manner of communicating disease risk and involving the general public are given in Chapter 16, Section 16.3.

Risk management is the process of identifying and evaluating the efficacy and feasibility of measures to reduce the risk. The focus of the discussion in this chapter will be measures to prevent introduction, whereas Chapter 13 will focus on the implementation of various measures to reduce or eradicate a disease. In the absence of scientific certainty about what disease agents might be present, and who might be susceptible to these agents, preventative measures should follow the Precautionary Principle: "When human activities may lead to morally unacceptable harm that is scientifically plausible but uncertain, actions shall be taken to avoid or diminish that harm" (UNESCO 2005).

12.1.1 Anticipatory planning

Developing plans to address every potential introduction of exotic wildlife diseases is clearly not practical. There are, however, some cases where the risk is substantial enough for this to be worthwhile. A good example is given by Grant et al. (2017). *Batrachochytrium salamandrivorans* is a fungus closely related to *B. dendrobatidis* that has recently emerged in Europe, causing mass mortalities among European salamanders. At the time of writing, it has yet to be detected in the United States, but as the eastern

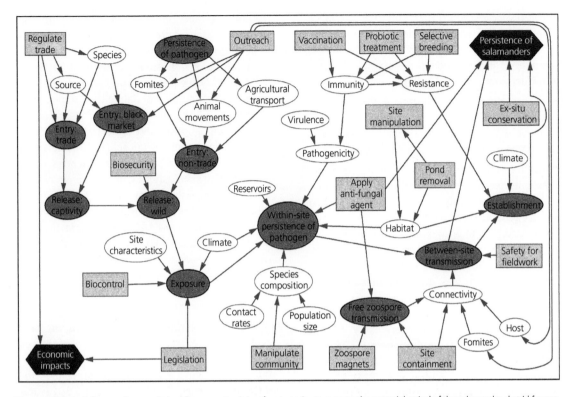

Figure 12.2 An influence diagram, derived from expert opinion, for strategies to manage the potential arrival of the salamander chytrid fungus in North America. The black hexagons represent two management objectives: maximizing the persistence of salamanders and minimizing economic impacts. Shaded ovals represent key stochastic events, and open ovals represent contributing processes. Rectangles represent potential management actions. (Reproduced from Grant et al. 2017.)

United States contains the highest diversity of salamanders in the world, the potential consequences of its introduction could be devastating. Grant et al. (2017) report the results of the workshop in which the advice of numerous experts was elicited to produce the influence diagram shown in Figure 12.2, which links potential management actions and biotic and abiotic factors to fundamental management objectives. The corresponding list of actions is given in Table 12.1.

12.1.2 Ecological and evolutionary considerations

Sometimes an exotic pathogen becomes established in a new population after even a brief first exposure, while other times, an epizootic fails to take place even after repeated contacts. Why is this? Whether

a pathogen can invade a new area is fundamentally determined by the biological characteristics of the system, including the life history of the invading organism, the ecology of the local host(s), as well as the prevailing environmental conditions (see also Section 2.3). The observation that not all types of pathogens can invade new systems with equal ease reflects the fact that commonly invasive pathogens have biological characteristics that are quite different from the traits of a noninvasive disease agent. Emerging pathogens tend to be direct life cycle microparasites (e.g., viruses, fungi, and bacteria) and only infrequently macroparasites (Lymbery et al. 2014; see Figure 12.3a–c). Similarly, they often have the ability to infect a broader range of host species, can be transmitted through a diversity of pathways, possess durable stages that allow them to survive outside the hosts for long periods

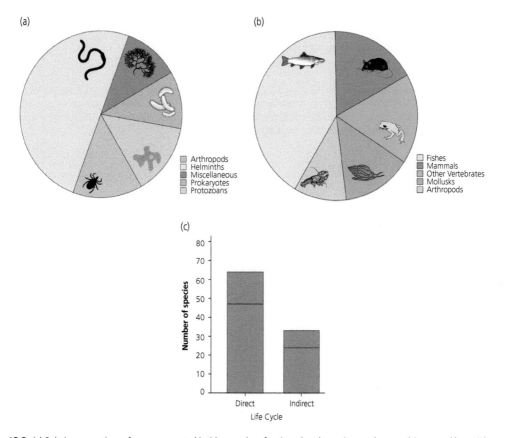

Figure 12.3 (a) Relative proportions of taxa represented in 98 examples of co-introduced parasites: prokaryotes (viruses and bacteria); protozoans; helminths (platyhelminths, nematodes, and acanthocephalans); arthropods (crustaceans, arachnids); and a miscellaneous group including fungi, myxozoans, annelids, mollusks, and pentasomids. (b) Relative proportions of alien hosts represented in 98 examples of co-introductions: mollusks; arthropods; fishes; mammals; and other vertebrates (amphibians, reptiles, and birds). (c) Number of co-introduced parasite species with direct and indirect life cycles that have switched (blue bars) or not switched (orange bars) from alien to native host species. (Modified from Lymbery et al. 2014.)

of time, and have an overall high R_0 (Dobson and Foufopoulos 2001). Understanding how pathogen characteristics are associated with higher invasion probabilities can allow conservation managers to prioritize those parasites that possess these particular traits in their prevention efforts. Similarly, one can focus prevention efforts on those specific pathogens that already have a track record of emergence either in other moments in time or in other parts of the world. In general, however, successful prevention of invasion hinges on a detailed understanding of the prevailing biological realities, so as to know where and how it is most profitable to disrupt transmission.

12.1.2.1 Parasite life cycle and introduction

Among the many biological characteristics of a pathogen, none is more important than the complexity of its life cycle, because it often determines whether an agent that has successfully crossed a species barrier will be able to maintain itself in the new population (May et al. 2001; Lymbery et al. 2014). Agents that are transmitted directly require only a population of susceptible hosts to complete their life cycle (Figure 12.3c). Once the agent makes contact at the destination with a new host species and manages to infect it, it will likely become established if the transmission rate is high enough (i.e. its $R_0 > 1$). In contrast, agents with a heteroxenous

Table 12.1 Preparatory planning is a key part of preventing a disease introduction. List of potential management actions for *Batrachochytrium salamandrivorans* (*Bsal*) generated during a preparatory conference in anticipation of the introduction of the pathogen into North America. Actions are arranged into nine broad categories (adapted from Grant et al. 2016).

1. Environmental and habitat actions

   Filter or remove of contaminated water; regulate water use or extraction; regulate and treat wastewater; apply antimicrobial agent to terrestrial/aquatic habitat; apply sterilization techniques (heat, UV, salt/copper treatment); use probiotics; manipulate local species assemblage (e.g., zooplankton) to impair transmission; modify vegetation or substrate to reduce pathogen spread; increase habitat fragmentation to reduce pathogen spread; contain site to restrict animal or pathogen dispersal; destroy affected habitat

2. Actions on animals

   Cull subset/all of animals or otherwise decrease population densities to reduce transmission; apply heat treatment; vaccinate; otherwise stimulate immune system; administer antimicrobial agents; administer probiotics; expose host to other less virulent strain (competition); translocate before/after treatment

3. Regulatory actions and policy

   Ban or regulate target species imports and enforce national import regulation and laws; enforce state laws and import restrictions; require health certificate; quarantine and/or decontaminate imported animals; require surveillance for pathogen in all traded animals, water, or fomites

4. Captive breeding of host

   Create assurance colonies; selective breeding (for resistance, tolerance); population augmentation; hybridize species for low pathogen virulence; captive breeding with immunization and reintroduction (to increase herd immunity); genetic modification of host or pathogen

5. Regulation and policy regarding commercial pet trade and zoos

   Require and enforce biosecurity protocols; enforce existing animal trade and health laws; require pre-trade health certificate; establish surveillance protocols for pathogen; treat or decontaminate animals prior to trade; require chain of custody documentation; quarantine and test animals; request/require pet marking

6. Authority and jurisdiction to act

   Lobby for research and surveillance funding; advocate for changes in relevant laws and regulations; request federal authority to act on wildlife diseases; participate in international control and coordination efforts; engage champions to raise public awareness; organize institutional response; engage decision makers and stakeholders (e.g., private land owners)

7. Human activities

   Regulate and enforce collection and movement of animals (at local, region, state level); limit spread of other hosts and vectors; regulate transport of fomites; create and enforce disinfection stations; establish decontamination protocols for all user groups (researchers, public, managers); reduce public access

8. Monitoring and research

   Create and implement surveillance program; develop rapid testing field protocols (enzyme-linked immunosorbent assay, environmental DNA); reduce diagnostic errors (centralize diagnostics, refine methodologies); identify origin and spread of disease; sequence pathogen genome and strains; develop risk assessment models; develop antimicrobial treatments; develop vaccinations; monitor pathogen effects on populations (population demography, nonlethal effects, survival); maintain database for information sharing

9. Public education and outreach

   Organize outreach and communication efforts; hand out detection and eradication kits to landowners/agencies; raise awareness to increase compliance with: (i) pet trade actions, (ii) local site-visit decontamination actions, and (iii) animal movement and fomite movement actions; prepare coloring books and student outreach materials, e.g., create a super hero (public figure/image—puppet, a voice) to alert public about disease problem

(i.e., complex, or indirect) life cycle require at least two suitable host species to reproduce successfully. For instance, the nematode *Parastrongylus cantonensis* requires both suitable vertebrate and invertebrate hosts for transmission. Following its introduction into New Orleans, *P. cantonensis* was able to become established in North American rodent populations only by also being able to infect native snails as intermediate hosts (Kim et al. 2002). If the required hosts are not present, they may be introduced, simultaneously or independently, and in any order. For instance, yellow fever virus was introduced into the Americas simultaneously with a suitable mosquito vector (Juliano and Lounibos 2005). In contrast, the vector mosquito *Culex quinquefasciatus* was introduced to Hawaii long before *Plasmodium relictum* (the agent of avian malaria) arrived in imported birds. Because a suitable vector was already present, the parasite was able to become established, and spread to indigenous birds (Van Riper et al. 1986). Conversely, while Arctic foxes (*Vulpes lagopus*) introduced the tapeworm

Echinococcus multilocularis to Svalbard, it did not become established there until a suitable intermediate host, the sibling vole (*Microtus levis*), was introduced inadvertently (Knapp et al. 2012, see also Section 3.2.3).

In the following sections, we discuss prevention of disease establishment under three distinct scenarios: (i) introduction of a pathogen into an area of susceptible hosts; (ii) introduction of susceptible hosts into an area where a disease is already present; and (iii) environmental modification that accelerates transmission of an existing disease through a susceptible population of resident hosts.

12.2 Introduction of pathogens into susceptible systems: Prevention and management

Sometimes disease agents move into a new setting through natural processes. For example, migrating birds introduced *Trichomonas gallinae* to the UK (Lawson et al. 2011) and West Nile virus to Israel (Malkinson et al. 2002); infected biting midges (*Culicoides* spp.), dispersed by wind, spread bluetongue virus to ruminants in new locations (Ander et al. 2012; Maclachlan and Mayo 2013). Similarly, ringed seals (*Phoca hispida*)—displaced from the Arctic by the collapse of a fish population—introduced phocine morbillivirus to southern harbor seal (*Phoca vitulina*) populations (Härkönen et al. 2006).

However, most disease agents do not arrive in novel situations by natural means; typically they are moved directly or indirectly by humans (Dobson and Foufopoulos 2001; Wikelski et al. 2004; Hayes and Gubler 2006; see Table 12.1). Many such introductions could have been prevented if appropriate care had been taken early on and the focus of this section is to discuss ways to achieve this. Prevention begins with the recognition that, when animals are moved, disease agents living in or on the animals are also likely to be translocated. Several distinct stages exist during the introduction and establishment of an invasive organism including pathogens, and distinct preventative activities are appropriate for each stage (Dunn and Hatcher 2015).

When discussing the prevention of the introduction of a pathogen in a new region, it is useful to distinguish between introductions as the result of accidental transfer of hosts, versus introductions that are the result of intentional movements of hosts by humans (e.g., during reintroduction or translocation campaigns).

12.2.1 Preventing disease introduction after accidental movement of hosts

In the majority of cases, pathogens are introduced into a new host population as the inadvertent result of human activities. For example, pathogens may arrive in an area through fomites, such as contaminated natural products. However, more often, parasites are introduced via their host, so the challenge of control often defaults to managing the unintended introduction of infected hosts. As a result, many of the approaches to prevent the introduction of infected hosts are similar to the approaches used to prevent the introduction of invasive species (Dunn and Hatcher 2015; Figure 12.4). While there are a variety of diverse circumstances where pathogens and parasites can be introduced accidentally (see Table 12.2), we discuss some of the most important cases in the following sections.

12.2.1.1 Exotic pets

Because exotic pets originate from a variety of far-flung localities and represent a broad spectrum of phylogenetic diversity they have the potential to host a very wide range of parasites or pathogens. Increasing numbers of escaped or released pets can become invasive and disrupt whole species communities. For example, in the US state of Florida, 56 species of exotic reptiles and amphibians have already established self-maintaining populations and are in various degrees of expansion (Krysko et al. 2011). Beyond their immediate ecological impacts, such invasive pets represent a significant disease risk, as they can act as conduits for the introduction of exotic pathogens into native reptile and amphibian populations (Pasmans et al. 2008). There is increasing evidence that invasive pathogens originating from exotic pets precipitate epizootics in native species communities (Ariel 2005; Nowak 2010; Can et al. 2019). Under the right circumstances, such pathogens can directly facilitate their host's invasion, if they have a relatively higher detrimental effect on resident species that

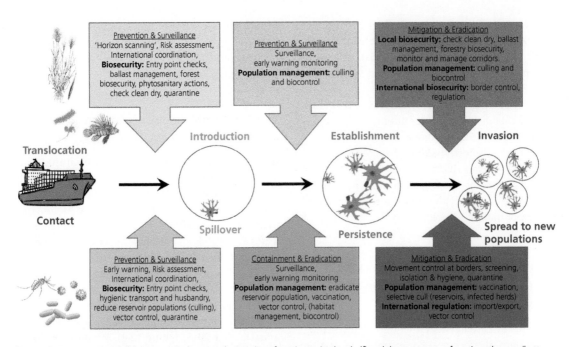

Figure 12.4 Numerous parallel aspects exist between the invasion of exotic species (top half) and the emergence of exotic pathogens (bottom half). The presence of parallel distinct stages for species invasion (translocation, introduction, establishment, invasion) and pathogens (contact, spillover, persistence, spread into new populations) also allows the implementation of stage-specific control strategies (boxes; top row control strategies for invasive species, bottom row control strategies for emerging pathogens). Circles represent recipient habitat/invaded host. (Adapted and modified from Dunn and Hatcher 2015.)

lack immunity (Prenter et al. 2004). Such disease-mediated invasions (DMIs) appear to become an increasingly common problem (Tompkins et al. 2003; Strauss et al. 2012; Schrimpf et al. 2013). Given the impacts that such pathogens can have, it is particularly important to educate pet importers and owners on the dangers that pets may pose, as well as the merits of preventing exotic animal escape.

Two specific periods are particularly important for the prevention of the introduction of exotic pet pathogens. The first is during the initial importation of the animals into a new country and the second occurs at the end, when an animal either dies and the body needs to be decontaminated, or when it cannot be cared for anymore and the temptation exists to release it into the wild. During initial importation, and especially with wild-caught individuals, it is most pressing to implement comprehensive health surveillance of all imported animals together with appropriate quarantine and disinfection protocols to avoid infection of other

pets (Pasmans et al. 2008). Also important is the availability of reliable tests that can be used to rapidly assess infection status of all animals imported (Grant et al. 2016). Various additional tools such as pre-trade health certificates or pet marking can be used to further prevent entry of infected hosts. Alternatively, if the risk is too high, it is possible to prohibit—as has been already done by some countries—the importation of particularly invasive taxa into a country (García de Lomas and Vilà 2015). Once a species has been imported, and to avoid potential spillover of the infection into wild animals, it is important that pet owners are educated both about the ecological risks associated with the escape or deliberate release of any pets into the wild, *and* of the key ways to prevent this from happening.

12.2.1.2 Feral wildlife

Pathogens are frequently introduced into wildlife populations through the secondary spread into

Table 12.2 Examples of infectious diseases of wild species translocated to new locations through movement of infected animals.

Disease agent	Wild species affected	Method of introduction	References
Rabbit hemorrhagic disease virus	European rabbit (*Oryctolagus cuniculus*)	Infected domestic rabbits	Lenghaus et al. (2001)
Pilchard herpesvirus	Pilchard (*Sardinops sagax neopilchardus*) (Australia)	Infected frozen pilchards imported to feed tuna	Gaughan (2002)
Aleutian mink disease virus	European mink (*Mustela lutreola*) and other mustelids	Escaped North American mink (*Neovison vison*) from fur farms	Mānas et al. (2001); Knuuttila et al. (2009)
Squirrel parapoxvirus	Eurasian red squirrel (*Sciurus vulgaris*)	Infected eastern gray squirrels (*Sciurus carolinensis*)	Tompkins et al. (2002)
West Nile virus	Dozens of North American bird species especially songbirds	Infected pet of vector introduced from Israel	Hayes and Gubler (2006)
Canine distemper virus	Lions (*Panthera leo*), African hunting dogs (*Lycaon pictus*)	Infected dogs, both domestic and feral	Anderson (1995); Prager et al. (2012)
Mycobacterium bovis	Wood bison (*Bison b. athabascae*) (Canada)	Translocation of infected plains bison (*B.b. bison*)	Tessaro (1986)
Yersinia pestis	Various indigenous rodents (North America)	Infected rats and rat fleas by ship from the Orient	Gasper and Watson (2001)
Aphanomyces astaci (Oomycete)	Native European crayfish	Introduced North American crayfish species	Vrålstad et al. (2011); Schrimpf et al. (2013)
Trichomonas gallinae	Galápagos dove (*Zenaida galapagoensis*)	Infected rock doves (*Columba livia*)	Harmon et al. (1987)
Myxosoma cerebralis	Native trout (North America)	Infected frozen fish from Europe	Hoffmann (1990)
Echinococcus multilocularis	Arctic fox (*Vulpes lagopus*)	Infected sibling vole (*Microtus laevis*)	Knapp et al. (2012)
Fascioloides magna	European cervids	Infected North American elk	Pybus (2001)
Anguillicola spp.	Eel (*A. anguilla*) (Europe)	Imported eels from Asia and New Zealand	Koops and Hartmann (1989)
Parastrongylus cantonensis	Cockatoos (*Callocephalon fimbriatum*) (Australia) Wood rat (*Neotoma floridanus*), opossum (*Didelphis virginia*) (USA)	Infected rats (*Rattus rattus*, *R. norvegicus*)	Reece et al. (2013); Kim et al. (2002)

natural ecosystems of feral livestock, peridomestic affiliates, or other human-associated taxa. Such species, including rats, pigs, dogs, cats, and pigeons, carry a well-known community of micro- and macroparasites that can spill into free-ranging wildlife populations when regular contact is established between free-ranging domestics and wildlife (Harmon et al. 1987; Gasper and Watson 2001; Lenghaus et al. 2001; Mānas et al. 2001; Knuuttila et al. 2009; Deem et al. 2012; Chalkowski et al. 2018). In domestic animals, the risk of disease incursion can be mitigated through legally enforced measures, including tests, inspections, and veterinary certifications (OIE 2013). However, preventing the expansion of such hosts and their associated pathogens into wildlife populations can be challenging, and

often involves one or a combination of some of the following approaches.

In the most high-risk situations it may be advisable to completely ban the importation of certain species known to be especially dangerous; this approach can be especially successful in well-circumscribed, isolated ecosystems, such as the Galápagos Islands (Wikelski et al. 2004). Second, in the vicinity of protected areas, human-associated hosts that are known to be able to introduce a disease may be vaccinated to reduce the risk of spillover into wildlife (see also Chapter 13, Section 13.5). For example, extensive field investigations combined with computer modeling have shown that the outbreak of rabies virus epidemics and perhaps of canine distemper virus in African lions can be prevented using a comprehensive dog vaccination campaign (Craft 2010; Lembo et al. 2010; Prager et al. 2012). Improved techniques in remote vaccine deployment have now made it possible to treat otherwise difficult-to-reach populations (Rosatte 2013). An alternative measure to this may be depopulation of feral animals: this can either be achieved through traditional culling methods, or via more novel techniques such as immunocontraception (Carroll et al. 2011; Naz and Saver 2016). This approach has the disadvantage that it is labor-intensive and may have to be sustained in the long term. As a result, elimination of invasive hosts and their pathogens has the best chances of success if conducted in isolated systems of relatively small size, such as island locations. Lastly, because many invasive species depend on degraded or human-modified habitats, it is possible to limit their contact with resident wildlife through appropriate environmental management, for example, via the maintenance and enhancement of native habitat buffer zones around protected areas.

12.2.1.3 Disease vectors

Vectors of infectious agents are often adaptable, mobile organisms that can easily be introduced into previously isolated regions and transmit their pathogens into native wildlife. They are often transported as stowaways and are one of the key targets for the government agencies carrying out preventative biosecurity activities at, and beyond a nation's borders (García-Diaz et al. 2017). Vector surveillance and biosecurity measures are most successful if they are implemented based on relevant risk analysis (MacLeod 2015). The importance of such routine measures, typically conducted at airports and naval ports of entry, becomes most obvious in the cases where they fail, and dangerous vectors become established as a result. Such was the case with two Asian mosquitoes (*Aedes albopictus*, *A. j. japonicus*) that were introduced recently into North America and which are suitable vectors for several arboviruses circulating in wildlife (Armstrong et al. 2013; Kampen and Werner 2014). Other examples include the well-known mosquito *Culex quinquefasciatus*, which now transmits diseases not only in Hawaii (Van Riper III et al. 1986; see Chapter 3, Section 3.2.3), but also in the Galápagos Islands

(Whiteman et al. 2005) and New Zealand (Tompkins and Gleeson 2006). Because vectors may be introduced through the transport of water that contains larvae (as was the case for the introduction of *A. albopictus* into North America with a shipment of rain-filled tires; O'Meara 1997), prevention also needs to focus on the interception and sterilization of infested water containers, including ship ballast.

Other vectors are introduced while attached to their hosts, as is the case with ticks and mites, which transmit a variety of pathogens across a range of domestic, wildlife, and human hosts (Gray 1985; Jongejan and Uilenberg 2004; Hansford et al. 2018). Ticks already constitute a big focus of livestock veterinary management: just a single species of tick (*Rhipicephalus microplus*) is responsible for extensive economic losses totaling over US$13.9–18.7 billion annually in cattle alone (Betancur Hurtado and Giraldo-Rios 2018). As a result of this concern, numerous methods exist to control and prevent tick infestations, including livestock dipping, use of oral acaricides, and even vaccination with recombinant *Bm86* antitick vaccines which have been developed recently, as an environmentally friendly alternative to toxic chemical agents (de la Fuente et al. 2007; Valle and Guerrero 2018). Whether applied to livestock or to appropriate free-ranging vertebrates, many of these tools can help prevent the spread of tick-transmitted pathogens into wildlife populations. Nonetheless, true prevention needs to aim at the root of the problem, which is the international dissemination of exotic tick and mite species via the global animal trade. The two main avenues that facilitate the introduction of ticks and the colonization of new host species are, first, the worldwide trade of domesticated animals and especially livestock (Barré et al. 1987; Barré and Uilenberg 2010) and, second, the booming international pet trade (Keirans and Durden 2001; Burridge and Simmons 2003; Mihalca 2015). Often, while little is known about many of the more obscure taxa of ticks detected at international points of entry, their control and management remains straightforward, as it only requires a thorough visual examination of all animals imported combined with well-established ectoparasite treatments (Pasmans et al. 2008).

12.2.1.4 Fomites

Sometimes animals do not even need to be released into the wild to introduce a novel disease. Tiger salamanders (*Ambystoma tigrinum*) have been used as bait by fishermen and have been transported and sold across the US West, therefore introducing ranavirus and chytrid infection into previously uninfected populations (Jancovich et al. 2005). Similar problems have arisen from the introduction of invasive pathogens via frozen pilchards and eels that were used as animal feed (Koops and Hartman 1989; Gaughan 2002).

12.2.2 Preventing disease introduction during intentional host movement: Wildlife translocations and reintroductions

While the previous section discussed the prevention of disease outbreaks during accidental host movement, here we focus on preventing emergence of pathogens following intentional host introductions. Animal translocations and (re)introductions have historically taken place for hunting and fish managing purposes and are increasingly becoming important tools in conservation operations. There are, however, significant risks associated not just with the ecological impacts of the translocated animals, but also with the exotic diseases that they may carry along (Tessaro 1986; Pybus 2001). Such microbial "hitchhikers" may threaten both the translocated individuals, as well as the recipient ecological community (Sainsbury and Vaughan-Higgins 2012). Irrespective of the ultimate purpose, it is therefore necessary to conduct a careful assessment of the disease introduction risk associated with animal translocations. This type of assessment, called disease risk analysis, has been formalized recently and is discussed in Chapter 11, Section 11.4 (MacDiarmid 1997; Jakob-Hoff et al. 2014; Hartley and Sainsbury 2017). The exact risk tends to be difficult to determine, so introduction of exotic individuals should always be considered as an inherently risky option to be conducted with great care.

Identifying potential hazards—before animals are translocated—requires answering several questions:

What diseases or agents are known to occur in the species to be translocated? Preparing a comprehensive list of potential diseases requires thorough review of available information about specific disease agents reported in the species anywhere, including case histories of disease events of unknown causation (Gaughan 2002). Given the generally poor knowledge that we presently have of the world's parasite communities, this list is likely to be incomplete.

Are any of these disease agents present at the site from which animals are to be moved? Structured population-based surveys may be required to determine whether specific diseases occur in the group or population from which animals will be selected for translocation.

Are disease agents present in the individuals selected for translocation? This requires examination to determine the health of individuals selected for translocation, including testing for diseases for which validated tests with high sensitivity are available.

Are any of the indigenous species at the destination susceptible to agents that might be moved? This requires a thorough review and cross-matching of all information available about specific agents, and the species present at the destination. Since there are sometimes only a limited number of potentially competent native hosts occurring at the site of the introduction (e.g., on species-poor islands), this process can sometimes be more circumscribed than expected. If the susceptibility of an indigenous species to an agent that might be introduced is unknown, experimental infection trials should be conducted, as was done prior to the introduction of rabbit hemorrhagic disease virus to Australia and New Zealand (Cooke and Fenner 2002).

Unfortunately, there is neither a "universal test" that will detect all disease agents, nor a general technique for predicting which, if any, of the species at the destination might be susceptible to a specific agent. For this reason, obtaining sufficient information to ensure that a translocation is safe is always going to be a challenging task and translocations should always be viewed with concern. Ignorance of either the hazards or the potential consequences of a translocation is alone sufficient reason to delay this action until better understanding of the risks is achieved.

If a pathogen *is* identified in the source population, it may be possible to mitigate the risk of translocation by testing and/or treating individual animals before they are moved. This requires a test with known sensitivity (see Chapter 8 Section 8.2) and/or a treatment with proven efficacy in the particular species. Beyond repeated testing of the animals to be introduced, quarantine and vaccination of the imported individuals can also be of importance.

Experience with *Elaphostrongylus cervi*, a nematode that causes neurological disease in deer, illustrates the difficulty in preventing movement of a known disease agent—even when testing and treatment are done—and the value of effective quarantine together with multiple tests, applied in series, in preventing establishment of a disease agent at a new site. Prior to the initial discovery in red deer (*Cervus elaphus*) in Scotland (Cameron, 1931), *E. cervi* had already been exported with red deer and become established in New Zealand (Mason et al. 1976). Subsequently, despite repeated testing and anthelmintic treatment, *E. cervi* was moved in red deer from New Zealand to Australia (Presidente 1986) and Canada (Gajadhar et al. 1994). Fortunately, at each destination, the imported deer were held in quarantine and retested, during which the parasite was detected, so that the deer, and the parasite, were not released.

A special subcategory of disease introduction problems involves fish hatcheries and farms, which, in the worst case, aggregate fish (and their pathogens) from diverse locations, cross-breed them, and then release potentially cross-infected offspring into a variety of sites. Hatcheries have emerged in the past as focal places of dissemination of various fish pathogens, including whirling disease, rosette agent, salmon sarcoma virus, and furunculosis (Koops and Hartman 1989; Hoffmann 1990; Kennedy 1994; Dobson and Foufopoulos 2001; Combe and Gozlan 2018). Both state and private hatchery managers tend to be aware of these issues and take precautions to avoid cross-contamination; nonetheless, the large numbers of animals moving through such facilities leave a small but not inconsiderable residual risk of disease transmission. To prevent pathogen dissemination at sites like this, not mixing fish from different locations is important, and strict adherence to veterinary guidelines and sanitary protocols is an absolute necessity.

12.2.3 Preventative protection of individual hosts

The aforementioned strategies, all of which aim at anticipating and preventing the introduction of novel pathogens into a new area, are clearly the preferred approach to protect species of conservation concern. However, sometimes a pathogen has already been introduced into a previously isolated area, and great risk exists of it spreading into a wildlife population with high conservation value. In this case, if managers have access to the target wildlife population, it may be appropriate to employ last-ditch individual protection strategies to protect a limited number of individuals. Most of these approaches (such as vaccination) are discussed in other chapters (e.g. Chapter 13) as disease elimination techniques, though sometimes there is a place for them as anticipatory or preventative strategies. For example, high-priority species that can be accessed for treatment may be vaccinated against a disease known to have already invaded, in order to protect an important breeding population. This was the case with an island population of scrub jays that needed to be protected against West Nile virus (see Boyce et al. 2011). Similarly, active infections may be treated with antimicrobial agents to protect hosts (Fix et al. 1988). Ecologists are also increasingly considering approaches that manipulate an individual's gut microbiota to promote conditions that prevent the establishment of an invasive pathogen (De Schryver and Vadstein 2014). Such approaches are labor-intensive and expensive but may provide longstanding protection and prevent the establishment of invasive pathogens in a population of conservation concern.

12.3 Introduction of susceptible hosts to a disease nidus: Realities and prevention

The mirror situation of the introduction of a novel pathogen into a susceptible host population is the introduction of a susceptible host into an existing disease nidus. Spillover into the newly arrived host may result in the failure of the translocation. The introduced species may also alter the ecology of an existing indigenous disease. An example of the first situation has been the introduction of *Parelaphostrongylus tenuis*, the meningeal worm of deer. This nematode is not usually harmful to its normal host, the white-tailed deer (*Odocoileus virginianus*), but often causes fatal neurologic injury in many other ungulates and is the reason for the failure of numerous attempts to introduce caribou and reindeer (*Rangifer tarandus*) into areas where the parasite is present.

An example of the second outcome where normal transmission patterns were disrupted following the arrival of a new host comes from Omsk hemorrhagic fever virus. When muskrats (*Ondatra zibethicus*) were introduced to Siberia from Canada (Růžek et al. 2010) they became infected with this Siberian flavivirus. The virus was originally transmitted by local ticks (*Dermacentor marginatus* and *D.*

reticulatus) among various local vertebrate species, including narrow-skulled voles (*Microtus gregalis*); after becoming infected, the introduced muskrats also became bridge (or amplification) hosts that transmitted the infection to humans (Lebedev 1955; Figure 12.5). Further examples include whooping cranes (*Grus americana*), which developed fatal infections after being moved into an area where eastern equine encephalitis virus was present in indigenous birds (Carpenter et al. 1989); and New Zealand keas (*Nestor notabilis*), which suffered high mortality from avian malaria infections when moved to Malaysia (Bennett et al. 1993).

Bovine tuberculosis in New Zealand is a striking example of an introduced species changing the ecology of an existing disease. *Mycobacterium bovis* was likely introduced to New Zealand with infected cattle at some unknown time in the past, but transmission was then enhanced by introduction of another host species, the brushtail possum (*Trichosurus vulpecula*), which became the major reservoir of bovine tuberculosis (O'Neil and Pharo

Figure 12.5 Transmission of Omsk hemorrhagic disease virus. This is an example of a pathogen that emerged as a significant disease in human populations in southwestern Siberia, after humans introduced a new, highly susceptible host (muskrats, *Ondatra zibethicus*, center) into the region. The virus normally circulates among a variety of small native mammals in the region (*Left*) and appears to be mostly transmitted by two local species of ticks (*Dermacentor reticulatus* and *D. marginatus*) (heavy arrow, *Left*). After the introduction of muskrats, which acted as an amplification, or bridge host, the virus started infecting humans, both trappers who were directly exposed to the virus while hunting muskrats (*Right*, middle arrow), but also locals who contracted the disease via infected tick bites (*Right*, top arrow). Under lower temperatures the virus can persist for long periods in the environment suggesting a possible additional environmental infection pathway (*Right*, lower arrow). Thickness of arrows indicates relative frequency of each pathway. (Based on Lebedev 1955, and Růžek et al. 2010.)

1995). Similarly, nutria (*Myocastor coypus*), introduced from South America to France, became a reservoir for two existing diseases, liver flukes (*Fasciola hepatica*) (Ménard et al. 2001) and leptospirosis (Vein et al. 2014).

Lastly, Karateyev et al. (2012) found that six of seven exotic mollusk species introduced into the Great Lakes region of North America became infected with intermediate stages of trematodes native to this continent. The introduced mollusks became established in dense populations, with a high prevalence of infection, resulting in "high risk of parasite transmission to the subsequent vertebrate hosts." In addition, one introduced snail, *Bithynia tentaculata*, was also infected with three trematode species not native to North America, which have since caused the death of tens of thousands of waterbirds (Cole and Franson 2006; Sauer et al. 2007).

Preventing the introduction of susceptible animals into an area in which an indigenous disease is present is even more difficult than simply preventing movement of disease agents. The reason for this is that not only is the number of agents and potential sources of infection that need to be considered much larger, but that it is difficult to predict which of the many agents present at the site might actually infect the introduced species. For example, it is unlikely that anyone could have predicted that an introduced arboreal marsupial would become important in the epidemiology of tuberculosis among cattle in New Zealand, particularly as brushtail possums do not naturally become infected with bovine tuberculosis in their Australian native range (Radunz 2006). Factors to consider during risk assessment include: the basic ecology and life history traits of the species to be introduced; the diversity and phylogenetic distance between members of the native fauna and the species to be introduced; and properties of the local parasite fauna, including diversity and phylogenetic distance to parasites of the species to be introduced in its area of origin (Poulin et al. 2011).

Nonetheless, there is sometimes opportunity to learn from previous mistakes, if previous introductions in other regions or related species have resulted in disease outbreaks. For example, after reviewing failed attempts to introduce other cervid species into areas where *P. tenuis* was present in white-tailed deer, Lankester (2001) concluded that the "complete failure of past attempts to introduce caribou, reindeer and black-tailed deer into enzootic areas should clearly discourage any future efforts." Hence, a careful review of the translocation literature needs to be a first step when evaluating infection risk for introduced species. If the susceptibility of a species to a disease present at the proposed introduction site is unknown, it can be resolved through infection trials prior to introduction.

12.4 Preventing environmental changes that allow a disease to establish or increase in prevalence

The last scenario to discuss involves prevention of environmental changes that permit or accelerate the transmission of an existing disease. A comprehensive analysis of disease outbreaks in wildlife revealed that habitat modification and environmental degradation are the most common drivers of pathogen emergence (Dobson and Foufopoulos 2001; Figure 12.6). Environmental alterations that affect wildlife disease dynamics often involve human activities that breach disease barriers or the distribution and abundance of species involved in disease transmission or persistence. Because of these indirect links, an alteration that may appear unlikely to have any relationship to disease can have important negative effects. For instance, dumping ballast water from ships would seem to have no obvious relationship with disease in wild birds. Yet, as mentioned earlier, this is how several species of wildlife pathogens were introduced into North America.

Transmission of an existing disease may be enhanced by any environmental change that promotes contact between species or aggregates hosts, such as supplementary feeding or creation of watering holes (see Figure 12.6). For example, supplementary feeding is important in maintaining *Brucella abortus* infection of elk (*Cervus canadensis*) (Scurlock and Edwards 2010) and bovine tuberculosis in white-tailed deer in North America (O'Brien et al. 2002). Similarly, spread of an introduced mycoplasmal conjunctivitis into North American passerine birds was accelerated through widespread winter feeding, which led to congregation of birds at the

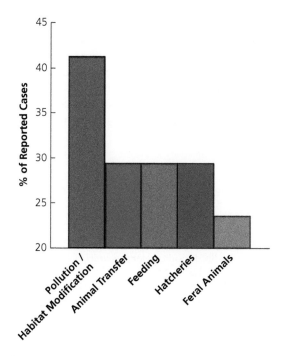

Figure 12.6 Factors associated with wildlife disease emergence. (From Dobson and Foufopoulos, 2001.)

feeders and enhanced transmission (Hartup et al. 1998; Moyers et al. 2018).

Many emerging pathogens have been shown to be tied to environmental degradation; maintaining healthy habitats through environmental protection, can be critical in maintaining health of wildlife, domesticated animal, and human populations. For example, Ebola virus can have devastating effects on wildlife and humans alike, and outbreaks are associated with recent deforestation—"preventing the loss of forests could reduce the likelihood of future outbreaks" (Olivero et al. 2017). Similarly, deforestation that leads to the collapse of freshwater food webs in Africa is associated with the emergence of the highly virulent pathogen *Mycobacterium ulcerans* (Morris et al. 2016).

Manipulation of surface water has had a greater effect on the occurrence, distribution, and prevalence of infectious diseases than any other anthropogenic environmental change. This is because water not only influences distribution and abundance of both animals and vectors, but also enhances the survival of many infectious agents in

the environment. Water may also transport disease agents between different habitats. For example, *Toxoplasma gondii*, *Sarcocystis neurona*, and fecal bacteria from terrestrial animals are washed with freshwater into the marine environment, where they likely have contributed to disease in southern sea otter (*Enhydra lutris nereis*) (Kreuder et al. 2003; Miller et al. 2010). The scale of recent human-caused surface water manipulations, such as river dams, is simply staggering. As stated by Keiser et al. (2005), "an estimated 40,000 large dams and 800,000 small dams have been built, and 272 million hectares of land are currently under irrigation worldwide." In the past, potential risks that might result from such water developments were rarely assessed prior to construction, and even today are only investigated in regard to human and not wildlife disease, despite the fact that such developments have frequently been found to have intensified existing diseases or permitted the establishment of introduced novel pathogens (Steinmann et al. 2006).

Often several, apparently unrelated, environmental changes combine to create a situation in which an infectious disease thrives. The emergence of two zoonotic diseases illustrates how a chain of actions—none of which was perceived as objectionable or dangerous in itself—can create a novel disease outbreak. For example, Japanese encephalitis virus causes an estimated 45,000 cases of human disease annually in Asia (Van den Hurk et al. 2009). Herons are the primary vertebrate hosts for the virus and transmission is by opportunistic mosquitoes of the genus *Culex*. Human disease is strongly associated with irrigated rice culture, a land use that provides habitat for both herons and paddy-breeding mosquitoes. Herons have come to rely upon rice fields because natural wetlands have been greatly reduced (Czech and Parsons 2002). Villages are built near irrigated areas, and the virus is spread by mosquitoes from herons to domestic pigs and humans in the villages. There is a high prevalence of infection in pigs, providing large amounts of circulating virus to infect additional mosquitoes, which, in turn, also infect humans.

In another example, an outbreak of disease among domestic pigs and humans in Malaysia was initially suspected to be caused by Japanese encephalitis virus (Paton et al. 1999); however, a pre-viously unknown virus, with features characteristic of the family Paramyxoviridae, was subsequently identified in human patients (Chua et al. 2000). The agent was called Nipah virus, after the Malaysian location of the initial outbreak. Disease spread to other areas of Malaysia and to Singapore through the movement of pigs. While the original source of the Nipah virus remained unknown, the virus cross-reacted serologically with a similar virus carried by fruit bats (*Pteropus* spp.) in Australia. This led to examination of fruit bats in Malaysia, which were found to be infected with the new virus (Johara et al. 2001). In retrospect, it was recognized that defor-estation for agriculture, commercial logging, and urban development had resulted in loss of natural food resources (Ambat et al. 2019). This, together with harassment of fruit bats, led to abandonment of roosting sites and greater dependence by bats on commercial fruit crops, including fruit trees that had been planted close to, and overhanging, pens on pig farms. Nipah virus was recovered from bat urine and partially eaten fruit dropped by bats into the pens. The "intersection of commercial fruit produc-tion with intensive pig-farming and resident fruit bat populations" (Epstein et al. 2006) produced "a pathway for a virus circulating in flying foxes to infect an intensely managed commercial pig popu-lation" (Pulliam et al. 2011) and humans.

The emergence of Japanese encephalitis and of Nipah virus had several features in common, including changes in land use and agricultural intensification that altered the distribution of reser-voir species and promoted close contact between wild and domestic animals, leading to the spillover of an infectious agent, which was then amplified in a domestic animal and passed to humans.

12.5 Conclusion

The basic method of hazard identification and risk assessment, described earlier, can be applied to any proposed environmental change. Biotic and abiotic factors, including changes in land use, water management, and distribution and abundance of species, should be considered. In the field of pub-lic health, "health impact assessment" has become a part of environmental impact assessment poli-cy, programs, and projects (Scott-Samuel 1998; Lock 2000; McCarthy et al. 2002; Dannenberg et al. 2006), with the goal of ensuring that proposed changes do not cause deterioration in health status. Diseases of wild animals should receive similar considera-tion during evaluation of proposed environmental changes.

References

Ambat, A.S., Zubair, S.M., Prasad, N., Pundir, P., Rajwar, E., Patil, D.S., et al. (2019). Nipah virus: A review on epidemiological characteristics and outbreaks to inform public health decision making. *Journal of Infection and Public Health*, **12**:634–39.

Ander, M., Meiswinkel, R., and Chirico, J. (2012). Seasonal dynamics of biting midges (Diptera: Ceratopogonidae: Culicoides), the potential vectors of bluetongue virus in Sweden. *Veterinary Parasitology*, **284**:59–67.

Anderson, E.C. (1995). Morbillivirus infections in wildlife in relation to their population biology and disease control in domestic animals, *Veterinary Microbiology*, **44**:319–32.

Ariel, E. (2005). Ornamental fish as trans-boundary vectors of viral diseases. In: Walker, P., Lester, R., and Bondad-Reantaso, M.G. (eds.), *Diseases in Asian Aquaculture V*, pp. 103–12. Fish Health Section, Asian Fisheries Society, Manila.

Armstrong, P.M., Anderson, J.F., Farajollahi, A., Healy, S.P., Unlu, I., Crepeau, T.N., et al. (2013). Isolations of Cache Valley virus from *Aedes albopictus* in New Jersey and evaluation of its role as a regional arbovirus vector. *Journal of Medical Entomology*, **50**: 1310–14.

Arthur, J.R. (1995). Efforts to prevent the international spread of diseases of aquatic organisms, with emphasis on the southeast Asian region. In: Shariff, M., Arthur, J.R., and Subasinghe, R.P. (eds.), *Diseases in Asian Aqua-culture II*, pp. 9–25. Fish Health Section, Asian Fisheries Society, Manila.

Audy, J.R. (1958). The localization of disease with special reference to the zoonoses. *Transactions of the Royal Society of Tropical Medicine and Hygiene*, **52**:308–34.

Barré, N. and Uilenberg, G. (2010). Spread of para-sites transported with their hosts: Case study of two species of cattle tick. *OIE Revue Scientifique et Technique*, **29**:149–60.

Barré, N., Uilenberg, G., Morel, P.C., and Camus, E. (1987). Danger of introducing heartwater onto the Ameri-can mainland: Potential role of indigenous and exotic

Amblyomma ticks. *Onderstepoort J. Veterinary Research*, **54**:405–17.

Bennett, G.F., Peirce, M.A., and Ashford, R.W. (1993). Avian haematazoa: Mortality and pathogenicity. *Journal of Natural History*, **27**:993–1001.

Betancur Hurtado, O.J. and Giraldo-Ríos, C. (2018). Economic and health impact of the ticks in production animals. IntechOpen. doi:10.5772/intechopen.81167. https://www.intechopen.com/online-first/economic-and-health-impact-of-the-ticks-in-production-animals (accessed December 7, 2018).

Boyce, W.M., Vickers, W., Morrison, S.A., Sillett, T.S., Caldwell, L., Wheeler, S.S., et al. (2011). Surveillance for West Nile virus and vaccination of free-ranging island scrub-jays (*Aphelocoma insularis*) on Santa Cruz Island, California. *Vector-Borne and Zoonotic Diseases*, **11**:1063–68.

Burridge, M.J. and Simmons, L.A. (2003). Exotic ticks introduced into the United States on imported reptiles from 1962 to 2001 and their potential roles in international dissemination of diseases. *Veterinary Parasitology*, **113**:289–320.

Cameron, T.W.M. (1931). On two new species of nematodes from the Scottish red deer. *Journal of Helminthology*, **9**:213–16.

Can, Ö.E., D'Cruze, N., and Macdonald, D.W. (2019). Dealing in deadly pathogens: Taking stock of the legal trade in live wildlife and potential risks to human health. *Global Ecology and Conservation*, 17:p.e00515.

Carpenter, J.W., Clark, G.G., and Watts, D.M. (1989). The impact of eastern equine encephalitis virus on efforts to recover the endangered whooping crane. In: Cooper, J.E. (ed.), *Disease and Threatened Birds*, pp. 115–120. Proceedings of a symposium held at the XIX World Conference of the International Council for Bird Preservation, Technical Publication 10. International Council for Bird Preservation, Queens University, Kingston, Canada..

Carroll, M.J., Singer, A., Smith, G.C., Cowan, D.P., and Massei, G. (2011). The use of immunocontraception to improve rabies eradication in urban dog populations. *Wildlife Research*, **37**:676–87.

Chalkowski, K., Lepczyk, C.A., and Zohdy, S. (2018). Parasite ecology of invasive species: Conceptual framework and new hypotheses. *Trends in Parasitology*, **34**:655–63.

Chua, K.B., Bellini, W.J., Rota, P.A., Harcourt, B.H., Tamin, A., Lam, S.K., et al. (2000). Nipah virus: A recently emergent deadly paramyxovirus. *Science*, **288**:1432–35.

Cole, R.A. and Franson, J.C. (2006). Recurring waterbird mortalities of unusual etiologies. In: Boere, C., Galbraith, C.A., and Stroud, D.A. (eds.), *Waterbirds Around the World: A Global Overview of the Conservation, Management and Research of the World's Waterbird Flyways*, pp. 439–440. The Stationery Office, Edinburgh.

Combe, M. and Gozlan, R.E. (2018). The rise of the rosette agent in Europe: An epidemiological enigma. *Transboundary and Emerging Diseases*, **65**:1474–81.

Cooke, B.D. and Fenner, F. (2002). Rabbit haemorrhagic disease and biological control of wild rabbits, *Oryctolagus cuniculus*, in Australia and New Zealand. *Wildlife Research*, **29**:689–706.

Covello, V.T. and Merkhoher, M.W. (1993). Risk Assessment Methods: Approaches for Assessing Health and Environmental Risks. Plenum Publishing, New York.

Craft, M.E. (2010). Ecology of infectious diseases in Serengeti lions. In: Macdonald, D. and Loveridge, A. (eds.), *The Biology and Conservation of Wild Felids*, pp. 263–81. Oxford University Press, Oxford.

Czech, H.A. and Parsons, K.C. (2002). Agricultural wetlands and waterbirds: A review. *Waterbirds*, **25**:56–65.

Dannenberg, A.L., Bhatia, R., Cole, B.L., Dora, C., Fielding, J.E., Kraft, K., et al. (2006). Growing the field of health impact assessment in the United States: An agenda for research and practice. *American Journal of Public Health*, **96**:262–70.

Daszak, P., Cunningham, A.A., and Hyatt, A.D. (2000). Emerging infectious diseases of wildlife—threats to biodiversity and human health. *Science*, **287**:443–49.

De La Fuente, J., Almazán, C., Canales, M., De La Lastra, J.M.P., Kocan, K.M., and Willadsen, P. (2007). A ten-year review of commercial vaccine performance for control of tick infestations on cattle. *Animal Health Research Reviews*, **8**:23–28.

De Schryver, P. and Vadstein, O. (2014). Ecological theory as a foundation to control pathogenic invasion in aquaculture. *The ISME Journal*, **8**:2360–68.

Deem, S.L., Cruz, M.B., Higashiguchi, J.M., and Parker, P.G. (2012). Diseases of poultry and endemic birds in Galapagos: Implications for the reintroduction of native species. *Animal Conservation*, **15**:73–82.

Deem, S.L., Karesh, W.B., and Weisman, W. (2001). Putting theory into practice: Wildlife health in conservation. *Conservation Biology*, **15**:1224–33.

Dobson, A.P. and Foufopoulos, J. (2001). Emerging infectious pathogens in wildlife. *Philosophical Transactions of the Royal Society of London B: Biological Sciences*, 356: 1001–12.

Dunn, A.M. and Hatcher, M.J. (2015). Parasites and biological invasions: Parallels, interactions, and control. *Trends in Parasitology*, **31**:189–99.

Epstein, J.H., Field, H.E., Luby, S., Pulliam, J.R., and Daszak, P. (2006). Nipah virus: Impact, origins, and causes of emergence. *Current Infectious Disease Reports*, 8:59–65.

Fix, A.S., Waterhouse, C., Greiner, E.C., and Stoskopf, M.K. (1988). *Plasmodium relictum* as a cause of avian malaria in wild-caught Magellanic penguins (*Spheniscus magellanicus*). *Journal of Wildlife Diseases*, 24:610–19.

Gajadhar, A.A., Tessaro, S.V., and Yates, W.D. (1994). Diagnosis of *Elaphostrongylus cervi* infection in New Zealand red deer (*Cervus elaphus*) quarantined in Canada, and experimental determination of a new extended prepatent period. *Canadian Veterinary Journal*, 35:433–37.

García de Lomas, J. and Vilà, M. (2015). Lists of harmful alien organisms: Are the national regulations adapted to the global world? *Biological Invasions*, 17:3081–91.

García-Díaz, P., Ross, J.V., Woolnough, A.P., and Cassey, P. (2017). Managing the risk of wildlife disease introduction: Pathway-level biosecurity for preventing the introduction of alien ranaviruses. *Journal of Applied Ecology*, 54:234–41.

Gasper, P.W. and Watson, R.R. (2001). Plague and yersiniosis. In: Williams, E.S. and Barker, I.K. (eds.), *Infectious Diseases of Wild Mammals*, 3rd edn, pp. 313–29. Iowa State University Press, Ames, IA.

Gaughan, D.J. (2002). Disease translocation across geographical boundaries must be recognized as a risk even in the absence of disease identification: the case with Australian Sardinops. *Reviews in Fish Biology and Fisheries*, 11:113–23.

Grant, E.H., Muths, E., Katz, R.A., Canessa, S., Adam, M.J., Ballard, J.R., et al. (2016). Salamander chytrid fungus (*Batrachochytrium salamandrivorans*) in the United States—Developing research, monitoring, and management strategies: US Geological Survey Open-File Report 2015-1233. http://dx.doi.org/10.3133/ofr20151233 (accessed February 13, 2019).

Grant, E.H., Muths, E., Katz, R.A., Canessa, S., Adams, M.J., Ballard, J.R., et al. (2017). Using decision analysis to support proactive management of emerging infectious wildlife diseases. *Frontiers in Ecology and the Environment*, 15:214–21.

Gray, J.S. (1985). Ticks: Their economic importance and methods of control. *Outlook on Agriculture*, 14:136–42.

Hansford, K.M., Pietzsch, M.E., Cull, B., Gillingham, E.L., and Medlock, J.M. (2018). Potential risk posed by the importation of ticks into the UK on animals: Records from the Tick Surveillance Scheme. *Veterinary Record*, 182:107–107.

Härkönen, T., Dietz, R., Reijnders, P., Teilmann, J., Harding, K., Hall, A., et al. (2006). A review of the 1988 and 2002 phocine distemper virus epidemics in European harbour seals. *Diseases of Aquatic Organisms*, 68:115–30.

Harmon, W.M., Clark, W.A., Hawbecker, A.C., and Stafford, M. (1987). *Trichomonas gallinae* in columbiform birds from the Galapagos Islands. *Journal of Wildlife Diseases*, 23:492–94.

Hartley, M. and Sainsbury, A. (2017). Methods of disease risk analysis in wildlife translocations for conservation purposes. *EcoHealth*, 14:16–29.

Hartup, B.K., Mohammed, H.O., Kollias, G.V., and Dhondt, A.A. (1998). Risk factors associated with mycoplasmal conjunctivitis in house finches. *Journal of Wildlife Diseases*, 34: 281–88.

Hayes, E.B. and Gubler, D.J. (2006). West Nile virus: Epidemiology and clinical features of an emerging epidemic in the United States. *Annual Review of Medicine*, 57:181–94.

Hoffmann, G.L. (1990). *Myxosoma cerebralis*, a worldwide cause of salmonid whirling disease. *Journal of Aquatic Animal Health*, 2:30–37.

Huang, A., Drake, J.M., Gittleman, J.L., and Altizer, S. (2015). Parasite diversity declines with host evolutionary distinctiveness: A global analysis of carnivores. *Evolution*, 69:621–30.

Jakob-Hoff, R.M., MacDiarmid, S.C., Lees, C., Miller, P.S., Travis, D. and Kock, R. (2014). Manual of Procedures for *Wildlife Disease Risk Analysis*. World Organisation for Animal Health, and the International Union for Conservation of Nature and The Species Survival Commission, Paris.

Jancovich, J.K., Davidson, E.W., Parameswaran, N., Mao, J., Chinchar, V.G., Collins, J.P., et al. (2005). Evidence for emergence of an amphibian iridoviral disease because of human-enhanced spread. *Molecular Ecology*, 14:213–24.

Johara, M.Y., Field, H., Rashdi, A.M., Morrissy, C., Van Der Heide, B., Rota, P., et al. (2001). Nipah virus infection in bats (Order Chiroptera) in peninsular Malaysia. *Emerging Infectious Diseases*, 7:439–41.

Jones, J.B., Hyatt, A.D., Hine, P.M., Whittington, R.J., Griffin, D.A., and Bax, N.J. (1997). Special topic review: Australasian pilchard mortalities. *World Journal of Microbiology and Biotechnology*, 13:383–92.

Jongejan, F. and Uilenberg, G. (2004). The global importance of ticks. *Parasitology*, 129:3–14.

Juliano, S.A. and Lounibos, L.P. (2005). Ecology of invasive mosquitoes: Effect on resident species and on human health. *Ecology Letters*, 8:558–74.

Kampen, H. and Werner, D. (2014). Out of the bush: The Asian bush mosquito *Aedes japonicus japonicus* (Theobald, 1901) (Diptera, Culicidae) becomes invasive. *Parasites & Vectors*, 7:59–68.

Karatayev, A.Y., Mastitsky, S.E., Burlakova, L.E., Karateyev, V.A., Hajduk, M.M., and Conn, D.B. (2012). Exotic molluscs in the Great Lakes host epizootiologically important trematodes. *Journal of Shellfish Research*, **31**:885–94.

Keirans, J.E. and Durden, L.A. (2001). Invasion: Exotic ticks (Acari: Argasidae, Ixodidae) imported into the United States. A review and new records. *Journal of Medical Entomology*, **38**:850–61.

Keiser, J., De Castro, M.C., Maltese, M.F., Bos, R., Tanner, M., Singer, B.H., et al. (2005). Effect of irrigation and large dams on the burden of malaria on a global and regional scale. *American Journal of Tropical Medicine and Hygiene*, **72**:392–406.

Kennedy, C.R. (1994). Introductions, spread and colonization of new localities by fish helminth and crustacean parasites in the British Isles: A perspective and appraisal. *Journal of Fish Biology*, **43**:287–301.

Kilpatrick, A.M. (2011). Globalization, land use and the invasion of West Nile virus. *Science*, **334**:323–27.

Kim, D.Y., Stewart, T.B., Bauer, R.W., and Mitchell, M. (2002). *Parastrongylus (=Angiostrongylus) cantonensis* now endemic in Louisiana wildlife. *Journal of Parasitology*, **88**:1024–26.

Knapp, J., Staebler, S., Bart, J.M., Stien, A., Yoccoz, N.G., Drögemüller, C., et al. (2012). *Echinococcus multilocularis* in Svalbard, Norway: Microsatellite genotyping to investigate the origin of a highly focal contamination. *Infection, Genetics and Evolution*, **12**:1270–74.

Knuuttila, A., Uzcategui, N., Kankkonen, J., Vapalahti, O., and Kinnunen, P. (2009). Molecular epidemiology of Aleutian mink disease virus in Finland. *Veterinary Microbiology*, **133**:229–38.

Koops, H. and Hartmann, F. (1989). Anguillicola—infestations in Germany and German eel imports. *Journal of Applied Ichthyology*, **1**:41–45.

Kreuder, C., Miller, M.A., Jessup, D.A., Lowenstein, L.J., Harris, M.D., Ames, J.A., et al. (2003). Patterns of mortality in southern Sea otters (*Enhydra lutris nereis*) from 1998–2001. *Journal of Wildlife Diseases*, **39**:495–509.

Krysko, K.L., Burgess, J.P., Rochford, M.R., Gillette, C.R., Cueva, D., Enge, K.M., et al. (2011). Verified nonindigenous amphibians and reptiles in Florida from 1863 through 2010: Outlining the invasion process and identifying invasion pathways and stages. *Zootaxa*, **3028**:1–64.

Lankester, M.W. (2001). Extrapulmonary lungworms of cervids. In: Samuel, W.M., Pybus, M.J., and Kocan, A.A. (eds.), *Parasitic Diseases of Wild Mammals*, 2nd edn, pp. 228–78. Iowa State University Press, Ames, IA.

Lawson, B., Robinson, R.A., Neimanis, A., Handeland, K., Isomursu, M., Angren, E.O., et al. (2011). Evidence of spread of the emerging infectious disease, finch trichomonosis, by migrating birds. *Ecohealth*, **8**:143–53.

Lebedev, A.D. (1955). Role of small mammals as carriers of Omsk haemorrhagic fever. *Zoologicheskii Zhurnal*, **34**:605–08.

Lembo, T., Hampson, K., Kaare, M.T., Ernest, E., Knobel, D., Kazwala, R.R., et al. (2010). The feasibility of canine rabies elimination in Africa: Dispelling doubts with data. *PLoS Neglected Tropical Diseases*, **4**:p.e626.

Lenghaus, C., Studdert, M.J., and Gavier-Widén, D. (2001). Calicivirus infections. In: Williams, E.S. and Barker, I.K. (eds.), *Infectious Diseases of Wild Mammals*, 3rd edn, pp. 280–291, Iowa State University Press, Ames, IA.

Lock, K. (2000). Health impact assessment. *British Medical Journal*, **7246**:1395–98.

Longdon, B., Brockhurst, M.A., Russell, C.A., Welch, J.J., and Jiggins, F.M. (2014). The evolution and genetics of virus host shifts. *PLoS Pathogens*, **10**:e1004395.

Lymbery, A.J., Morine, M., Kanani, H.G., Beatty, S.J., and Morgan, D.L. (2014). Co-invaders: The effects of alien parasites on native hosts. *International Journal for Parasitology: Parasites and Wildlife*, **3**:171–77.

MacDiarmid, S.C. (1997). Risk analysis, international trade, and animal health. In: Molak, V. (ed.), *Fundamentals of Risk Analysis and Risk Management*, pp. 377–87. CRC Press, Boca Raton, FL.

MacLachlan, N.J. and Mayo, C.E. (2013). Potential strategies for control of bluetongue, a globally emerging, *Culicoides*-transmitted viral disease of ruminant livestock and wildlife. *Antiviral Research*, **99**:79–90.

MacLeod, A. (2015). The relationship between biosecurity surveillance and risk analysis. In: Jarrad, F., LowChoy, S., and Mengersen, K. (eds.), *Biosecurity Surveillance: Quantitative Approaches*, pp. 109–22. CAB International, Wallingford.

Malkinson, M., Banet, C., Weisman, Y., Pokamunski, S., King, R., Drouet, M.-T. et al. (2002). Introduction of West Nile Virus in the Middle East by migrating white storks. *Emerging Infectious Diseases*, **8**:392–97.

Mañas, S., Ceña, J.C., Ruiz-Olmo, J., Palazón, S., Domingo, M., Wolfinbarger, J.B., et al. (2001). Aleutian mink disease parvovirus in wild riparian carnivores in Spain. *Journal of Wildlife Diseases*, **37**:138–44.

Mason, P.C., Kiddey, N.R., Sutherland, R.J., Rutherford, D.M., and Green, A.G. (1976). *Elaphostrongylus cervi* in red deer. *New Zealand Veterinary Journal*, **24**:22–23.

May, R.M., Gupta, S., and McLean, A.R. (2001). Infectious disease dynamics: What characterizes a successful invader? *Philosophical Transactions of the Royal Society London B: Biological Sciences*, **356**:901–10.

McCarthy, M., Biddulph, J.P., Utley, M., Ferguson, J., and Gallivan, S. (2002). A health impact assessment model for environmental changes attributable to developmental projects. *Journal of Epidemiology and Community Health*, **56**:611–16.

Ménard, A., Agoulon, A., L'Hostis, M., Rondelaud, D., Collard, S., and Chauvin, A. (2001). *Myocastor coypus* as a reservoir host *of Fasciola hepatica* in France. *Veterinary Research*, **32**:499–508.

Mihalca, A.D. (2015). Ticks imported to Europe with exotic reptiles. *Veterinary Parasitology*, **213**:67–71.

Miller, M.A., Byrne, B.A., Jang, S.S., Dodd, E.M., Dorfmeier, E., Harris, M.D., et al. (2010). Enteric bacterial pathogen detection in southern sea otters (*Enhydra lutris nereis*) is associated with coastal urbanization and freshwater runoff. *Veterinary Research*, **41**:1–13.

Morris, A.L., Guégan, J.F., Andreou, D., Marsollier, L., Carolan, K., Le Croller, M., et al. (2016). Deforestation-driven food-web collapse linked to emerging tropical infectious disease, Mycobacterium ulcerans. *Science Advances*, **2**:p.e1600387.

Moyers, S.C., Adelman, J.S., Farine, D.R., Thomason, C.A., and Hawley, D.M. (2018). Feeder density enhances house finch disease transmission in experimental epidemics. *Philosophical Transactions of the Royal Society B: Biological Sciences*, **373**:20170090.

Naz, R.K. and Saver, A.E. (2016). Immunocontraception for animals: Current status and future perspective. *American Journal of Reproductive Immunology*, **75**:426–39.

Nowak, M. (2010). The international trade in reptiles (Reptilia)—The cause of the transfer of exotic ticks (Acari: Ixodida) to Poland. *Veterinary Parasitology*, **169**:373–81.

O'Brien, D.J., Schmitt, S.M., Fierke, J.S., Hogle, S.A., Winterstein, S.R., Cooley, T.M., et al. (2002). Epidemiology of *Mycobacterium bovis* in free-ranging white-tailed deer, Michigan, USA, 1995–2000. *Preventive Veterinary Medicine*, **54**:47–63.

O'Meara, G.F. (1997). The Asian tiger mosquito in Florida. Publication #ENY632, *EDIS*. http://edis.ifas.ufl.edu/MG339 (accessed February 4, 2019).

O'Neill, B.D. and Pharo, H.J. (1995). The control of bovine tuberculosis in New Zealand. *New Zealand Veterinary Journal*, **43**:249–55.

OIE. (2013). *Terrestrial Animal Health Code*, 22nd edn, 2 vols. World Organisation for Animal Health, Paris.

Olivero, J., Fa, J.E., Real, R., Márquez, A.L., Farfán, M.A., Vargas, J.M., et al. (2017). Recent loss of closed forests is associated with Ebola virus disease outbreaks. *Scientific Reports*, **7**:1–9.

Parrish, C.R., Holmes, E.C., Morens, D.M., Park, E.C., Burke, D.S., Calisher, C.H., et al. (2008). Cross-species virus transmission and emergence of new epidemic diseases. *Microbiology and Molecular Biology Reviews*, **72**:457–70.

Pasmans, F., Blahak, S., Martel, A., and Pantchev, N. (2008). Introducing reptiles into a captive collection: The role of the veterinarian. *The Veterinary Journal*, **175**:53–68.

Paton, N.I., Leo, Y.S., Zaki, S.R., Auchus, A.P., Lee, K.E., Ling, A.E., et al. (1999). Outbreak of Nipah-virus infection among abattoir workers in Singapore. *The Lancet*, **354**:1253–56.

Pavlovsky, E.N. (1966). *Natural with Special Reference to the Landscape Epidemiology of Zooanthroponoses*. University of Illinois Press, Urbana, IL.

Peeler, E.J., Murray, A.G., Thebault, A., Brun, E., Giovaninni, A., and Thrush, M.A. (2007). The application of risk analysis in aquatic animal health management. *Preventive Veterinary Medicine*, **81**:3–20.

Poulin, R., Paterson, R.A., Townsend, C.R., Tompkins, D.M., and Kelly, D.W. (2011). Biological invasions and the dynamics of endemic diseases in freshwater ecosystems. *Freshwater Biology*, **56**:676–88.

Prager, K.C., Mazet, J.A., Dubovi, E.J., Frank, L.G., Munson, L., Wagner, A.P., et al. (2012). Rabies virus and canine distemper virus in wild and domestic carnivores in Northern Kenya: Are domestic dogs the reservoir? EcoHealth, **9**:483–98.

Prenter, J., MacNeil, C., Dick, J.T., and Dunn, A.M. (2004). Roles of parasites in animal invasions. *Trends in Ecology & Evolution*, **19**:385–90.

Presidente, P. (1986). Tissue worm: Implications for live deer exports. In: Owen, P. (ed.), *Deer Farming into the Nineties*, pp. 194–202. Owen Art & Publishing, Brisbane.

Pulliam, J.R., Epstein, J.H., Dushoff, J., Rahman, S.A., Bunning, M., Jamaluddin, A.A., et al. (2011). Agricultural intensification, priming for persistence and emergence of Nipah virus: A lethal bat-borne zoonosis. *Journal of the Royal Society Interface*, **9**:89–101.

Pybus, M.J. (2001). Liver flukes. In: Samuel, W.M., Pybus, M.J., and Kocan, A.A. (eds.), *Parasitic Diseases of Wild Mammals*, 2nd edn, pp. 121–49. Iowa State University Press, Ames, IA.

Radunz, B. (2006). Surveillance and risk management during the latter stages of eradication: Experiences from Australia. *Veterinary Microbiology*, **112**:283–90.

Reece, R.L., Perry, R.A., and Spratt, D.M. (2013). Neurangiostrongylosis due to *Angiostrongylus cantonensis* in gang-gang cockatoos (*Callocephalon fimbriatum*). *Australian Veterinary Journal*, **91**:477–81.

Rosatte, R.C. (2013). Rabies control in wild carnivores. In: Jackson A. (ed.), *Rabies: Scientific Base of the Disease and Its Management*, 3rd edn, pp. 617–70. Elsevier, Amsterdam.

Růžek, D., Yakimenko, V.Y., Karan, L.S., and Tkachev, S.E. (2010). Omsk haemorrhagic fever. *Lancet*, **376**:2104–13.

Sainsbury, A.W. and Vaughan-Higgins, R.J. (2012). Analyzing disease risks associated with translocations. *Conservation Biology*, **26**:442–52.

Sauer, J.S., Cole, R.A., and Nissen, J.M. (2007). Finding the exotic snail (*Bithynia tentaculata*): Investigation of waterbird die-offs on the Upper Mississippi River National Wildlife and Fish Refuge., US Geological Survey Open-File Report 2007–1065. http://pubs.er.usgs.gov/publication/ofr20071065 (accessed January 26, 2019).

Schrimpf, A., Chucholl, C., Schmidt, T., and Schulz, R. (2013). Crayfish plague agent detected in populations of the invasive North American crayfish *Orconectes immunis* (Hagen, 1870) in the Rhine River, Germany. *Aquatic Invasions*, **8**:103–09.

Scott-Samuel, A. (1998). Heath impact assessment: Theory into practice. *Journal of Epidemiology and Community Health*, **52**:704–05.

Scurlock, B.M. and Edwards, W.H. (2010). Status of brucellosis in free-ranging elk and bison in Wyoming. *Journal of Wildlife Diseases*, **46**:442–49.

Steinmann, P., Keiser, J., Bos, R., Tanner, M., and Utzinger, J. (2006). Schistosomiasis and water resources development: Systematic review, meta-analysis, and estimates of people at risk. *Lancet Infectious Diseases*, **6**:411–25.

Strauss, A., White, A., and Boots, M. (2012). Invading with biological weapons: The importance of disease-mediated invasions. *Functional Ecology*, **26**:1249–61.

Tessaro, S.V. (1986). The existing and potential importance of brucellosis and tuberculosis in Canadian wildlife: A review. *Canadian Veterinary Journal*, **27**:119–24.

Tompkins, D.M. and Gleasson, D.M. (2006). Relationship between avian malaria distribution and an exotic invasive mosquito in New Zealand. *Journal of the Royal Society of New Zealand*, **36**:51–62.

Tompkins, D.M., Sainsbury, A.W., Nettleton, P., Buxton, D., and Gurnell, J. (2002). Parapoxvirus causes a deleterious disease in red squirrels associated with UK population declines. *Proceedings of the Royal Society London B: Biological Sciences*, **269**:529–33.

Tompkins, D.M., White, A.R., and Boots, M. (2003). Ecological replacement of native red squirrels by invasive greys driven by disease. *Ecology Letters*, **6**:189–96.

UNESCO. (2005). The Precautionary Principle. World Commission on the Ethics of Scientific Knowledge and Technology (COMEST)-United Nations Educational, Scientific and Cultural Organization, Paris. https://unesdoc.unesco.org/ark:/48223/pf0000139578) (accessed January 25, 2019).

Valle, M.R. and Guerrero, F.D. (2018). Anti-tick vaccines in the omics era. *Frontiers in Bioscience* **10**:122–36.

Van Den Hurk, A.F., Ritchie, S.A., and Mackenzie, J.S. (2009). Ecology and geographical expansion of Japanese encephalitis virus. *Annual Reviews of Entomology*, **54**:17–35.

Van Riper III, C., Van Riper, S.G., Goff, M.L., and Laird, M. (1986). The epizootiology and ecological significance of malaria in Hawaiian land birds. *Ecological Monographs*, **56**:327–44.

Vein, J., Leblond, A., Belli, P., Kodjo, A., and Berny, P.J. (2014). The role of the coypu (*Myocaster coypus*), an invasive aquatic rodent species, in the epidemiological cycle of leptospirosis: A study in two wetlands in the east of France. *European Journal of Wildlife Research*, **60**:125–33.

Vrålstad, T., Johnsen, S.I., and Taugbøl, T. (2011). NOBANIS – Invasive Alien Species Fact Sheet– *Aphanomyces astaci*. Online Database of the European Network on Invasive Alien Species—NOBANIS. www.nobanis.org (accessed January 25, 2019).

Webby, R., Hoffmann, E., and Webster, R. (2004). Molecular constraints to interspecies transmission of viral pathogens. *Nature Medicine*, **10**:77–109.

Whiteman, N.K., Goodman, S.J., Sinclair, B.J., Walsh, T.I.M., Cunningham, A.A., Kramer, L.D., et al. (2005). Establishment of the avian disease vector *Culex quinquefasciatus* Say, 1823 (Diptera: Culicidae) on the Galápagos Islands, Ecuador. *Ibis*, **147**:844–47.

Wikelski, M., Foufopoulos, J., Vargas, H., and Snell, H. (2004). Galápagos birds and diseases: Invasive pathogens as threats for island species. *Ecology and Society*, **9**:5.

Woolhouse, M.E.J., Haydon, D.T., and Antia, R. (2005). Emerging pathogens: The epidemiology and evolution of species jumps. *Trends in Ecology and Evolution*, **20**:238–44.

CHAPTER 13

Disease Elimination and Eradication

"policy-makers can easily set ambitious disease management objectives when they have little conception of what may be practically required to attain them."

Ramsey et al. (2014)

13.1 Introduction

The first step when trying to affect the occurrence of a disease in a wild population is to have clear management goals and a good understanding of what is possible. Managers for example, need to determine whether a disease needs to be controlled, eliminated, or eradicated, as these aims are by definition very different.

Disease control involves reducing disease incidence to a defined level, as a result of deliberate efforts. For example, control of fox rabies in Europe was defined as reduction in annual cases by >90%, relative to endemic levels (Freuling et al. 2013). Programs to control disease have no foreseeable endpoint, as infections continue to occur.

Disease elimination is reduction to zero of the incidences of infection caused by a specified agent, in a defined geographical area, as a result of deliberate efforts. For example, Freuling et al. (2013) reported the elimination of rabies from nine countries in western and central Europe by 2010 through the use of bait-delivered vaccine. *Batrachochytrium dendrobatidis*, the causative agent of amphibian chytridiomycosis, was eliminated from four of five ponds on an island through a combination of antifungal treatment of tadpoles and draining and chemical disinfection of ponds (Bosch et al. 2015). Programs to eliminate disease may cease if there is no threat of

reinfection from outside the area. If a threat exists, intervention changes to prevention of reinfection. Elimination of a disease in wild animals is only likely to be attempted in extreme situations, such as incursion of a foreign livestock disease into wild species; a disease of humans or domestic animals for which wild animals are maintenance hosts; or a disease that threatens extinction of a valued species.

Disease eradication is the reduction to zero of the worldwide incidence of infection caused by a specific agent, as the result of deliberate effort. Further intervention is unnecessary when a disease has been eradicated. Disease eradication is difficult and rare-only two infectious diseases, smallpox and rinderpest, have been eradicated to date (Figure 13.1). Attributes contributing to their eradication included a single pathogen immunotype, few inapparent infections, lack of a chronic carrier state, and a vaccine that induced long-lasting protective immunity (Morens et al. 2011); in addition, smallpox had only one host species. Attempts to eradicate other diseases, including several in humans, have failed (Aylward et al. 2000), although many diseases found in humans and domestic animals have been eliminated from whole continents.

13.2 Preintervention considerations

The "eradicability" of a disease is critically determined by the ability to interrupt transmission of the

Infectious Disease Ecology and Conservation. Johannes Foufopoulos, Gary A. Wobeser and Hamish McCallum, Oxford University Press.
© Johannes Foufopoulos, Gary A. Wobeser and Hamish McCallum (2022). DOI: 10.1093/oso/9780199583508.003.0013

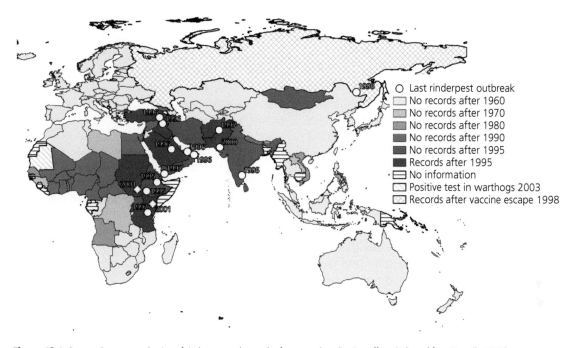

Figure 13.1 Progressive range reduction of rinderpest as the result of concerted eradication efforts (Adapted from Normile 2008.)

agent. Equally important is the availability of diagnostic tools with sufficient sensitivity and specificity
to detect levels of infection that can lead to transmission (Ottesen et al. 1998). Such tools include both
methods to identify the disease status of individual animals accurately, and also methods to measure
host population features, including distribution and
density.

Before attempting eradication, it is important to
identify the target population. This target population is not necessarily the main host population but
rather the host population that is expected to benefit
primarily from intervention. Humans or domestic
animals have been the target population in most
past attempts to eliminate disease from wild animals. Success of intervention is measured in the target population, for example, effectiveness of culling
Eurasian badgers (*Meles meles*) in order to eliminate bovine tuberculosis (bTB) is measured in cattle
rather than in badgers.

For successful transmission, susceptible individuals must come into suitable contact with the
causative agent, which must be in a stage and in
sufficient numbers to represent an infective dose

(Corner 2006). To interrupt transmission, it is necessary to identify, and then reduce or eliminate either
the source of infection, or the means of transmission, or the availability of susceptible hosts, or some
combination of these. The source of infection may be
within the target population, or come from a reservoir. The latter is to be understood as one, or more,
epidemiologically connected populations or environments in which the agent can be permanently
maintained and from which infection is transmitted to the target population (Haydon et al. 2002).
The "mere presence of infection does not of itself
provide evidence" that a species is a significant
reservoir (Corner 2006) or is important in a specific situation (Caley and Hone 2005). In some cases,
the source of infection for the target population can
be identified by using genotyping to discriminate
between slightly different pathogen strains circulating in different host populations. For example,
where brucellosis occurred in elk (*Cervus elaphus*),
bison (*Bison bison*), and cattle, infection in cattle
(the target population) was linked to the elk, rather
than the bison strain, by genotyping (Beja-Periera
et al. 2009). A species may have dissimilar roles in

transmission in different situations. For example, wild boars/feral pigs (*Sus scrofa*) are a maintenance host for bTB in Spain (Naranjo et al. 2008), but not in Australia (Corner 2006) or New Zealand (Nugent 2011). Within a species, the source of infection may be a particular segment of the population, for example, young, reproductively active females are the source for transmission of brucellosis among bison (Ebinger et al. 2011).

Transmission occurs at the level of individuals and involves a series of steps (Figure 13.2, also see McCallum et al. 2017). To intervene most effectively, it is necessary to understand how agents move from infectious to susceptible individuals, as well as the functional relationship between the rate of disease transmission, host population density, and social organization. Multiple transmission routes may be relevant, in which case it is important to identify and evaluate their relative importance, particularly if the agent infects multiple host species. It may not be necessary to entirely interrupt transmission to the target population, because elimination simply requires reducing R_0 (the basic reproductive number of the pathogen—see Chapter 2, Section 2.2.3) to below one and maintaining $R_0 < 1$ for a sufficient length of time for infection levels to decline to zero. This must be achieved for the whole community of hosts infected the by pathogen in question. Methods to calculate R_0 for multispecies infections are discussed in Chapter 9, and methods to estimate it from field data are further discussed in Chapter 10.

Effective surveillance—to detect disease, and determine its frequency, spatial distribution, and progression—is fundamental to management. Surveillance that begins before intervention can provide guidance as to where to intervene; furthermore, it also provides baseline information, against which progress can be evaluated. Surveillance during intervention monitors effectiveness and allows the program to be adjusted as needed; surveillance after apparent success confirms that disease has, in fact, been eliminated. Sampling may focus on specific geographic areas, or on segments within the population, for which there is a greater probability of disease presence (i.e., "targeted," or "risk-based surveillance"). Sampling units may be chosen according to habitat type (Rees et al. 2011),

species composition and density (Snow et al. 2007), or particular demographic features (Walsh and Miller 2010). Data should be georeferenced, so that spatial heterogeneity and patterns of spread can be identified, and intervention can be targeted at high-risk areas (see Chapter 4 for a more detailed description of sampling strategies). Wild animals usually must be captured for surveillance (see Chapters 5 and 7), and sentinel species (often a predator or scavenger) may be used to detect disease. For example, coyotes (*Canis latrans*) have been used as a sentinel for detection of bTB in deer (Berentsen et al. 2011).

"Situation-based surveillance" may be used for diseases that cause recognizable morbidity or mortality (Thulke et al. 2009). Depending upon the situation, sampling may focus on "indicator animals" (IA), which are individuals suspected of having disease, because of clinical signs, being found dead, or belonging to high-risk species. Alternatively, investigators may focus on "hunted animals" (HA), suspected of *not* having disease, and assumed to be representative of the healthy population. When disease is thought *not* to be present, sampling IA, for example, animals with appropriate clinical signs, is more efficient than sampling HA to determine if disease is present. When disease is known to be present, HA can be sampled systematically to measure effectiveness of intervention (e.g., the proportion vaccinated), and IA can be sampled to determine distribution and prevalence of disease.

The scope of an epidemiological intervention is dependent on the interrelated factors of disease distribution, timing, and intensity of infection. To eliminate or eradicate a disease, intervention must occur everywhere disease occurs. Because size of the area, distribution of the disease, and spatial structure of the host population influence disease persistence (Rossi et al. 2005; Kramer-Schadt et al. 2007), they also dictate the type, intensity, and cost of intervention. When the disease is confined to a small area, it may be possible to capture a significant fraction of the host population for treatment, vaccination, or culling. When the disease is widespread, or occurs in inaccessible areas, intervention may be feasible only by remote means, such as through distribution of baits containing the "active" compound. Examples of remote delivery include distribution of

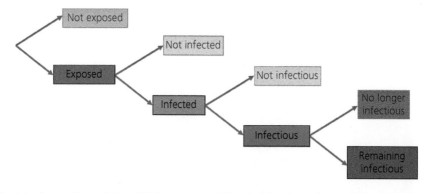

Figure 13.2 The chain of events ("transmission axis") that may occur within a single host in an infectious disease. Possible interventions can focus on reducing or preventing exposure to the causative agent, preventing infection in exposed individuals, or reducing the infectiousness of infected individuals. In some diseases (e.g., West Nile Virus in humans), infected individuals do not become infectious and play no further role in transmission (dead-end hosts). Frequently, there is also a substantial latent period between exposure resulting in infection and the host becoming infectious. Depending on the pathogen, individuals may quickly cease being infectious, or remain chronically infectious.

anthelminthic drugs to control lungworms in wild sheep (Schmidt et al. 1979), rabies vaccine to foxes (*Vulpes vulpes*) (Freuling et al. 2013), and poison to reduce the abundance of brushtail possums (*Trichosurus vulpecula*) in management of bTB in New Zealand (Caley et al. 1999). In a spatially structured population, precise targeting of the intervention may reduce the size of the area within which intervention is necessary, and the associated cost (Haydon et al. 2006; Eisinger and Thulke 2008; Cowled et al. 2012). Intensity of intervention depends in part on frequency and consistency of coverage (proportion of the area or population influenced by intervention). Timing includes when intervention begins; when and how often it is repeated; and how long it should continue. Ideally, intervention would begin when disease is first detected so as to reduce disease spread. Disease detection is often delayed in wild animals, especially if the prevalence is low, if infection causes no obvious disease, or because detection often depends on the incidental discovery of infected animals. When disease is detected, pressure for a rapid response must be balanced against the need for proper preparation, particularly if aggressive action is to be acceptable to the public. Initially, intervention might be designed as a management experiment that provides disease control, while simultaneously increasing understanding of the relevant epidemiologic processes (Gordejo and Vermeesch 2006; Wasserberg et al. 2009).

The relationships among distribution, intensity, and time were evident in a program to eliminate rabies from red foxes (*V. vulpes*) in Europe (Freuling et al. 2013). While it was predicted that rabies could be eliminated within 5–6 years, the actual time ranged from 5 to 26 years (median 14 years) in different countries. Figure 13.3 shows the spatial distribution of vaccination campaigns and compares the number of reported rabies cases in 1983 and 2010. The number of individual vaccination campaigns required varied with size of the affected area, and more campaigns were required where coverage was incomplete and nonoverlapping than where it was consistent and overlapping. Modeling and limited trials can help predict the probable length of a campaign. However, programs must include regular monitoring, to determine whether intervention is having a positive effect, and so that supporters can be reassured, and adjustments can be made. If there is uncertainty about the optimal intensity of intervention, it is "preferable to err towards too much rather than too little" (Woolhouse 2011, p2046).

Intervention may involve manipulating the agent, the vertebrate host(s), the environment (any factor that is not an attribute of the agent or the host, including vectors), or combinations of these. Programs to eliminate or eradicate disease in wild animals are not only expensive, controversial, and may require extensive societal justification (see Chapter 16), but also face the challenge of having

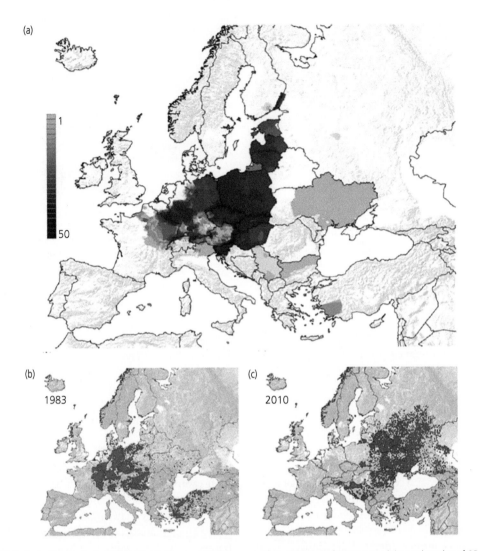

Figure 13.3 Oral rabies vaccination (ORV) effort in Europe and rabies cases. (a) Spatial extent of ORV area and the total number of ORV campaigns conducted in each country between 1978 and 2010. Reported rabies cases in (b) 1983 and (c) 2010. The vaccination campaign succeeded in eliminating rabies from most of W. Europe. (Reproduced from Freuling et al. 2013.)

to predict the effects of various management approaches when there is no precedent for their use in the species/situation (Peterson et al. 1991). Such programs are not "turn-key" operations; they are complex, management-intensive efforts, and success cannot be taken for granted (Vreysen et al. 2013). Modeling helps to predict probable effects, but, before attempting elimination or eradication, there should be a reasonably thorough plan, an established research program, success in

a significantly large demonstration site, and a firm commitment of resources to support the program (Henderson and Klepac 2013).

13.3 Manipulating the causative agent

The most straightforward way to manage a disease is to directly eliminate the infectious agent itself. The ease with which this can be done is influenced by the ability of the agent to persist without transmission.

Persistence may occur in the environment, or in infected hosts. Decontamination (disinfection) of "infected" environments is important in eliminating outbreaks of livestock disease, but is seldom feasible in the wild, unless the agent is concentrated in an identifiable focus (see Bosch et al. 2015). Lungworms (*Protostrongylus* spp.) were eliminated from an island population of arctic hares (*Lepus arcticus*) by liming soil—to destroy parasite eggs—where hares concentrated about feeding sites (Skrjabin 1970). Destruction of carcasses containing infectious agents is routine in domestic livestock, and can be effective in wildlife, but carcasses of wild animals may be difficult to find. For example, during botulism outbreaks, carcasses of dead waterfowl support large numbers of fly maggots that concentrate the botulism toxin. Because additional birds are exposed to the toxin after feeding on these maggots, targeted collection and destruction of waterfowl carcasses by wildlife managers walking along the water's edge can break the cycle of transmission and end an outbreak (Rocke and Friend 1999). If the agent reproduces outside the host, elimination may be possible by interfering with its replication. For example, the "sterile-fly technique" has been used to eliminate the screwworm fly (*Cochliomyia hominivorax*) from the United States (Wyss 2000).

Treatment of hosts with drugs, that is, chemotherapy, to reduce infectiousness is used in human and domestic animal diseases. For example, treatment of humans, to reduce infection of blackfly (*Simulium* spp.) vectors, eliminated onchocerciasis (river blindness) from an area of Guatemala (Lindblade et al. 2007). Pharmaceuticals may be equally effective in wild animals—the difficulty lies in delivery. Wild animals may have to be captured for treatment (e.g., Foufopoulos 1999; León-Vizcaíno et al. 2001), or treated remotely, for example, with treated grain to eliminate ticks from wild cervids (Pound et al. 2009). Oral baits, containing anthelminthic, were used in attempts to eliminate *Echinococcus multilocularis* from foxes (Hegglin and Deplazes 2008) and *Baylisascaris procyonis* from raccoons (*Procyon lotor*) (Page et al. 2011). Treatment reduced the prevalence of infection in foxes and raccoons, but did not eliminate the parasites, because treated animals became reinfected from infective stages in the environment. A critical issue is that

when chemotherapy, vaccine, or poison is delivered remotely, there is limited control over access. For example, over 50% of anthelminthic baits intended for raccoons were consumed by Virginia opossums (*Didelphis virginianus*) (Smyser et al. 2013). A further problem is that administration of a drug over long periods and large areas may lead to the evolution of drug resistance in the infectious agent.

13.4 Manipulating the host(s)

Culling and vaccination have been the most common interventions used in attempts to eliminate disease in wild animals. Culling is controversial and unpopular, but more acceptable when the species concerned is a damaging, invasive species, rather than a valued endemic species (Cowled et al. 2012). The relative features of culling and vaccination, for a variety of diseases, have been discussed by Barlow (1996), Smith and Cheeseman (2002), and Kramer-Schadt et al. (2007).

Selective culling ("test-and-cull") is the least controversial form of culling, because only infected individuals are removed. However, it is challenging because it requires both the identification of infected individuals under field conditions, and the ability to test a substantial proportion of the population. Selective culling of the aforementioned infected European badgers (*Meles meles*) could not be used to eliminate bTB, because "there is no adequately sensitive or reliable field-based test available to enable detection and selection of individual infected badgers" (Defra 2011, p15). Wild animals usually must be captured for testing. If test results are not immediately available, the animals must be held (risking infection or injury in captivity), or released, with positive individuals recaptured when results are available. If the test lacks sensitivity, infected ("false-negative") individuals may be returned to the population; if it does not distinguish between infected and recovered individuals (as is the case with many serological tests), more resistant than infected individuals may be removed (Treanor et al. 2011), and if it lacks specificity, culling "false-positive" individuals may be unacceptable. Selective culling requires considerable effort to find infected individuals, which becomes increasingly difficult as the prevalence declines. If infection is

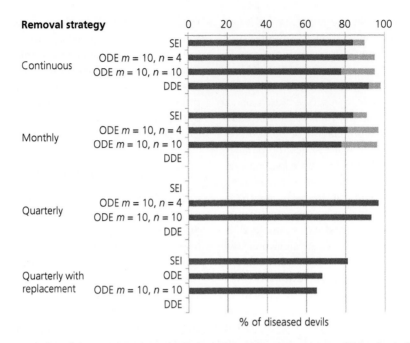

Figure 13.4 Tasmanian devils are facing an eminent threat of extinction because of the recent emergence of Tasmanian devil facial tumor disease (DFTD), a lethal infectious cancer. Diverse strategies have been proposed to control DFTD and prevent devil extinction all of which necessitate removal of animals from the wild. Tasmanian devils rarely become infected with DFTD until they are sexually mature, and once infected, there is a latent period (the length of which is poorly known) until they are able to transmit infection on to other devils. It is important to include these time delays when modeling the removal rate necessary to control the disease, and the estimated removal rate may be sensitive to the way in which these time delays are modeled. Four ways of modeling the time delays are shown:
- SEI—simple, exponentially distributed delays.
- ODE $m = 10$, $n = 4$—ordinary differential equations with tightly distributed sexual maturation of devils ($m = 10$), but a broadly distributed latent period of infection ($n = 4$).
- ODE $m = 10$, $n = 10$—ordinary differential equations with tightly distributed maturation ($m = 10$) and a tightly distributed latent period ($n = 10$).
- DDE—delay differential equations.

In addition, four removal strategies are shown: continuous, where trapping occurs continually; monthly; quarterly, the strategy used in practice; and quarterly with replacement, where diseased animals are replaced with captive-reared disease-free individuals. Some removal may be sufficient to control the disease so that it does not wipe out the devils; however, higher removal rates will be needed to completely eradicate the disease. They are given respectively in the dark and light grey bars. Relative to most wild mammals, Tasmanian devils are very trappable: mark–recapture analysis suggested that given a standard trapping effort, the typical devil in a target population would have a 57–94% chance of being trapped over a 10-day trapping session (Lachish et al. 2010). Nevertheless, the removal rates necessary to eliminate disease are at the very top of this range, and are unachievable with quarterly trapping. (From Beeton and McCallum 2011; McCallum 2016.)

confined to, or concentrated in, an identifiable segment of the population, then that segment might be culled without testing. For instance, cattle ≥5 years of age were culled during elimination of bTB in Australia (Cousins and Roberts 2001). Although the primary impetus for eliminating bTB in Australia was to protect the cattle industry, bTB was also prevalent among introduced feral water buffalo in northern Australia, which had the potential to act as a reservoir, so in addition to selective culling

of cattle, broadscale nonselective culling of buffalo also took place (More et al. 2015). Selective culling will be more effective, if infected individuals can be identified and culled, *before* they become infectious. For example, it was proposed that seropositive, prereproductive, female bison (*B. bison*), which are likely to become infectious at their first pregnancy, could be culled to manage brucellosis (Ebinger et al. 2011). Selective culling will also be particularly effective if individuals with the potential to infect a disproportionately large number of contacts ("*superspreaders*") can be identified and culled early on (Woolhouse et al. 1997; Lloyd-Smith et al. 2005; Paull et al. 2012).

Selective culling has been used to eliminate diseases in domestic animals, but we are not aware of a disease that has been eliminated from a wild species by this means. The difference between the two situations relates to the relative intensity of testing possible. Complete groups of domestic animals, such as all the cattle on a farm, can be gathered relatively easily for repeated testing, and movement of animals between tests can be controlled. In contrast, entire groups of wild animals, such as all the deer in a large park, can seldom be gathered for simultaneous testing. As a consequence, testing has to continue over the prolonged period needed to capture animals, and movement of individuals between tests is difficult to control. An attempt was made to eliminate Tasmanian devil facial tumor disease (DFTD), a lethal infectious cancer of the Tasmanian devil (*Sarcophilus harrisii*), by selective culling on a nearly isolated peninsula, approximately 100 km^2 in area. All animals with clinical signs of disease were removed in an intensive trapping campaign undertaken quarterly between 2004 and 2010. Not only did the culling fail to arrest population decline, but the rate at which uninfected individuals became infected did not differ between the peninsula in which culling was undertaken and a similar peninsula where no culling took place (Lachish et al. 2010). Models used were sensitive to different epidemiological assumptions, as well as the frequency of trapping (see Figure 13.4). Still, they predicted that the proportion of the Tasmanian devil population that would need to be tested and culled to eliminate disease was so high as not to be feasible (Beeton and McCallum 2011).

The fraction of a population tested at a given time can be very important in the success of a disease control campaign. When all animals within a group cannot be tested, or testing is extended over time, prevalence of disease will decline only if the rate of removal of infected animals exceeds the rate at which new animals become infected. For instance, Gross and Miller (2001) predicted that chronic wasting disease (CWD) could be eliminated by annual culling of >10–20% of infected deer from a population, although this might require 80 years of testing. In contrast, selective culling, based on testing of more than 50% of a mule deer (*Odocoileus hemionus*) herd annually, was predicted to reduce CWD prevalence from 8 to <2% within 5–10 years (although the time to eliminate the disease was not predicted) (Wolfe et al. 2004). Ultimately, the success of a selective culling campaign depends critically on the transmission particularities of the disease in the situation; for example, while this approach was impractical for eliminating bTB from white-tailed deer (*Odocoileus virginianus*) in Michigan (Cosgrove et al. 2012), selective culling of African buffalo (*Syncerus caffer*) (Renwick et al. 2006) and elk (*C. elaphus*) (Shury and Bergeson 2011) reduced the apparent incidence of the disease to zero in animals born after initiation of the campaign.

Focal culling ("reactive culling," "point infection control") is used to eliminate disease in livestock. The entire herd on an "infected" farm is removed and culling may extend to adjacent premises, to remove potential sources of infection and reduce availability of susceptible animals. The technique requires rapid identification of cases, prompt intervention, and restriction of movement of animals into and out of the cull site. For example, in a foot and mouth disease (FMD) outbreak, the aim was to cull animals on infected farms within 24 hours of disease identification (Ferguson et al. 2001).These requirements are problematic in wild animals, because (i) detection is likely to be delayed; (ii) it is difficult to maintain low density of wild animals even in small areas, because of immigration (Warburton and Thomson 2002); and (iii) culling may increase dispersal and contact (Vial and Donnelly 2012). Intense, focal culling may be appropriate when a newly discovered disease is limited to a small geographic area, such as culling raccoons and striped skunks (*Mephitis mephitis*) within a 5-km radius of the first confirmed cases of raccoon rabies in Canada (Rosatte et al. 2001). Culling all Tasmanian devils in the region where DFTD was first recognized "could have removed the disease threat to the entire species" but could not have been justified with the information available at the time (McCallum 2008, p635). Aggressive culling in a small area prevented the spread of bTB and establishment of deer as a maintenance host in Minnesota (Carstensen and Don Carlos 2011). Focal nonselective culling in areas with a high prevalence of infection, or removal of entire groups of female deer within which an infected individual has been found, have been suggested as potential management strategies for CWD (Storm et al. 2013).

Focal culling—within a larger area where disease is present—is unlikely to be effective in eliminating disease, even locally, if animal movements cannot be controlled. When deer mice (*Peromyscus maniculatus*) were removed from ranch buildings to reduce risk of human exposure to hantavirus, immigrants quickly replaced the mice that were removed and there was constant turnover. This resulted in more mice in the buildings than prior to removal and increased the probability that an infected mouse would enter the buildings (Douglass et al. 2003). Reactive culling of badgers had "a beneficial impact" on bTB in cattle herds in the Republic of Ireland (Olea-Popelka et al. 2009), although there was no consistent trend in disease prevalence in badgers (Corner et al. 2008). In contrast to that, in Great Britain, reactive culling resulted in increased ranging behavior by badgers, potentially greater contact with cattle and other badgers, and increased incidence of bTB in cattle (Woodroffe et al. 2006a, 2006b; Pope et al. 2007; Karolemeas et al. 2012).

Barrier culling (to create a "cordon sanitaire," in which the density of potential hosts has been reduced) has been used to prevent entry of disease into an area where it does not occur, or has been eliminated, for example, local depopulation of foxes (using gassing of dens) apparently prevented rabies entering areas of Switzerland (Wandeler et al. 1974) and Denmark (Muller 1971).

Area-wide, nonselective culling ("proactive culling," "host population reduction") involves removal of hosts, without regard to disease status, on the assumption that transmission is density-dependent and that a density threshold exists, below which disease will not persist. If the rate of transmission is independent of population density, and no threshold exists, nonselective culling is unlikely to be successful in reducing transmission (McCallum et al. 2001; Lloyd-Smith et al. 2006; Storm et al. 2013). Although proactive culling has been used

widely, it seldom has been credited with eliminating disease from a wild species, and the effect of culling appears questionable even in some of the few apparent successes. For instance, culling of 20,698 mule deer was credited with ending an outbreak of FMD in cattle in California (Anonymous 1921). However, culled deer were not tested for FMD, and lesions attributed to FMD, may instead have been the result of infection with the epizootic hemorrhagic disease virus (Weaver et al. 2013). The culling of >158,000 carnivores was credited with eliminating a rabies epidemic in Alberta, Canada (Ballantyne and O'Donoghue 1954), but rabies also disappeared from neighboring provinces, without extensive predator culling. When rabies disappeared from Corsica, it was unclear whether this resulted from culling, or mortality from the disease (Aubert 1994). Conversely, population reduction failed to eliminate rabies from foxes in Europe (Selhorst et al. 2005), vampire bats (*Desmodus rotundus*) in Latin America (Streicker et al. 2012), or classical swine fever (CSF) from wild boars (*Sus scrofa*) in Europe (Laddomada 2000).

The minimum host population size/density that will allow disease persistence is generally unknown, and "Very few plans to reduce host population density include an evaluation of the desired level of population decrease" (Artois et al. 2001, p143). In some situations, disease may persist in a very small population, for example, brucellosis and bTB might persist in as few as 200 bison (Dobson and Meagher 1996; Joly and Messier 2004). A single ("one-off") cull is unlikely to eliminate disease, especially if the prevalence of disease is high, unless almost the entire host population is removed, but repeated culling (that maintains the population at a reduced level) may be effective. A single cull of >96% of a rabbit (*Oryctolagus cuniculus*) population would be required "to have even a small probability" of eliminating *Mycobacterium avium spp. paratuberculosis* infection; however, culling 45% of the population annually for 33 years had a high probability of eliminating the disease (Davidson et al. 2009). Intense annual culling of brushtail possums (*T. vulpecula*), in which density was kept at <40% precontrol for 6–10 years, resulted in "major reduction" in incidence of BTB in cattle herds in New Zealand, suggesting that the disease might be eliminated from possums by intense culling (Caley et al. 1999). Intense culling (50% reduction of abundance) reduced the prevalence of bTB but had no effect on the prevalence of Aujeszky's disease in wild boars in an area of Spain (Boadella et al. 2012). In another study, also in Spain, intense culling of wild boar failed to reduce the prevalence of bTB in boars but reduced the prevalence in sympatric fallow deer (*Dama dama*) (García-Jiménez et al. 2013). Proactive culling of badgers had different effects on bTB in the Republic of Ireland and the UK. These variations may be attributable to differences in cattle and badger populations, the history of prior culling, intensity of culling (proportion of land on which culling occurred, and method of capture of badgers), length of time over which culling occurred, and barriers to movement of badgers into and out of trial sites (Woodroffe et al. 2006a, 2006b; O'Connor et al. 2012).

In summary, any proposal to use culling of wildlife as a strategy to control infectious disease requires extensive and careful evaluation. The effectiveness of culling strategies is limited by a range of ecological, practical, and sociological constraints, and culling can lead to counterintuitive and detrimental outcomes (Miguel et al. 2020). In the case of fox rabies, efficient culling can only be sustained on relatively small, localized areas (Sterner and Smith 2006). Culling should not be considered without a good understanding of the transmission cycle of the disease and response of the population to culling (Harrison et al. 2010). Culling may cause changes in reproduction or recruitment that result in an increased frequency of susceptible individuals in the population, increased dispersal and effective contact among remaining animals, and interference with establishment of herd immunity (Prentice et al. 2019; Miguel et al. 2020). Repeated nonselective culling was associated with increasing prevalence of BTB in badgers in an area in the UK (Woodroffe et al. 2006b). Models predict that some levels of nonselective culling may increase, rather than decrease, disease incidence (Choisy and Rohani 2006; Bolzoni et al. 2007). There can be a "Goldilocks zone" in which culling is sufficient to control disease but not so intense that the host population is threatened, but whether and under what conditions this zone exists depends on a good understanding of the nature of density-dependent dispersal in the focal host and the spatial and temporal distribution of the culling effort relative to the spatiotemporal dynamics of the host population (Prentice et al. 2019).

13.5 Vaccination

Mass immunization was the major technique used to eradicate smallpox and rinderpest and has led to the control of many diseases of humans and domestic animals. Vaccination generally does not disrupt territorial behavior (although see Munoz et al. 2010) or density-dependent population dynamics, and is acceptable to the public; however, suitable vaccines are not available for all diseases. The type and degree of protection induced by vaccination depends on features of the vaccine, dose, method of delivery, and host. A vaccine may not even need to prevent infection to be effective in a disease elimination or eradication program, so long as it reduces the infectiousness of vaccinated individuals below that required to sustain infection in the population (Buddle et al. 2000; Nol et al. 2008).

A disease can be most easily eliminated if a safe, effective, vaccine exists, as well as a method for delivering it to a large proportion of the host population. Safety of the vaccine and delivery method for the target species, and for other species that might be exposed, should be determined prior to field studies. Although a perfect vaccine would provide lifelong protection for every vaccinated individual, in reality many vaccines protect only a proportion of those who receive vaccine, that is, they reduce, but do not eliminate the probability of infection; furthermore, they often provide protection that wanes over time. Such imperfections should be identified before use in the field and considered in planning a campaign (McLean and Blower 1993).

Several basic questions apply to any vaccination program:

Who should be vaccinated? To eliminate or eradicate disease, vaccination must be applied to those specific hosts that maintain infection in an area. For instance, where rabies is maintained in dogs and spills over into wild animals, vaccination of dogs might eliminate the disease, but vaccination of wild hosts will not. Where rabies is maintained in wild species, vaccination of dogs will protect immunized dogs (and, indirectly, their owners), but will not eliminate the disease. In humans and domestic animals, vaccine administration can target specific individuals, such as those of a certain sex or age. In wild animals, this is only possible if they are captured ("trap–vaccinate–release"), or otherwise vaccinated selectively, such as by a dart fired from an aircraft. When vaccine is distributed in oral baits, there is limited control over who is exposed, for example, only 22% of baits intended for feral pigs, were taken by pigs, and 51% were taken by raccoons (Campbell et al. 2006). As a result, it may be difficult to reach focal segments of the population. For instance, it is important to immunize young wild boars to reduce transmission of CSF, but piglets <4–5 months old tend to not take up vaccine bait (Kramer-Schadt et al. 2009), and if they do, they must compete for baits with older animals, who may already be immune (Bolzoni et al. 2007). Special bait-delivery methods may therefore be required to vaccinate particular segments of a population, and to reduce the number of baits "wasted" by other species.

What proportion of the population must be vaccinated? It is not necessary to vaccinate every individual in a population to eliminate or eradicate a disease, because each infection prevented reduces further transmission. This "reduction in infection or disease in the unimmunized segment as a result of immunising a proportion of the population" has been termed "herd effect" by John and Samuel (2000) p601, who differentiated it from "herd immunity," which they defined as "the proportion of subjects with immunity in a given population." If there is a single host species, whose population is homogeneous and well-mixed, the "critical proportion" (p_c) that must be immunized to eliminate disease is approximated by $1 - 1/R_0$ (Anderson and May 1982). This will not prevent infection in unvaccinated individuals, but, if they do become infected, they will on average infect <1 further individuals and any possible epizootic outbreak will quickly fade out. If R_0 is large, p_c approaches the entire population. If more than one host type is involved, alternative methods are needed to estimate the effort required (Roberts and Heesterbeek 2003). Vaccine imperfections

must be considered in calculating p_c. Such imperfections are often included under the summary variable "vaccine impact" (ϕ), which is based on take, mean duration of protection, and mean life expectancy of hosts, and can be used to adjust for vaccine imperfections: $p_c = 1/\phi (1 - 1/R_0)$ (Woolhouse et al. 1997).

When a vaccine is delivered in baits, p_c translates into the density and distribution of baits that must be placed. During elimination of fox rabies in Europe, the goal was to immunize ≥70% of foxes. Baits were distributed via aircraft in spring and autumn each year, at an average density of 20–25 baits/km, along parallel flight lines spaced 1,000 m apart (Freuling et al. 2013). This was generally successful in eliminating rabies, and the disease was eliminated even in some locations where <70% of foxes were immunized (Eisinger and Thulke 2008). Vaccination was less successful in these latter settings because baits were not dropped within the home range of some fox family groups. This can be alleviated by delivering the same number of baits but reducing the distance between flight lines (Thulke et al. 2004). Nonuniform distribution of vaccine baits may be required where host density is spatially clumped (Bohrer et al. 2002).

When should vaccination occur? Vaccination only protects against new infection, so vaccination should occur at a young age, to protect animals before exposure. However, antibodies transferred from the mother may interfere with immunization of young individuals, for example, fox cubs <8 weeks old, born to immune dams, have an impaired immune response, because of maternally transferred immunity (Müller et al. 2001). Thus, there may be a limited time window, between waning of maternal immunity and the average age of infection, during which vaccination is optimal.

Vaccination of wild animals usually occurs in "pulses," in which the population is exposed to vaccine at fixed intervals. This allows vaccination of individuals missed in previous pulses, and new individuals, including immigrants from nonvaccinated populations, as well as "boosting" of previously vaccinated animals. In seasonally breeding hosts, the initial annual pulse should be timed to coincide with waning of maternal immunity, and when young animals are able to access baits effectively. A single annual pulse may be adequate, if a single vaccination confers long-lasting effective immunity, relative to the lifespan of the host, and there is sufficient coverage to reach p_c. Vaccination at more frequent intervals is necessary to generate and maintain p_c if coverage is low, protection is of short duration relative to lifespan, R_0 is high, ø is low, or there is rapid recruitment of susceptible hosts. For example, Cross and Getz (2006) predicted that, if the half-life of protection against bTB was <5 years, vaccination of every African buffalo calf for 50 years would not eliminate tuberculosis from this long-lived species. Timing of vaccination also must consider seasonal factors, such as how long baits remain effective in the environment, and food supply, that may reduce, or augment, bait uptake.

Can vaccinated individuals be differentiated from infected individuals? For effective surveillance, it is important to differentiate between naturally infected and vaccinated individuals, and to identify vaccinated individuals that have become infected. Strategies to allow differentiation of infected and vaccinated individuals have been reviewed by Henderson (2005).

Will vaccine-resistant agents develop? If there is genetic diversity in the infectious agent population, vaccination may select for resistant organisms. Vaccine-driven selection has not been reported in wild animals, but experience in humans demonstrates a potential for this to occur, for example, whooping cough was controlled through mass immunization of humans, but reemerged, despite high levels of vaccination coverage, as a result of clonal expansion of resistant forms (Schouls et al. 2004; Octavia et al. 2011).

13.6 Manipulating environmental factors

The environmental "stage" upon which agent and hosts interact is critically important for disease transmission and includes both biotic (e.g., vegetation) and abiotic factors (topography, climate, weather, water, soils). Many diseases have a more or less sharply defined spatial and temporal "niche"

(See Chapter 12, Section 12.1). Identifying the features of the niche that are required for disease occurrence and transmission is an important component of disease intervention. This is typically done first through a systematic comparison of conditions prevailing when and where disease occurs versus when and where it is absent; this can then be tested through experimental manipulations of those factors. Once such specific features have been identified, they might be modified on a large scale to facilitate disease elimination, particularly if transmission is temporally or spatially concentrated. Such manipulations can focus on: (i) creating changes in the spatial distribution or size of habitat patches and host populations; or (ii) on modifications of habitat quality that undermine transmission. In regard to the first approach, recent investigations have demonstrated that experimental habitat fragmentation can impact transmission of parasitic nematodes in Australian lizards (Resasco et al. 2019). Habitat fragmentation has even been proposed as an approach to control pathogen spread, although the benefits of controlling disease need to be weighed against the damage done by habitat degradation to the wildlife (Bozzuto et al. 2019).

Two examples illustrate the process of identifying, and then manipulating, an environmental driver to reduce disease. A study of tick-borne disease in wild ungulates in Tanzania (Fyumagwa et al. 2007) found that changes in abiotic (fire frequency, climate) and biotic (vegetation, abundance of herbivore species) factors over 30 years produced "cumulative positive effects" on the ticks that transmitted the causative protozoan. Fire suppression was identified as an important driver that resulted in expansion of the area occupied by tall grasses ("ideal habitat for free-living stages of ticks"), increased abundance of African buffalo ("a superb host for ticks"), and decreased abundance of short grass-grazers ("with relatively high innate resistance to ticks"). Intervention, in the form of prescribed burning, lowered tick numbers significantly, maintained these tick populations at low levels, and changed the relative abundance of herbivore species. Similarly, Allan et al. (2010) found that areas invaded by the exotic shrub Amur honeysuckle (*Lonicera maacki*) had fivefold greater density of white-tailed deer, and 10-fold greater

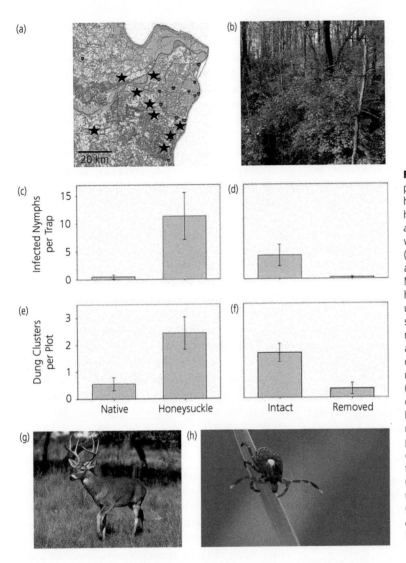

Figure 13.5 Invasive plants species can promote wildlife diseases. Invading Amur honeysuckle (*Lonicera maackii*) provides habitat for lone star ticks (*Amblyomma americanum*) that transmit ehrlichiosis to white-tailed deer (*Odocoileus virginianus*). (a) Map of the nine study sites in natural areas located the vicinity of St. Louis, Missouri, USA. (b) Example of Amur honeysuckle bushes invading the understory of a local forest. Density of lone star ticks was higher in invasive versus in native vegetation (c), and declined in plots after honeysuckle removal (d). (e) Density of dung clusters of white-tailed deer in native forest vs. in honeysuckle plots. (f) Density of dung clusters of white-tailed deer honeysuckle-intact vs. honeysuckle-eradicated plots. Error bars reflect 1 SE. Both deer and tick nymphs prefer the invasive honeysuckle thickets creating excellent conditions for *Ehrlichia* transmission. (Source: Allan et al. 2010.) (g) White-tailed deer (*Odocoileus virginianus*). (Source: Scott Bauer, USDA.) (h) Lone star tick (*Amblyomma americanum*) questing for hosts. (Source: Michael Dryden, Kansas State University.)

density of nymphal ticks, infected with *Ehrlichia chaffeensis* (a zoonotic parasite), than areas with native vegetation. The relationship was confirmed by experimental removal of honeysuckle, resulting in decreased use by deer and density of infected ticks (Figure 13.5).

Even if it is impossible to manipulate many abiotic factors, it is still important to understand how they influence disease dynamics, because such factors may assist or compromise intervention efforts. For example, geophysical features may limit movement of animals and prevent the spread of disease. Alpine areas above 1,500 m prevented spread

of rabies between valleys in Switzerland (Steck et al. 1982); fenced motorways and large rivers limit movement of wild boars and the spread of CSF (European Food Safety Authority 2009), and large rivers slow spread of raccoon rabies (Smith et al. 2002). Geophysical barriers can be used to improve disease intervention campaigns, for example, by distributing vaccine baits only in valleys to control fox rabies or using barriers to reduce immigration into areas where animals were culled or vaccinated.

The most direct form of environmental intervention is the elimination of a factor required for transmission, such as an invertebrate vector or intermediate host. Tsetse flies (*Glossinia* spp.) are the vector of human and animal trypanosomiasis in sub-Saharan Africa. Early methods to control tsetse flies included environmental manipulations that are no longer acceptable, including removal of native vegetation preferred by the flies, elimination of wildlife hosts, and widespread spraying of persistent insecticides. More recently, integrated pest management, including release of sterile male flies, have eliminated tsetse flies from the Island of Unguja, Zanzibar (Vreysen et al. 2000). Blackflies (*Simulium* spp.), the vector of human onchocerciasis (river blindness), were eliminated from an island in the Republic of Equatorial Guinea, by using insecticide to kill larvae in streams (Traoré et al. 2009). Success in the latter situation was attributed to the distribution of the vector being "isolated, well-defined and accessible," features that may not be present when a disease and vector are widespread.

Many environmental features that affect disease result from human activities, including alterations of the landscape that influence the distribution, abundance, and behavior of wild animals (Farnsworth et al. 2005). An important example is supplemental feeding, which congregates wild animals, increasing opportunities for direct transmission between individuals, and indirect transmission by exposure to a contaminated environment. Feeding also may promote exchange between otherwise isolated groups and maintain populations at a density beyond normal carrying capacity of the habitat. Feeding is important in many diseases, including CSF in wild boar (Kramer-Schadt et al. 2007), bTB in deer (Miller et al. 2003), brucellosis in elk (Scurlock and Edwards 2010), and conjunctivitis caused by *Mycoplasma gallisepticum* in house finches (*Carpodacus mexicanus*) (Dhondt et al. 2005; Becker et al. 2015). While stopping feeding seems an obvious remedy, there may be strong social and political support for artificially inflated populations maintained by supplemental feeding (Scurlock and Edwards 2010). If feeding cannot be stopped, its impact might be reduced by changing the method and timing of feeding: for instance, brucellosis is transmitted by aborting females; reducing congregation of elk at feeding sites in early spring could result in decreased transmission of the pathogen (Cross et al. 2007; Creech et al.

2012). In contrast, Thompson et al. (2008) found that restricting the quantity of feed did not reduce bTB transmission among deer substantially. In any case, any management based on changing human behavior requires public support to be successful (see Chapter 16).

13.7 Combined techniques

A single method may be inadequate to eliminate or eradicate a disease, particularly in wild animals, where movement and contact cannot be controlled. Combinations of methods, applied either simultaneously or consecutively, may be more effective than a single method. As an example, Kramer-Schadt et al. (2007) suggested that different techniques are appropriate for management of CSF in wild boars, depending on the phase of disease. In an acute outbreak, artificial feeding should be stopped to limit contact and population reduction by hunting "should only be carried out if it can be guaranteed that no infected boar escape the outbreak zone" or if vaccination is not useful. Where CSF is endemic, vaccination is appropriate and supplementary feeding should be banned. In areas proximal to an outbreak or endemic focus, hunting should be reduced to avoid movement of boar into the area affected by disease, and vaccination is recommended. Different methods may be applied consecutively. For example, it was suggested that selective culling for a limited period, followed by vaccination, might be used to eliminate brucellosis from bison (Ebinger et al. 2011). In the aforementioned example of tsetse fly elimination on Zanzibar, initial suppression of the flies by limited insecticide use was followed by the release of sterile male flies (Vreysen et al. 2000).

13.8 The endgame

Challenges emerge during "the final stages of an elimination or eradication programme, when the disease is still circulating, although at much reduced levels" (Klepac et al. 2013, p2), because:

- Intervention becomes increasingly difficult as remaining foci of infection are likely to be in remote areas, the cost per case prevented increases, while awareness and skill of investigators, support for the program, and incentive to participate may decline.

- The proportion of susceptible individuals in the population increases and herd immunity wanes, resulting in increased risk of a major outbreak.
- Elimination of one disease may allow another to become problematic if cross-reacting immunity suppressed another infection. For example, eradication of rinderpest may have allowed the *peste de petits ruminants* to increase in range (Libeau et al. 2011).

Deciding when disease has been eliminated is problematic and "some sort of proof of the final success is required" (Tischendorf et al. 1998, p839), so that intervention can cease. Stopping prematurely may allow disease reemergence, while continuing intervention after disease has been eliminated is wasteful. During the endgame, ever more sensitive detection methods are required; thus, if vaccination was used to eliminate disease, vaccination may have to cease to allow surveillance. Structured, representative surveys are used in human and domestic animal populations to assess the likelihood that the population is free of a disease. This type of survey is difficult in wild animals, because information on population size, appropriate sample size, and spatial distribution is often inadequate (Anderson et al. 2013). It also may be difficult to sample sufficient animals, for example, if the prevalence of CWD is <0.1%, it would be necessary to sample >25,000 deer to have >50% probability of detecting a diseased individual (Diefenbach et al. 2004). If resources are limited, it may be appropriate to focus surveillance on areas or individuals (e.g., older animals) that have a higher probability of disease occurrence. Surveillance systems that use data from multiple sources to quantitatively assess the probability of disease and vector elimination have been described (Barclay and Hargrove 2005; Martin et al. 2007; Anderson et al. 2013).

Programs to eliminate or eradicate disease need an "exit strategy." For example, Thulke et al. (2000) suggested that "terminating but observing" was an appropriate strategy for use after rabies appeared to have disappeared from a fox population. As part of the strategy, surveillance would be continued after vaccination ended, at a level designed to detect a 2% prevalence of infection with 95% confidence. Modeling predicted that any new outbreak of rabies would be detected within 2 years with this level of surveillance. The strategy included maintenance of emergency preparedness for intensive vaccination to surround any potential detection of a rabid animal.

13.9 Establishing an "ark" or "insurance" population

If disease threatens to cause extirpation of a species, and no effective technique is available to eliminate the disease, a protected population of disease-free animals can be established, which can then be used to reintroduce a free-living population should the species, and the disease, become extinct in the wild. Such an "insurance" population of Tasmanian devils, threatened by DFTD, could include both captive animals and free-ranging individuals on offshore islands (Jones et al. 2007). The "salvage" of disease-free individuals from a diseased population requires a highly sensitive technique for detecting infection (to prevent infected individuals from being included in the insurance group) and long-term isolation facilities. Attempted salvage of disease-free animals from a wood bison (*B. b. athabascae*) population infected with BTB and brucellosis illustrates the difficulties of this approach (Nishi et al. 2006). Neonatal calves, captured from the wild, were tested, treated intensively, raised in isolation, and retested annually, with the aim of producing a captive herd of disease-free animals that could be reintroduced to an area from which infected wild bison would be eliminated. Unfortunately, bTB occurred within the salvaged bison, 9 years into the project, resulting in herd depopulation. The likely source of infection was a single animal, infected prior to capture, that remained undetected, despite rigorous annual testing (Himsworth et al. 2010).

References

Allan, B.F., Dutra, H.P., Goessling, L.S., Barnett, K., Chase, J.M., Marquis, R.J., et al. (2010). Invasive honeysuckle eradication reduces tick-borne disease risk by altering host dynamics. *Proceedings of the National Academy of Sciences*, **107**:18523–27.

Anderson, D.P., Ramsey, D.S.L., Nugent, G., Bosson, M., Livingstone, P., Martin, P.A.J., et al. (2013). A novel approach to assess the probability of disease eradication from a wild-animal reservoir host. *Epidemiology Infection*, **141**:1509–21.

Anderson, R.M. and May, R.M. (1982). Directly transmitted infectious diseases: Control by vaccination. *Science*, **215**:1053–60.

Anonymous. (1921). Foot and mouth disease in deer. *North American Veterinarian*, **7**:53.

Artois, M., Delahay R., Guberti, V., and Cheeseman, C. (2001). Control of infectious diseases of wildlife in Europe. *Veterinary Journal*, **162**:141–52.

Aubert, M. (1994). Control of rabies in foxes: What are the appropriate measures? *Veterinary Record*, **134**:55–59.

Aylward, B., Henney, K.A., Zagaria, N., and Olivé, J.-M. (2000). When is a disease eradicable? 100 years of lessons learned. *American Journal of Public Health*, **90**:1515–20.

Ballantyne, E.E. and O'Donahue, J.G. (1954). Rabies control in Alberta. *Journal American Veterinary Medical Association*, **125**:316–26.

Barclay, H.J. and Hargrove, J.W. (2005). Probability models to facilitate a declaration of pest-free status, with special reference to tsetse (Diptera: Glossinidae). *Bulletin Entomological Research*, **95**:1–11.

Barlow, N.D. (1996). The ecology of wildlife disease control: Simple models revisited. *Journal Applied Ecology*, **33**:303–14.

Becker, D.J., Streicker, D.G., and Altizer, S. (2015). Linking anthropogenic resources to wildlife–pathogen dynamics: A review and meta-analysis. *Ecology Letters*, **18**:483–95.

Beeton, N. and McCallum, H. (2011). Models predict that culling is not a feasible strategy to prevent extinction of Tasmanian devils from facial tumour disease. *Journal of Applied Ecology*, **48**:1315–23.

Beja-Pereira, A., Bricker, B., Chen, S., Almendra, C., White, P.J., and Luikart G. (2009). DNA genotyping suggests that recent brucellosis outbreaks in the Greater Yellowstone Area originated from elk. *Journal Wildlife Diseases*, **45**:1174–77.

Berentsen, A.R., Dunbar, M.R., Johnson, S.R., Robbe-Austerman, S., Martinez, L.L., and Jones, R.L. (2011). Active use of coyotes (*Canis latrans*) to detect bovine tuberculosis in northeastern Michigan, USA. *Veterinary Microbiology*, **151**:126–32.

Boadella, M., Vicente, J., Ruiz-Fons, F., de la Fuente, J., and Gortázar, C. (2012). Effects of culling Eurasian wild boar on the prevalence of *Mycobacterium bovis* and Aujeszky's disease virus. *Preventive Veterinary Medicine*, **107**: 214–21.

Bohrer, G., Shem-Tov, S., Summer, E., Or, K., and Saltz, D. (2002). The effectiveness of various rabies spatial vaccination patterns in a simulated host population with clumped distribution. *Ecological Modelling*, **152**: 205–11.

Bolzoni, L., Real, L., and De Leo, G. (2007). Transmission heterogeneity and control strategies for infectious disease emergence. *PLoS ONE*, **2** (8):e747, doi:10.1371/jourbnal.pone.0000747.

Bosch, J., Sanchez-Tomé, E., Fernández-Loras, A., Oliver, J.A., Fisher, M.C., and Garner, T.W.J. (2015). Successful elimination of a lethal wildlife infectious disease in nature. Biology Letters, **11**:20150874. http://dx.doi.org/10.1098/rsbl.2015.0874.

Bozzuto, C., Canessa, S., and Koella, J.C. (2019). Planning artificial habitat fragmentation to mitigate invasion by infectious wildlife diseases. Preprint on Research Gate: License CC BY-NC-ND 4.0. doi:10.13140/RG.2.2.31616.89606.

Buddle, B.M., Skinner, M.A., and Chambers, M.A. (2000). Immunological approaches to the control of tuberculosis in wildlife reservoirs. *Veterinary Immunology Immunopathology*, **74**:1–16.

Caley, P., Hickling, G.J., and Cowan, P.E. (1999). Effects of sustained control of brushtail possums on levels of *Mycobacterium bovis* infection in cattle and brushtail possum populations from Hohotaka, New Zealand. *New Zealand Veterinary Journal*, **47**:133–42.

Caley, P. and Hone, J. (2005). Assessing the host status of wildlife and the implications for disease control: *Mycobacterium bovis* infection in feral ferrets. *Journal Applied Ecology*, **42**:708–19.

Campbell, T.A., Lapidge, S.J., and Long, D.B. (2006). Using baits to deliver pharmaceuticals to feral swine in southern Texas. *Wildlife Society Bulletin*, **34**:1184–89.

Cartensen, M. and Don Carlos, M.W. (2011). Preventing the establishment of a wildlife disease reservoir: A case study of wild deer in Minnesota. *Veterinary Medicine International*, 2011: 413240 doi:10.4061/2011/413240.

Choisy, M. and Rohani, P. (2006). Harvesting can increase the severity of wildlife disease epidemics. *Proceedings of the Royal Society London B: Biological Sciences*, **273**:2025–34.

Corner, L.A. (2006). The role of wild animal populations in the epidemiology of tuberculosis in domestic animals: How to assess the risk. *Veterinary Microbiology*, **112**:303–12.

Corner, L.A.L., Clegg, T.A., More, S.J., Williams, D.H., O'Boyle, I., Costello, E., et al. (2008). The effect of varying levels of population control on the prevalence of tuberculosis in badgers in Ireland. *Research Veterinary Science*, **85**:238–49.

Cosgrove, M.K., Campa, H., III, Schmidt, S.M., Marks, D.R., Wilson, A.S., and O'Brien D.J. (2012). Live-trapping and bovine tuberculosis testing of free-ranging white-tailed deer for targeted removal. *Wildlife Research*, **39**:104–11.

Cousins, D.V. and Roberts, J.L. (2001). Australia's campaign to eradicate bovine tuberculosis: The battle for freedom and beyond. *Tuberculosis*, **81**:5–15.

Cowled, B.D., Garner, M.G., Negus, K., and Ward M.P. (2012). Controlling disease outbreaks in wildlife using limited culling: Modelling classical swine fever incursions in wild pigs in Australia. *Veterinary Research*, 43:3 https://doi.org/10.1186/1297-9716-43-3 (accessed August 12, 2021).

Creech, T.G., Cross, P.C., Scurlock, B.M., Miachak, E.J., Rogerson, J.D., Henningsen, J.C., et al. (2012). Effects of low-density feeding on elk-fetus contact rates on Wyoming feedgrounds. *Journal Wildlife Management*, **76**:877–86.

Cross, P.C., Edwards, W.H., Scurlock, B.M., Maichek, E.J., and Rogerson, E.D. (2007). Effects of management and climate on elk brucellosis in the Greater Yellowstone Ecosystem. *Ecological Applications*, **17**:957–64.

Cross, P.C. and Getz, W.M. (2006). Assessing vaccination as a control strategy in an on-going epidemic: Bovine tuberculosis in African buffalo. *Ecological Modelling*, **196**:494–504.

Davidson, R.S., Marion, G., White, P.C.L., and Hutchings, M.R. (2009). Use of host population to control wildlife infection: Rabbits and paratuberculosis. *Epidemiology Infection*, **137**:131–38.

Defra (2011). The Government's policy on Bovine TB and badger control in England. https://www.gov.uk/government/publications/the-government-s-policy-on-bovine-tb-and-badger-control-in-england (accessed August 12, 2021).

Dhondt, A.A., Altizer, S., Cooch, E.G., Davis, A.K., Dobson, A., Driscoll, M.J., et al. (2005). Dynamics of a novel pathogen in an avian host: Mycoplasmal conjunctivitis in house finches. *Acta Tropica*, **94**:77–93.

Diefenbach, D.R., Rosenberry, C.S., and Boyd, R.C. (2004). From the field: Efficacy of detecting chronic wasting disease via sampling hunter-killed deer. *Wildlife Society Bulletin*, **32**:267–72.

Dobson, A. and Meagher, M. (1996). The population dynamics of brucellosis in the Yellowstone National Park. *Ecology*, **77**:1026–36.

Douglass, R.J., Kuenzi, A.J., Williams, C.Y., Douglass, S.J., and Mills, J.N. (2003). Removing deer mice from buildings and the risk for human exposure to Sin Nombre virus. *Emerging Infectious Diseases*, **9**:390–92.

Ebinger, M., Cross, P., Wallen, R., White, P.J., and Treanor, J. (2011). Simulating sterilization, vaccination, and test-and-remove as brucellosis control measures in bison. *Ecological Applications*, **21**:2944–59.

Eisinger, D. and Thulke, H.-H. (2008). Spatial pattern formation facilitates eradication of infectious diseases. *Journal of Applied Ecology*, **45**:415–23.

European Food Safety Authority. (2009). Scientific Opinion of the Panel on AHAW on a request from Commission on "Control and eradication of classical swine fever in wild boar." *EFSA Journal*, **932**:1–18.

Farnsworth, M.L., Wolfe, L.L., Hobbs, N.T., Burnham, K.P., Williams, E.S. Theobald, D.M., et al. (2005). Human land use influences chronic wasting disease prevalence in mule deer. *Ecological Applications*, **15**:119–26.

Ferguson, N.M., Donnelly, C.A., and Anderson, R.M. (2001). Transmission intensity and impact of control policies on the foot and mouth epidemic in Great Britain. *Nature*, **413**:542–48.

Foufopoulos, J. (1999). Host-parasite interactions in the mountain spiny lizard *Sceloporus jarrovii*. Unpublished dissertation, University of Wisconsin, Madison, WI.

Freuling, C.M., Hampson, K., Selhorst, T., Schroeder, R., Meslin, F.X., Mettenleiter, T.C., et al., (2013). The elimination of fox rabies from Europe: Determinants of success and lessons for the future. *Philosophical Transactions of the Royal Society of London B: Biological Sciences*, **368**:20120142. http://dx.doi.org/10.1098/rstb.2012.0142.

Fyumagwa, R.D., Runyoro, V., Horak, I.G., and Hoare, R. (2007). Ecology and control of ticks as disease vectors in wildlife of the Ngorongoro Crater, Tanzania. *South African Journal Wildlife Research*, **37**:79–90.

García-Jiménez, W.L., Fernández-Llario, P., Benítez-Medina, J.M., Cerrato, R., Cuesta, J., García-Sánchez, A., et al. (2013). Reducing Eurasian wild boar (*Sus scrofa*) population density as a measure of bovine tuberculosis control: Effects in wild boar and a sympatric fallow deer (*Dama dama*) population in Spain. *Preventive Veterinary Medicine*, **110**:435–46.

Gordejo, F.J.R. and Vermeesch, J.P. (2006). Towards eradication of bovine tuberculosis in the European Union. *Veterinary Microbiology*, **112**:101–09.

Gross, J.E. and Miller, M.W. (2001). Chronic wasting disease in mule deer: Dynamics and control. *Journal Wildlife Management*, **65**:205–15.

Harrison, A., Newey, S., Gilbert, L., Haydon, D.T., and Thirgood, S. (2010). Culling wildlife hosts to control disease: Mountain hares, red grouse and louping ill virus. *Journal Applied Ecology*, **47**:926–30.

Haydon, D.T., Cleaveland, S., Taylor, L.H., and Laurenson, M.K. (2002). Identifying reservoirs of infection: A conceptual and practical challenge. *Emerging Infectious Diseases*, 8:1468–73.

Haydon, D.T., Randall, D.A., Matthews, L., Knobel, D.L., Tallents, L.A., Gravenor, M.B., et al. (2006). Low-coverage vaccination strategies for the conservation of endangered species. *Nature*, 443:692–95.

Hegglin, D. and Deplazes, P. (2008). Control strategy for Echinococcus multilocularis. *Emerging Infectious Diseases*, 14:1626–28.

Henderson, D.A. and Klepac, P. (2013). Lessons from the eradication of smallpox: An interview with D.A. Henderson. *Philosophical Transactions of the Royal Society of London B: Biological Sciences*, 368:20130113. http://dx.doi.org?10.1098/rstb.2013.0113.

Henderson, L.M. (2005). Overview of marker vaccine and differential diagnostic test technology. *Biologicals*, 33:203–09.

Himsworth, C.G., Elkin, B.T., Nishi, J.S., Neimanis, A.S., Wobeser, G.A., Turcotte, C., et al. (2010). An outbreak of bovine tuberculosis in an intensively managed conservation herd of wild bison in the Northwest Territories. *Canadian Veterinary Journal*, 51:593–97.

John, T.J. and Samuel, R. (2000). Herd immunity and herd effect; new insights and definitions. *European Journal of Epidemiology*, 16:601–06.

Joly, D.O. and Messier, F. (2004). Factors affecting apparent prevalence of tuberculosis and brucellosis in wood bison. *Journal Applied Ecology*, 73:623–31.

Jones, M.E., Jarman, P.J., Lees, C.M., Hesterman, H., Hamede, R.K., Mooney, N.J., et al. (2007). Conservation management of Tasmanian devils in the context of an emerging extinction-threatening disease: Devil facial tumor disease. *Ecohealth*, 4:326–37.

Karolemeas, K., Donnelly, C.A., Conlan, A.J., Mitchell, A.P., Clifton-Hadley, R.S., Upton, P., et al. (2012). The effect of badger culling on breakdown prolongation and recurrence of bovine tuberculosis in cattle herds in Great Britain. *PLoS ONE*, 7:e51342, doi:0.1371/journal.pone.0051342.

Klepac, P., Metcalf, J.E., McLean, A.R., and Hampson, K. (2013). Toward the endgame and beyond: Complexities and challenges for the elimination of infectious diseases. *Philosophical Transactions of the Royal Society of London B: Biological Sciences*, 368:20120137. http://dx.doi.org/10.1098/restb.2012.0137.

Kramer-Schadt, S., Fernández, N., Eisinger, D., Grimm, V., and Thulke, H.-H. (2009). Individual variations in infectiousness explain long-term persistence in wildlife populations. *Oikos*, 118:199–208.

Kramer-Schadt, S., Fernández, N., and Thulke, H. (2007). Potential ecological and epidemiological factors affecting the persistence of classical swine fever in wild boar *Sus scrofa* populations. *Mammal Review*, 37:1–20.

Lachish, S., McCallum, H., Mann, D., Pukk, C., and Jones, M. (2010). Evaluation of selective culling of infected individuals to control Tasmanian devil facial tumour disease. *Conservation Biology*, 24:841–51.

Laddomada, A. (2000). Incidence and control of CSF in wild boar in Europe. *Veterinary Microbiology*, 73:121–30.

León-Vizcaíno, L., Cubero, M.J., González-Capitel, E., Simón, M.A., Pérez, L., De Ybáñez, M.R.R., et al. (2001). Experimental Ivermectin treatment of sarcoptic mange and establishment of a mange-free population of Spanish ibex. *Journal Wildlife Diseases*, 37:775–85.

Libeau, G., Kwiatek, O., Lancelot, R., and Albino, E. (2011). Pest des petit ruminants, growing incidence. *OIE Bulletin*, 52:4.

Lindblade, K.A., Arana, B., Zea-Flores, G., Rizzo, N., Porter, C.H., Dominguez, A., et al. (2007). Elimination of Onchocerca volvulus transmission in the Santa Rosa focus of Guatemala. *American Journal Tropical Medicine Hygiene*, 77:334–41.

Lloyd-Smith, J.O., Cross, P.C., Briggs, C.J., Daugherty, M., Getz, W.M., Latto, J., et al. (2006). Should we expect population thresholds for wildlife diseases? *Trends in Ecology Evolution*, 20:511–19.

Lloyd-Smith, J.O., Schreiber, S.J., Kopp, P.E., and Getz, W.M. (2005). Superspreading and the effect of individual variation on disease emergence. *Nature*, 438:355–59.

Martin, P.A.J., Cameron, A.R., and Greiner, M. (2007). Demonstrating freedom from disease using multiple complex data sources 1: A new method based on scenario trees. *Preventive Veterinary Medicine*, 79:71–97.

McCallum, H. (2008). Tasmanian devil facial tumour disease: Lessons for conservation biology. *Trends in Ecology & Evolution*, 23:631–37.

McCallum, H. (2016). Models for managing wildlife disease. *Parasitology*, 143:805–20.

McCallum, H., Barlow, N., and Hone, J. (2001). How should pathogen transmission be modelled? *Trends in Ecology & Evolution*, 16:295–300.

McCallum, H., Fenton, A., Hudson, P.J., Lee, B., Levick, B., Norman, R., et al. (2017). Breaking beta: Deconstructing the parasite transmission function. *Philosophical Transactions of the Royal Society of London B: Biological Sciences*, 372: 20160084.

McLean, A.R. and Blower, S.M. (1993). Imperfect vaccines and herd immunity to HIV. *Proceedings of the Royal Society of London B: Biological Sciences*, 253: 9–13.

Miguel, E., Grosbois, V., Caron, A., Pople, D., Roche, B., and Donnelly C.A. (2020). A systemic approach to assess the potential and risks of wildlife culling for infectious disease control. *Communications Biology*, 3:14.

Miller, R., Kaneene, J.B., Fitzgerald, S.D., and Schmitt, S.M. (2003). Evaluation of the influence of supplementary feeding of white-tailed deer (*Odocoileus virginianus*) on the prevalence of bovine tuberculosis in the Michigan wild deer population. *Journal Wildlife Diseases*, **39**: 84–95.

More, S.J., Radunz, B., and Glanville, R.J. (2015). Lessons learned during the successful eradication of bovine tuberculosis from Australia. *Veterinary Record*, **177**:224–32.

Morens, D.M., Holmes, E.C., Davis, A.S., and Taubenburger, J.K. (2011). Global rinderpest eradication: Lessons learned and why humans should celebrate too. *Journal Infectious Diseases*, **204**:502–05.

Muller, J. (1971). The effect of fox reduction on the occurrence of rabies. Observations from two outbreaks of rabies in Denmark. *Bulletin de l'Office International Épizooties*, **75**:763–76.

Müller, T.F., Schuster, P., Vos, A.C., Selhorst, T., Wenzel, U.D., and Neubert, A.M. (2001). Effect of maternal immunity of the immune response to oral vaccination against rabies in young foxes. *American Journal Veterinary Research*, **62**:1154–58.

Munoz, N.E., Blumstein, D.T., and Foufopoulos, J. (2010). Immune system activation affects song and territorial defense. *Behavioral Ecology*, **21**:788–93.

Naranjo, V., Gortazar, C., Vincente, J., and De La Fuente, J. (2008). Evidence of the role of the European wild boar as a reservoir of *Mycobacterium tuberculosis* complex. *Veterinary Microbiology*, **127**:1–9.

Nishi, J.S., Elkin, B.T., and Ellsworth, T.R. (2006). The Hook Lake Bison Recovery Project. *Annals New York Academy of Science*, **969**:229–35.

Nol, P., Palmer, M.V., Waters, W.R., Aldwell, F.E., Buddle, B.M., Triantis, J.M., et al. (2008). Efficacy of oral and parenteral routes of *Mycobacterium bovis* Bacille Calmette-Guerin vaccination against experimental tuberculosis in white-tailed deer (*Odocoileus virginianus*): a feasibility study. *Journal Wildlife Diseases*, **44**:247–58.

Normile, D. (2008). Driven to extinction. *Science*, **319**: 1606–09.

Nugent, G. (2011). Maintenance, spillover, and spillback transmission of bovine tuberculosis in multi-host complexes: A New Zealand case study. *Veterinary Microbiology*, **151**:34–42.

O'Connor, C.M., Haydon, D.T., and Kao, R.R. (2012). An ecological and comparative perspective on the control of bovine tuberculosis in Great Britain and the Republic of Ireland. *Preventive Veterinary Medicine*, **104**: 185–97.

Octavia, S., Maharjan, R.P., Sintchenko, V., Stevenson, G., Reeves, P.R., and Gilbert, G.L. (2011). Insight into evolution of *Bordetella pertussis* from comparative genomic analysis: Evidence of vaccine-driven selection. *Molecular Biology Evolution*, **28**:707–15.

Olea-Popelka, F.J., Fitzgerald, P., White, P., McGrath, G., Collins, J.D., O'Keeffe, J. et al. (2009). Targeted badger removal and subsequent risk of bovine tuberculosis in cattle herds in county Laois, Ireland. *Preventive Veterinary Medicine*, **88**:178–84.

Ottesen, E.A., Dowdle, W.R., Fenner, F., Habermehl, K.-O., John, T.J., Koch, M.A., et al. (1998). Group report: How is eradication defined and what are the biological criteria? In: Dowdle, W.R., and Hopkins, D.R. (eds.), *The Eradication of Infectious Diseases*, pp. 47–59. John Wiley & Sons, Chichester.

Page, K., Beasley, J.C., Olsen, Z.H., Smyser, T.J., Downey, M., Kellner, K.F., et al. (2011). Reducing *Baylisascaris procyonis* roundworm larvae in raccoon latrines. *Emerging Infectious Diseases*, **17**:90–93.

Paull, S.H., Song, S., McClure, K.M., Sackett, L.C., Kilpatrick, A.M., and Johnson, P.T.J. (2012). From superspreaders to disease hotspots: Linking transmission across hosts and space. *Frontiers in Ecology and Evolution*, **10**:75–82.

Peterson, M.J., Grant, W.E., and Davis, D.E. (1991). Bison-brucellosis management: Simulation of alternative strategies. *Journal Wildlife Management*, **55**:205–13.

Pope, L.C., Butlin, R.K., Wilson, G.J., Woodroffe, R., Erven, K., Conyers, C.M., et al. (2007). Genetic evidence that culling increases badger movement: Implications for the spread of bovine tuberculosis. *Molecular Ecology*, **16**:4919–29.

Pound, J.M., Miller, J.A., George, J.E., and Fish, D. (2009). The United States Department of Agriculture Northeast Area-wide Tick Control Project: History and protocol. *Vector-borne Zoonotic Diseases*, **9**:365–70.

Prentice, J.C., Fox, N.J., Hutchings, M.R., White, P.C.L., Davidson, R.S., and Marion, G. (2019). When to kill a cull: Factors affecting the success of culling wildlife for disease control. *Journal of the Royal Society Interface*, **16**:20180901.

Ramsey, D.S.L., O'Brien, D.J., Cosgrove, M.K., Rudolph, B.A., Locher, A.B., and Schmitt, S.M. (2014). Forecasting eradication of bovine tuberculosis in Michigan whitetailed deer. *Journal Wildlife Management*, **78**:240–54.

Rees, E.E., Bélanger, D., Lelièvre, F., Cotè, N., and Lambert, L. (2011). Targeted surveillance of raccoon rabies in Quebec, Canada. *Journal Wildlife Management*, **75**:1406–16.

Renwick, A.R., White, P.C.L., and Bengis, R.G. (2006). Bovine tuberculosis in southern African wildlife: A multi-species host-pathogen system. *Epidemiology Infection*, **135**:529–40.

Resasco, J., Bitters, M.E., Cunningham, S.A., Jones, H.I., McKenzie, V.J., and Davies, K.F. (2019). Experimental habitat fragmentation disrupts nematode infections in Australian skinks. *Ecology*, **100** (1–8):e02547.

Roberts, M.G. and Heesterbeek, J.A.P. (2003). A new method for estimating the effort required to control an infectious disease. *Proceedings of the Royal Society of London B: Biological Sciences*, **270**:1359–64.

Rocke T. and Friend, M. (1999). Botulism. In: Friend, M. and Franson, J.C. (eds.), *Field Manual of Wildlife Diseases. General Field Procedures and Diseases of Birds* (No. ITR-1999-001), pp. 271–281. Geological Survey, Biological Resources Division, Madison, WI.

Rosatte, R., Donovan, D., Allan, M., Howes, L.-A., Silver, A., Bennett, K., et al. (2001). Emergency response to raccoon rabies introduction into Ontario. *Journal Wildlife Diseases*, **37**:265–79.

Rossi, S., Artois, M., Pontier, D., Crucière, C., Hars, J., Barrat, J., et al. (2005). Long-term monitoring of classical swine fever in wild boar (*Sus scrofa* sp.) using serological data. *Veterinary Research*, **36**:27–42.

Schmidt, R.L., Hibler, C.P., Spraker, T.R., and Rutherford, W.H. (1979). An evaluation of drug treatment for lungworm in bighorn sheep. *The Journal of Wildlife Management*, **43**:461–67.

Schouls, L.M., Van Der Heide, H.G.J., Vauterin, L., Vauterin, P., and Mooi, F.R. (2004). Multiple-locus variable-number tandem repeat analysis of Dutch *Bordetella pertussis* strains reveals rapid genetic changes with clonal expansion during the late 1990s. *Journal Bacteriology*, **186**:5496–505.

Scurlock, B.M. and Edwards, W.H. (2010). Status of brucellosis in free-ranging elk and bison in Wyoming. *Journal Wildlife Diseases*, **46**:442–49.

Selhorst, T., Müller, T., Schwermer, H., Zuller, M., Schlütter, H., Beitenmoser, U., et al. (2005). Use of an area index to retrospectively analyze the elimination of fox rabies in European countries. *Environmental Management*, **35**: 292–302.

Shury, T.K. and Bergeson, D. (2011). Lesion distribution and epidemiology of *Mycobacterium bovis* in elk and white-tailed deer in south-western Manitoba, Canada. *Veterinary Medicine International*, 591980, doi:10.4061/2011/591980.

Skrjabin, K.I. (1970). Preventive measures against the spreading of helminthiasis among game animals. *Transactions International Congress Game Biologists*, **9**: 54–58.

Smith D.L., Lucey, B., Waller, I.A., Childs, J.E., and Real, L.A. (2002). Predicting the spatial dynamics of rabies epidemics on heterogeneous landscapes. *Proceedings of the National Academy of Sciences*, **99**: 3668–72.

Smith, G.C. and Cheeseman C.L. (2002). A mathematical model for the control of diseases in wildlife populations: Culling, vaccination and fertility control. *Ecological Modelling*, **150**:45–53.

Smyser, T.J., Page, L.K., Johnson, S.A., Hudson, C.M., Kellner, K.F., Swihart, R.K., et al. (2013). Management of raccoon roundworm in free-ranging raccoon populations via anthelmintic baiting. *Journal Wildlife Management*, **77**:1372–79.

Snow, L.C., Newson, S.E., Musgrove, A.J., Cranswick, P.A., Crick, H.P.Q., and Wilesmith, J.W. (2007). Risk-based surveillance for H5N1 avian influenza virus in wild birds in Great Britain. *Veterinary Record*, **161**:775–81.

Steck, F., Wandeler, A., Bichsel, P., Capt, S., Hafliger, U., and Schneider, L. (1982). Oral immunization of foxes against rabies, laboratory and field studies. *Comparative Immunology Microbiology Infectious Diseases*, **5**:165–71.

Sterner, R.T. and Smith, G.C. (2006). Modelling wildlife rabies: Transmission, economics, and conservation. *Biological Conservation*, **131**:163–79.

Storm, D.J., Samuel, M.D., Rolley, R.E., Shelton, P., Keuler, N.S., Richards, B.J., et al. (2013). Deer density and disease prevalence influence transmission of chronic wasting disease in white-tailed deer. *Ecosphere*, **4**:10. http://dx.doi.org/10.1890/ES12-00141.1.

Streicker, D.G., Recuenco, S., Valderrama, W., Benavides, J.G., Vargas, I., Pacheco, V., et al. (2012). Ecological and anthropogenic drivers of rabies exposure in vampire bats: Implications for transmission and control. *Proceedings of the Royal Society of London B: Biological Sciences*, **279**:3384–92.

Thompson, A.K., Samuel, M.D., and Van Deelen, T.R. (2008). Alternative feeding strategies and potential disease transmission in Wisconsin white-tailed deer. *Journal Wildlife Management*, **72**:416–21.

Thulke, H.-H., Eisinger, D., Freuling, C., Frölich, A., Globig, A., Grimm, V., et al. (2009). Situation-based surveillance: Adapting investigations to actual epidemic situations. *Journal Wildlife Diseases*, **45**:1089–100.

Thulke, H.-H., Selhorst, T., Müller, T., Wyszomirski, T., Müller, U., and Breitenmoser, U. (2004). Assessing anti-rabies baiting What happens of the ground. *BMC Infectious Diseases*, **4**:9. http://www.biomedcentral.co/1471-2334/4/9 (accessed August 3, 2018).

Thulke, H.-H., Tischendorf, L., Staubach, C., Selhorst, T., Jeltsch, F., Müller, T., et al. (2000). The spatio-temporal dynamics of a post-vaccination resurgence of rabies in foxes and emergency vaccination planning. *Preventive Veterinary Medicine*, **47**:1–21.

Tischendorf, L., Thulke, H.-H., Staubach, C., Müller, M.S., Jeltsch, F., Goretski, J., et al. (1998). Chance and risk of controlling rabies in large-scale and long-term immunized fox populations. *Proceedings of the Royal Society of London B: Biological Sciences*, **265**:839–46.

Traoré, S., Wilson, M.D., Sima, A., Barro, T., Diallo, A., Aké, A., et al. (2009). The elimination of the onchocerciasis vector from the island of Bioko as a result of larviciding by the WHO African Programme for Onchocerciasis Control. *Acta Tropica*, **111**:211–18.

Treanor, J.J., Geremia, C., Crowly, P.H., Cox, J.J., White, P.J., Wallen, R.L., et al. (2011). Estimating probabilities of active brucellosis infection through quantitative serology and culture. *Journal Applied Ecology*, **48**:1324–32.

Vial, F. and Donnelly, C.A. (2012). Localized reactive badger culling increases risk of bovine tuberculosis in nearby cattle herds. *Biology Letters*, **8**:50–53.

Vreysen, M.J.B., Saleh, K.M., Ali, M.Y., Abdulla, A.M., Zhu, Z.-R., Juma, K.G., et al. (2000). *Glossinia austeni* (Diptera: Glossinidae) eradicated on the Island of Unguja, Zanzibar, using the sterile insect technique. *Veterinary Entomology*, **93**:123–35.

Vreysen, M.J.B., Seck, M.S., Sall, B., and Bouyer, J. (2013). Tsetse flies: Their biology and control using area-wide integrated pest management approaches. *Journal Invertebrate Pathology*, **112**:S15–S25.

Walsh, D.P. and Miller, M.W. (2010). A weighted surveillance approach for detecting chronic wasting disease foci. *Journal of Wildlife Diseases*, **46**:118–35.

Wandeler, A., Muller, J., Wachendorfer, G., Schale, W., Forster, U., and Steck, F. (1974). Rabies in wild carnivores in central Europe III. Ecology and biology of the fox in relation to control operations. *Zentralblatt Veterinärmedizin*, **B21**:765–73.

Warburton, B. and Thomson, C. (2002). Comparison of three methods of maintaining possums at low density. *Science for Conservation*, **189**:20.

Wasserburg, G., Osnas, E.E., Rolley, R.E., and Samuel, M.D. (2009). Host culling as an adaptive management tool for chronic wasting disease in white-tailed deer: A modelling study. *Journal Applied Ecology*, **46**: 457–66.

Weaver, G.V., Domeniech, J., Thiermann, A.R., and Karesh, W.B. (2013). Foot and mouth disease: A look from the wild side. *Journal Wildlife Diseases*, **49**:759–85.

Wolfe, L.L., Miller, M.W., and Williams, E.S. (2004). Feasibility of "test-and-cull" for managing chronic wasting disease in urban mule deer. *Wildlife Society Bulletin*, **32**:500–05.

Woodroffe, R., Donnelly, C.A., Cox, D.R., Bourne, F.J., Cheeseman, C.L., Delahay, R.J., et al. (2006a). Effects of culling on badger *Meles meles* spatial organization: Implications for the control of bovine tuberculosis. *Journal Applied Ecology*, **43**:1–10.

Woodroffe, R., Donnelly, C.A., Jenkins, H.E., Johnston, W.T., Cox, D.R., Bourne, F.J., et al. (2006b). Culling and cattle controls influence tuberculosis risk for badgers. *PNAS*, **103**:14713–17.

Woolhouse, M. (2011). How to make predictions about future infectious disease risks. *Philosophical Transactions of the Royal Society of London B: Biological Sciences*, **366**:2045–54.

Woolhouse, M.E.J., Haydon, D.T., and Bundy, D.A.P. (1997). The design of veterinary vaccination programs. *Veterinary Journal*, **153**:41–47.

Wyss, J.H. (2000). Screwworm eradication in the Americas. *Annals New York Academy Sciences*, **916**:186–93.

CHAPTER 14

Disease Control: How to Live with Infection

You have to compromise all the way—the only thing that counts is the result.

Richard Widmark, US actor

14.1 Introduction

Much of this book's focus has been on the control and eradication of infectious agents that can have deleterious consequences for native biodiversity. Unfortunately, in the vast majority of cases where a wildlife disease is well established in a wildlife population, it will not be possible to eliminate it. Chapter 13 describes the handful of examples of widespread and well-established pathogens that were eliminated from wildlife populations on anything other than a local scale. In each of these cases, the pathogen concerned was a significant threat to either human or livestock populations, meaning that very considerable financial resources were available. We know of no example of successful elimination of a pathogen that is circulating in free-ranging wildlife species and that does not also infect humans or domesticated animals. The management strategy for "wildlife-only" pathogens must therefore be to enable the one or more species threatened by a pathogen to survive in the continued presence of the disease and to mitigate any consequent effects for the rest of the ecological community.

What remains less clear is how to proceed with managing an epizootic about which little is often known. A possible roadmap is provided by a deci-

sion tree showing possible management strategies when broad-scale elimination of a parasite or pathogen is not possible (Figure 14.1). The pathogen is not eradicated but through appropriate management the two organisms can continue coexisting without catastrophic impacts on the host species. This approach is based on the recognition of the fact that no host–parasite interaction exists in a uniform and completely connected environment—exploiting these heterogeneities in transmission means that it may be possible to control the disease in subsections of the range of the species. Even for well-established pathogens, there likely exist pockets of the host distribution that the pathogen has yet to reach, and there may also be relatively isolated subpopulations from which it may be possible to eliminate the pathogen and then prevent reestablishment. From a conservation perspective, such disease-free subpopulations can then serve as "insurance" populations to safeguard the survival of the host species. The strategies described in Chapter 12 to prevent disease arrival and/or those described in Chapter 13 to eliminate disease can therefore be applied in the appropriate subpopulations. Whether and where this is feasible will depend on the costs of eradication and continued

Infectious Disease Ecology and Conservation. Johannes Foufopoulos, Gary A. Wobeser and Hamish McCallum, Oxford University Press.
© Johannes Foufopoulos, Gary A. Wobeser and Hamish McCallum (2022). DOI: 10.1093/oso/9780199583508.003.0014

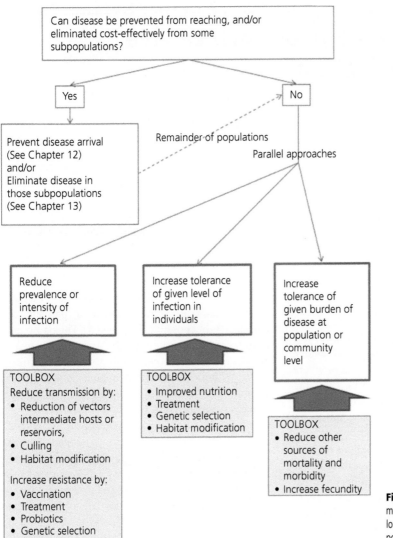

Figure 14.1 A decision tree for managing infectious disease once it can no longer be eradicated from a wildlife population.

isolation, and on the risk of pathogen reestablishment.

For the remainder of the host populations for which pathogen eradication is not feasible, the broad strategy is to increase either one, or both, of resistance or tolerance to infection, at the individual or population level. Although they are often conflated, *resistance* to infection (which is the ability to limit the level of infection) is different from *tolerance* of infection, which is the ability to survive, reproduce, and grow with a given level of infection (Råberg et al. 2009; Medzhitov et al. 2012). From the point of view of population dynamics, strategies to increase tolerance include any effort that diminishes the adverse consequences of infection on an individual's fecundity or survival. Similarly, at the population level, any strategy that increases population viability, at a given level of parasite or pathogen infection, can be regarded as increasing a species' tolerance.

There are three approaches that potentially can be taken to permit continued persistence of a threatened host species in the face of the ongoing presence of the disease. These are: (i) to reduce the *prevalence* or *intensity of infection*; (ii) to increase the *tolerance* of infected individuals to a given level of infection; and (iii) to increase the ability of the population to persist in the presence of a given *burden of disease* in individuals, by alleviating the impacts of other stressors or mortality sources it may be facing. The decision tree represents these as parallel approaches, as no approach has particular precedence over another and they can potentially be applied simultaneously, depending on the practicalities of the particular system under investigation.

14.2 Reducing prevalence or intensity of infection

The most obvious way to reduce the impact of a pathogen on a population is to reduce the prevalence or the intensity of the infection. Broadly, either one, or both of these can be achieved by two approaches: first, by reducing the infectious contact rate—the probability that an infective host or pathogen transmission stage encounters susceptible hosts; or, second, it may be possible to increase host *resistance*—to decrease the proportion of infective particles that, once arrived at an uninfected host, are actually able to progress into new infections (McCallum et al. 2017). Most of the approaches described in Chapters 12 and 13, even if they are unable to entirely eliminate disease from a particular population, can reduce disease intensity or prevalence and are therefore relevant to the "living with infection" approach described in this chapter.

14.2.1 Culling

Where a pathogen affects only one species, selectively culling a small number of highly infected individuals (superspreaders; Lloyd-Smith et al. 2005) could reduce the amount of transmission. Focusing on superspreaders only has the advantage of making a considerable dent in the epizootic while removing relatively few individuals, an aspect that is particularly attractive for endangered species. This approach relies, however, on three preconditions. First, there must indeed be individuals that are superspreaders. Second, it must be possible to identify these individuals in field conditions. Third,

it must be possible to remove them at a rate high enough to reduce prevalence while not otherwise compromising population viability. If these criteria can be met, then superspreader removal is an attractive choice, especially because it has the added benefit of removing genetically susceptible individuals and therefore selecting for resistance to the pathogen in the host population. On the other hand, superspreaders may be individuals that are particularly tolerant of infection (Gopinath et al. 2014) and able to persist with infection, so removal of superspreaders may select against tolerance.

Culling is often considered as a strategy for disease eradication. In principle, however, the objective may be to reduce prevalence or intensity of infection, recognizing that eradication is impractical (Prentice et al. 2019). There has only been limited modeling of when culling is sufficient to stave off the disease-caused extinction of an infected host population, yet not sufficient to eliminate disease. In the case of Tasmanian devil facial tumor disease, modeling suggested that there is a relatively small difference between the rate of culling sufficient to prevent devil extinction and that which would eliminate the disease itself (see Figure 13.4).

Disease reduction culling that does not result in elimination needs to be maintained indefinitely, reducing its attractiveness as a management option (Beeton and McCallum 2011; and Chapter 13). There is some evidence that hunting has potential to control chronic wasting disease in deer, but its effectiveness depends on the age classes targeted by hunters (Miller et al. 2020). In general, while there are several types of culling strategies (see also Chapter 13, Section 13.4), their efficiency and practicality depend greatly on the particularities of the transmission. Under most circumstances, however, culling appears to be a difficult management option to implement well, and present thinking is that it should only be utilized when population modeling indicates that it is likely to be successful (Beeton and McCallum 2011; Prentice et al. 2019).

14.2.2 Chemotherapy.

Reducing the population of parasites living in wild hosts through chemotherapy is—in principle—an

(a)

(b)

Figure 14.2 (a) Bare-nosed wombat (*Vombatus ursinus*) infected with sarcoptic mange (*Sarcoptes scabiei*). Note loss of fur and crusted skin on the rear flank. Infection with sarcoptic mange leads to increased metabolic rates, as well as high rates of heat loss through reduced fur insulation. Infected animals have a difficult time meeting their elevated energetic demands because they don't forage as much as healthy animals. (Photo: Scott Carver.) (b) Entrance of a wombat burrow. The treatment station installed above a burrow includes a moveable flap that is able to deliver a topical antiparasitic treatment to a passing animal. (Source: Old et al., 2018; Photo: Scott Carver.)

effective way to reduce overall disease transmission. Many efficacious antiparasitic agents have been developed to date, but the great challenge remains to deliver the drug to wild animals in a frequency and amount that will ensure parasite control. In situations like this, it is often useful to take advantage of a host's behavior and physiology. For example, several species of Galápagos Darwin's finches are presently threatened by an invasive ectoparasite, the fly *Philornis downsi*, whose larvae feed on finch nestlings (Fessl et al. 2006, 2010). Several approaches have been investigated to control the parasite; they include the use of microbial attractants (Cha et al. 2016), fly mating disruption using pheromones (Causton et al. 2011), and biological control agents (Bulgarella et al. 2017). Following the realization that the birds will utilize natural plant repellents to reduce their own ectoparasite load (Cimadom et al. 2016), researchers started offering the birds permethrin-treated cotton balls, which the birds incorporated into their nest structure, therefore significantly reducing the mortality of their nestlings (Knutie et al. 2014). Similarly, Alves et al. (2020) increased nestling survival 10-fold in the endangered forty-spotted pardalote (*Pardalotus quadragintus*) in Tasmania by providing insecticide-treated feathers for nest building.

Sarcoptic mange is a severe skin disease caused by a mite (*Sarcoptes scabiei*), which has been responsible for severe declines in a number of wild mammal populations (Rowe et al. 2019). While it is treatable with chemotherapeutic agents, delivered by injection, orally, or topically, administering these agents to free-ranging wildlife is difficult (Rowe et al. 2019). Bare-nosed wombats (*Vombatus ursinus*) are large burrowing marsupials endemic to south-eastern Australia (including Tasmania). The species is infected with sarcoptic mange (*Sarcoptes scabiei*), introduced with domestic mammals (Old et al. 2018, Figure 14.2a). The infection produces severe effects on wombat physiology and behavior and has resulted in catastrophic declines in some parts of the species' range (Martin et al. 2018a, 2018b). A relatively novel approach to treating the animals in the wild is to install plastic flaps that deliver an antiparasitic agent at the entrance of wombat burrows (see Figure 14.2b). Because wombats revisit their burrows on a regular basis, it is possible to ensure that infected animals receive regular treatment with antimange agents. This approach has the advantage of not necessitating capture. A trial of this method successfully reduced the level of sarcoptic mange infection within a population of wombats at Narawntapu National Park in northern Tasmania, but did not eradicate the mite infection (Martin et al. 2019). This may have been due to not all wombats being treated and also because of the likelihood that mites persist in the burrows. In general, treatment of wild animals with antiparasitic agents tends to miss a substantial fraction of the population and is therefore likely not to achieve parasite elimination. Another concern is that the long-term application of drugs is likely to cause the evolution of resistance in the parasite to the chemical agent. Still, it is possible that if implemented properly, it can become a stable, long-term support method helping to mitigate the impacts of a parasitic infection on the wildlife.

14.2.3 Managing species other than the focal host

In some cases, reducing contact rates in a focal species of conservation interest means managing other species in the area. Most parasites and pathogens that threaten to cause host extinction infect more than one host species and may have life cycles that involve vectors or intermediate hosts. The force of infection on a threatened species may therefore be reduced by managing populations of vectors, reservoir species, or intermediate hosts, or managing disease in those populations. For example, on the Hawaiian Islands, one way to reduce malaria in endemic birds is to control exotic mosquito (*Culex quinquefasciatus*) populations that transmit the disease. Because mosquitoes breed in nutrient-rich wallows created by feral pigs (Katahira et al. 1993), one out of several benefits of sustained pig control programs is thought to be the reduction in avian malaria infections (Stone and Loope 1987; Hess et al. 2006). Another well-investigated case involves lions (*Panthera leo*) in Serengeti National Park (Tanzania), which are threatened by irregular outbreaks of canine distemper and rabies (Roelke-Parker et al. 1996; Craft et al. 2009) that appear to occur by spillover from canids, including feral and domestic dogs around the park boundaries. Vaccination of these dog populations has therefore been proposed as a potential strategy to manage disease in lions and other carnivores in the Serengeti (Cleaveland et al. 2007). While rabies can apparently be controlled in this fashion, this is more difficult for canine distemper (see Chapter

12, Section 12.2.1.2). Even large-scale and intensive domestic dog vaccination programs appear to be less likely to eliminate canine distemper virus in lions in the Serengeti ecosystem, although vaccination campaigns can reduce the amplitude and severity of disease outbreaks in lions (Viana et al. 2015).

In an analogous situation, wildlife biologists have tried to protect prairie dogs (*Cynomys* sp.) and black-footed ferrets (*Mustela nigripes*)—an endangered, specialized prairie dog predator—from devastating sylvatic plague (*Yersinia*) outbreaks (see also Chapter 6, Section 6.4). While the infection is most noticeable during the epidemic phase with fulminating lethal outbreaks in prairie towns, the pathogen continues to persist both in prairie dog and ferret populations between outbreaks, posing therefore significant disease management challenges. Transmission reductions through flea vector control or vaccination were both shown to be clearly beneficial to prairie dog populations (Biggins et al. 2010; Abbott et al. 2012). A similar study in black-footed ferret populations had the same results (Matchett et al. 2010), suggesting that these are viable applications to help reduce the disease burden in these species.

14.3 Increasing resistance or tolerance

Reduction of other stressors on individuals is likely to increase their tolerance of a given level of infection (Scheele et al. 2014), but as described earlier, there may also be benefits in terms of increased resistance. Because of the similarities between these two approaches, they are discussed jointly in the text that follows.

14.3.1 Food supplementation.

Several strategies may be effective in increasing resistance of threatened species to infection by pathogens. Nutritional stress can suppress both the innate and adaptive immune systems, as seen in humans (Kau et al. 2011), as well as in wildlife (Kessler et al. 2018). Providing supplementary feed can therefore be an effective strategy to increase resistance in wildlife (Wiehn and Korpimäki 1998).

However, supplementary feeding has likely multiple, opposing effects on disease transmission and its impact is not straightforward. For example, providing artificial food sources often leads to animal aggregation, which in turn can intensify disease transmission. Although these increased contacts can be critically important in driving disease transmission, as shown in the case of mycoplasmal conjunctivitis (caused by *Mycoplasma gallisepticum*) in house finches (*Carpodacus mexicanus*) (Hochachka and Dhondt 2000; Dhondt et al. 2005), other factors, such as frequency of use and type of feeder, are also influential (Hartup et al. 1998; Adelman et al. 2015). The type of supplemental food is also significant, because provisioned food may on the one hand contain fewer infective stages than wild food (Hartup et al. 1998), but may lack micronutrients or protein, therefore leading to increased susceptibility to pathogens (Becker et al. 2015). As a result, supplemental feeding can have mixed effects on wildlife as was seen in the case of elk (*Cervus elaphus*) near Yellowstone National Park. Winter feeding stations resulted in more infectious contacts and elevated gastrointestinal nematode egg counts in food-supplemented elk (Hines et al. 2007). Intriguingly, however, these authors also found that artificially fed elk had lower burdens of gastrointestinal worm infection in early spring, possibly attributable to increased resistance to infection of animals with improved nutritional status or the lack of parasite eggs in the feed provided.

14.3.2 Vaccination.

Vaccination is perhaps the most obvious way to increase resistance in animals threatened by pathogens and has been employed with varying degrees of success in a number of situations. Examples include the bait-delivered oral vaccine that has eliminated fox rabies from Western Europe (see Chapter 13) and the controversial use of an injectable rabies vaccine to control rabies in African wild dogs (Burrows et al. 1995; Vial et al. 2006). While the objective of vaccination programs is often to reduce R_0 to below one and thus eliminate disease, much lower levels of vaccination still may still increase the probability that a wildlife population threatened by disease can persist. For example,

in a study of endangered Ethiopian wolves threatened by rabies, Haydon et al. (2006) used an epidemiological model to show that either (i) targeted vaccination of a limited number of select wolves residing in narrow habitat corridors connecting subpopulations, or (ii) reactive general vaccinations of relatively few individuals following small disease outbreaks, can be effective strategies to reduce the risk of disease-induced extinction (Figure 14.3). There have been proposals to vaccinate great apes against a number of human diseases, but as yet such vaccination programs have been small-scale and ad hoc (Ryan and Walsh 2011).

Vaccines can also be used with the objective of increasing tolerance—reducing the extent of disease in infected animals—rather than blocking transmission. For example, vaccines under development for chlamydial disease in koalas have not been shown clearly to reduce the probability of becoming infected, but there is some evidence that they reduce the extent of clinical disease when administered to infected koalas (Waugh et al. 2016; Nyari et al. 2019; Waugh and Timms 2020). However, "imperfect" vaccines that do not block transmission have the potential to select for increased virulence in pathogens, by relaxing the selection against highly virulent strains that occurs because the death of a host usually leads to cessation of transmission (Read et al. 2015).

14.3.3 Probiotics and related approaches.

An intriguing development is the possibility of using probiotics—other biological organisms—to increase resistance to infection. This has been proposed for the amphibian chytrid fungus (Woodhams et al. 2011; Bletz et al. 2013; Woodhams et al. 2016). Initially it was thought that bacteria had the most potential as probiotic agents to limit chytrid infection, but more recent work suggests that fungi may have even greater potential to inhibit amphibian chytrid infection (Kearns et al. 2017). A limited field trial in a high-altitude lake in the Sierra Nevada suggested that individual *Rana muscosa* treated with the bacterium *Janthinobacterium lividum* had lower chytrid burdens than control frogs (Vredenburg et al. 2011).

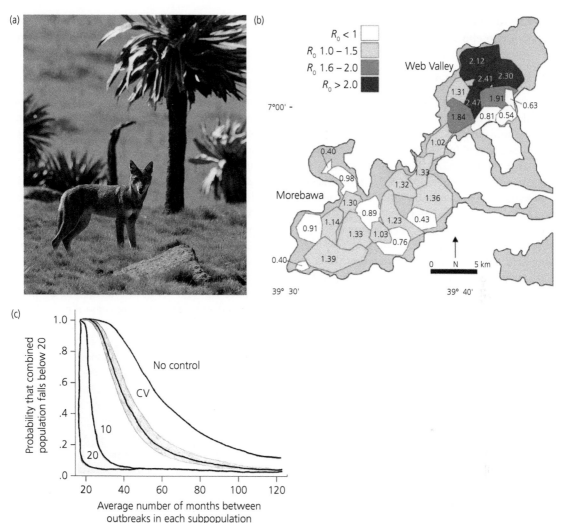

Figure 14.3 (a) Endangered Ethiopian wolf (*Canis simiensis*) in the Simien Mountains (Ethiopia). (Image by: Harri J.; photo cropped and modified from original.) (b) A map of estimated R_0 for rabies in Ethiopian wolves in the Bale Mountains of Ethiopia assuming intermediate disease transmission rates. Each wolf pack is shown by a closed polygon. Shade of filling indicates calculated R_0 for each pack. The Web Valley has higher local values of R_0 than Morebawa, with the two regions of suitable habitat connected by a relatively narrow corridor (center). (From Haydon et al. 2006.) (c) Model projections of the effect of diverse vaccination strategies on the overall probability of decline against frequency of outbreaks in Ethiopian wolves. The top black line shows the predicted poor outcome in the absence of any vaccination. The middle line with the associated gray error zones shows the effect of several corridor vaccination (CV) coverages in the habitat corridor connecting the two wolf subpopulations. The two innermost lines (10, 20) show the effect of reactive vaccination implemented after the death of 10 individuals with coverage of 10% and 20%. The CV approach offers good protection while only requiring the vaccination of few individuals. (Figure modified from Haydon et al. 2006.)

Interactions between parasites are increasingly being recognized as important and may be either synergistic or antagonistic (Sousa 1993). These may potentially be utilized to manipulate resistance. For example, it is known that intestinal nematodes increase susceptibility in African buffalo to *Mycobacterium tuberculosis*, which causes bovine tuberculosis (BTB). It may therefore be possible to confer increased resistance to BTB by controlling nematodes (Joseph et al. 2013).

14.3.4 Genetic management.

Pronounced differences exist among individual hosts (Hawley et al. 2005) as well as between different subpopulations (Adelman et al. 2013) in terms of their susceptibility to infection. As discussed in Chapter 2, Section 2.2.4, there is strong selective pressure on hosts to develop resistance to parasites or pathogens that have substantial effects on survival or fecundity. There are also selective pressures on pathogens, which usually, but not always, lead to highly pathogenic parasite or pathogen genotypes being selected against, because they are likely to kill their host before they are able to infect another (Delaney et al. 2012). The outcome of the introduction of the myxoma virus into rabbit populations in Australia and Europe is the best-known example of these host–pathogen coevolutionary processes and is discussed in detail in other chapters (Chapters 3, Section 3.2 and Chapter 15, Section 15.2). It is important to take such evolutionary processes into account when managing the emergence of infectious disease in a population of conservation significance. In some circumstances, it may be possible to accelerate the process of evolution of resistance in the host population, and in other circumstances it is important to ensure that management practices do not inhibit or reverse the natural evolution of resistance or tolerance.

Evolution of resistance or tolerance to pathogens that have threatened to cause extinction has occurred in a number of situations. Multiple lines of evidence indicate that Tasmanian devils may be rapidly evolving some resistance to Tasmanian devil facial tumor disease, despite very low levels of genetic diversity in the species (Epstein et al. 2016; Margres et al. 2018a, 2018b). Some frog populations in Panama have recovered following chytrid-induced declines (Figure 14.4). This appears to be a result of the evolution of increased resistance or tolerance in the frogs, rather than attenuation of the pathogen (Voyles et al. 2018). The native Hawaiian avifauna has been severely affected by avian malaria and bird pox (see also Chapter 2, Section 2.3.1, Chapter 2, Sections 3.2.3 and 3.5.1, and Chapter 7,

Section 7.3.2). There is, however, evidence of recovery of some populations in the presence of malaria infection. Some Hawaii Amakihi (*Hemignathus virens*) populations at low elevations on the island of Hawaii show evidence of recent population increases despite high prevalence of avian malaria (Woodworth et al. 2005). Despite high infection rates (24–40% by microscopy; 55–83% by serology), population densities at low altitudes were higher than those at disease-free high altitudes (Figure 14.5) and parasitemia at low altitudes remained relatively modest. Taken together, this evidence suggests that the Amakihi at low elevations are able to persist in the presence of the disease most likely because they are able to immunologically suppress the parasites to low infection burdens.

It may be possible to accelerate the evolution of resistance through artificial selection, or by managing wild populations to facilitate the evolution of resistance. For example, Scheele et al. (2014) have proposed that it may be possible to select for increased resistance to the amphibian chytrid fungus in captive-bred frogs and then release these resistant genotypes into wild populations. However, such artificial selection is expensive and would need to occur over multiple generations. Although it has been proposed regularly, we know of no case where artificial selection has been used successfully to reintroduce resistant genotypes into populations of wild animals threatened by disease.

A more realistic option may be to facilitate such evolution in the wild. For example, Kilpatrick (2006) has advocated that control of rodents—a major source of mortality for the Hawaiian avifauna—could facilitate the evolution of malaria resistance in several species of endangered Hawaiian birds. The proposed mechanism is that, by increasing the survival (and thus the reproductive output) of those birds with some resistance to avian malaria, these resistance alleles will spread more rapidly through the wild populations. Furthermore, Kilpatrick argues that better overall conservation outcomes would be achieved by targeting rodent control in areas of moderate malaria transmission, rather than at higher altitudes where malaria transmission is lower.

An alternative strategy to reduce the impact of avian malaria on native Hawaiian songbirds could

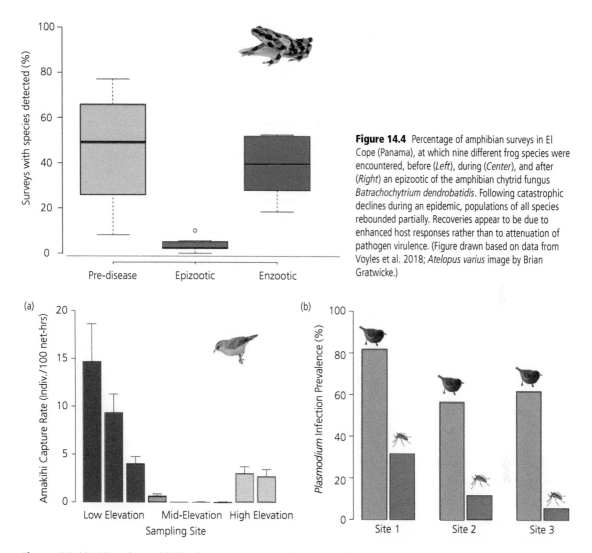

Figure 14.4 Percentage of amphibian surveys in El Cope (Panama), at which nine different frog species were encountered, before (*Left*), during (*Center*), and after (*Right*) an epizootic of the amphibian chytrid fungus *Batrachochytrium dendrobatidis*. Following catastrophic declines during an epidemic, populations of all species rebounded partially. Recoveries appear to be due to enhanced host responses rather than to attenuation of pathogen virulence. (Figure drawn based on data from Voyles et al. 2018; *Atelopus varius* image by Brian Gratwicke.)

Figure 14.5 (a) Evidence that amakihi (*Hemignathus virens*; a Hawaiian tree creeper) can survive in the presence of avian malaria. Relative abundance of amakihi at nine sites along and altitudinal gradient on the island of Hawaii. Mosquito vectors are present at the low- and mid-altitude sites but not at the high-attitude sites. The species is more common at lower elevations despite the presence of malaria. (Adapted and simplified from Woodworth et al. 2005; amakihi image by James Brennan; modified.) (b) Prevalence (% infected samples) of avian malaria (*Plasmodium relictum*) among amakihi (green bars) and mosquitoes (orange bars) at the three aforementioned low-altitude sites, showing that the birds persist despite the high prevalence of malaria in both vertebrate and invertebrate hosts. (Adapted and simplified from Woodworth et al. 2005; amakihi image by James Brennan; modified.)

be to translocate disease-resistant or tolerant individuals from low-altitude populations to highly susceptible mid-altitude populations. However, the effectiveness of the strategy depends on whether it is resistance or tolerance that is responsible for the recovery of some populations at low altitudes. Hobbelen et al. (2012) used a simulation model to show that translocating malaria-tolerant Amakihi from low to mid-altitudes would be likely to have a positive effect on the populations of Amakihi themselves, but tolerant hosts might also increase the prevalence of malaria infection in mosquitoes, leading to detrimental effects on other bird species. Nevertheless, provided that appropriate precautions are taken, facilitating the spread of resistance or tolerance alleles by translocating animals from locations where disease is endemic to those in which it has arrived more recently is a strategy worth considering.

If it is not possible to employ strategies to select actively for increased resistance in a wild population, it is important to avoid management strategies that may inadvertently slow down or arrest the evolution of resistance to a pathogen. For example, as mentioned earlier, Tasmanian devils appear to be rapidly evolving effective immune responses against Tasmanian devil facial tumor disease (Epstein et al. 2016, Margres et al. 2018a; 2018b), despite low overall genetic diversity (Jones et al. 2004; Lachish et al. 2011; Bruniche-Olsen et al. 2014). This raises issues important for genetic management, which are likely to be common to many situations where disease threatens an endangered species. Maintenance of genetic diversity by translocation is a key part of many management plans for small and isolated populations (Frankham 2015). However, translocation of animals from disease-free populations into diseased populations also dilutes the effect of resistant individuals in the genepool and is likely to delay, arrest, or even reverse evolution of resistance (Hohenlohe et al. 2019).

14.4 Increasing tolerance of infection at a population level

Based on first principles, a pathogen will cause the extinction of a given host species if the level of infection in the host population is sufficient to elevate the cumulative average host death rate above the average host birth rate. If it is not possible to eliminate a parasite or pathogen from a population, it may be possible to manage the host–parasite interaction so that the average host birth rate remains above the average host death rate. Alternatively, if this cannot be achieved, it may be possible to reduce the other nondisease sources of mortality, in which case the overall population growth rate will rise. For example, koalas (*Phascolarctos cinereus*) are known to be infected by bacteria in the family *Chlamydiacae*. These infections may lead to urogenital, respiratory, and ocular disease, resulting in increased mortality and decreased fecundity (Weigler et al. 1988; Polkinghorne et al. 2013). Although

some koala populations have few signs of clinical disease in the face of high levels of chlamydia (Weigler et al. 1988; Ellis et al. 1993; White and Timms 1994), therefore suggesting some tolerance of infection, it is generally thought that these chlamydial infections are a major agent of koala population declines (Rhodes et al. 2011). However, koalas are also subject to a range of other threatening processes, including habitat loss, urbanization, and edge effects, which manifest themselves in the form of high mortality from dog attacks and vehicular collisions. Reducing these additional sources of mortality will almost certainly enable koala populations to be more tolerant of reductions in fecundity and increases in mortality due to chlamydial infection (Grogan et al. 2018; McCallum et al. 2018).

Supplementation of a diseased host population by releasing additional animals, either from disease-free wild populations or with captive-reared animals, can be regarded as a means of increasing tolerance at a population level. It is, however, controversial. As discussed earlier, releasing naive individuals has the potential to slow down or reverse the evolution of resistance in the diseased population. The southern corroboree frog (*Pseudophryne corroboree*) is a brightly colored and charismatic frog species restricted to the alpine areas of southeastern Australia. It is threatened with extinction by the amphibian chytrid fungus, and its continued existence relies on an extensive captive breeding program. Some wild populations are currently maintained only by continual releases of captive-reared animals (Canessa et al. 2014). Aside from issues of cost-effectiveness, maintaining the presence of a wild population by continued introduction of captive-reared animals, which then succumb to disease, raises difficult ethical issues.

References

Abbott, R.C., Osorio, J.E., Bunck, C.M., and Rocke, T.E. (2012). Sylvatic plague vaccine: A new tool for conservation of threatened and endangered species? *EcoHealth*, 9:243–50.

Adelman, J.S., Kirkpatrick, L., Grodio, J.L., and Hawley, D.M. (2013). House finch populations differ in early inflammatory signaling and pathogen tolerance at the peak of *Mycoplasma gallisepticum* infection. *The American Naturalist*, 181:674–89.

Adelman, J.S., Moyers, S.C., Farine, D.R., and Hawley, D.M. (2015). Feeder use predicts both acquisition and transmission of a contagious pathogen in a North American songbird. *Proceedings of the Royal Society of London B: Biological Sciences*, **282**:20151429.

Alves, F., Langmore, N., Heinsohn, R., and Stojanovic, D. (2020). "Self-fumigation" of nests by an endangered avian host using insecticide-treated feathers increases reproductive success more than tenfold. *Animal Conservation*, **24**:239–45.

Becker, D.J., Streicker, D.G., and Altizer, S. (2015). Linking anthropogenic resources to wildlife-pathogen dynamics: A review and meta-analysis. *Ecology Letters*, **18**:483–95.

Beeton, N. and McCallum, H. (2011). Models predict that culling is not a feasible strategy to prevent extinction of Tasmanian devils from facial tumour disease. *Journal of Applied Ecology*, **48**:1315–23.

Biggins, D.E., Godbey, J.L., Gage, K.L., Carter, L.G., and Montenieri, J.A. (2010). Vector control improves survival of three species of prairie dogs (*Cynomys*) in areas considered enzootic for plague. *Vector Borne and Zoonotic Diseases*, **10**:17–26.

Bletz, M.C., Loudon, A.H., Becker, M.H., Bell, S.C., Woodhams, D.C., Minbiole, K.P., et al. (2013). Mitigating amphibian chytridiomycosis with bioaugmentation: Characteristics of effective probiotics and strategies for their selection and use. *Ecology Letters*, **16**:807–20.

Bruniche-Olsen, A., Jones, M.E., Austin, J.J., Burridge, C.P., and Holland, B.R. (2014). Extensive population decline in the Tasmanian devil predates European settlement and devil facial tumour disease. *Biology Letters*, **10**:20140619.

Bulgarella, M., Quiroga, M.A., Boulton, R.A., Ramírez, I.E., Moon, R.D., Causton, C.E., et al. (2017). Life cycle and host specificity of the parasitoid *Conura annulifera* (Hymenoptera: Chalcididae), a potential biological control agent of *Philornis downsi* (Diptera: Muscidae) in the Galápagos Islands. *Annals of the Entomological Society of America*, **110**:317–28.

Burrows, R., Hofer, H., and East, M.L. (1995). Population dynamics, intervention and survival in African wild dogs (*Lycaon pictus*). *Proceedings of the Royal Society of London B: Biological Sciences*, **262**:235–45.

Canessa, S., Hunter, D., McFadden, M., Marantelli, G., and McCarthy, M.A. (2014). Optimal release strategies for cost-effective reintroductions. *Journal of Applied Ecology*, **51**:1107–15.

Causton, C.E., Cunninghame, F., and Tapia, W. (2011). Management of the avian parasite *Philornis downsi* in the Galapagos Islands: A collaborative and strategic action plan. *Galapagos Report*, **2012**:167–73.

Cha, D.H., Mieles, A.E., Lahuatte, P.F., Cahuana, A., Lincango, M.P., Causton, C.E., et al. (2016). Identification and optimization of microbial attractants for *Philornis downsi*, an invasive fly parasitic on Galapagos birds. *Journal of Chemical Ecology*, **42**:1101–11.

Cimadom, A., Causton, C., Cha, D.H., Damiens, D., Fessl, B., Hood-Nowotny, R., et al. (2016). Darwin's finches treat their feathers with a natural repellent. *Scientific Reports*, **6**: 34559.

Cleaveland, S., Mlengeya, T., Kaare, M., Haydon, D., Lembo, T., Laurenson, M.K., et al. (2007). The conservation relevance of epidemiological research into carnivore viral diseases in the Serengeti. *Conservation Biology*, **21**:612–22.

Craft, M.E., Volz, E., Packer, C., and Meyers, L.A. (2009). Distinguishing epidemic waves from disease spillover in a wildlife population. *Proceedings of the Royal Society of London B: Biological Sciences*, **276**:1777–85.

Delaney, N.F., Balenger, S., Bonneaud, C., Marx, C.J., Hill, G.E., Ferguson-Noel, N., et al. (2012). Ultrafast evolution and loss of CRISPRs following a host shift in a novel wildlife pathogen, Mycoplasma gallisepticum . *PLoS Genetics*, **8**:e1002511.

Dhondt, A.A., Altizer, S., Cooch, E.G., Davis, A.K., Dobson, A., Driscoll, M.J., et al. (2005). Dynamics of a novel pathogen in an avian host: Mycoplasmal conjunctivitis in house finches. *Acta Tropica*, **94**:77–93.

Ellis, W.A., Girjes, A.A., Carrick, F.N., and Melzer, A. (1993). Chlamydial infection in koalas under relatively little alienation pressure. *Australian Veterinary Journal*, **70**:427–28.

Epstein, B., Jones, M., Hamede, R., Hendricks, S., McCallum, H., Murchison, E.P., et al. (2016). Rapid evolutionary response to a transmissible cancer in Tasmanian devils. *Nature Communications*, **7**:12684.

Fessl, B., Kleindorfer, S., and Tebbich, S. (2006). An experimental study on the effects of an introduced parasite in Darwin's finches. *Biological Conservation*, **127**:55–61.

Fessl, B., Young, G.H., Young, R.P., Rodríguez-Matamoros, J., Dvorak, M., Tebbich, S., et al. (2010). How to save the rarest Darwin's finch from extinction: The mangrove finch on Isabela Island. *Philosophical Transactions of the Royal Society of London B: Biological Sciences*, **365**:1019–30.

Frankham, R. (2015). Genetic rescue of small inbred populations: Meta-analysis reveals large and consistent benefits of gene flow. *Molecular Ecology*, **24**:2610–18.

Gopinath, S., Lichtman, J.S., Bouley, D.M., Elias, J.E., and Monack, D.M. (2014). Role of disease-associated tolerance in infectious superspreaders. *Proceedings of the National Academy of Sciences*, **111**:15780–85.

Grogan, L.F., Peel, A.J., Kerlin, D., Ellis, W., Jones, D., Hero, J.-M., et al. (2018). Is disease a major causal factor in declines? An evidence framework and case study on koala chlamydiosis. *Biological Conservation*, **221**:334–44.

Hartup, B.K., Mohammed, H.O., Kollias, G.V., and Dhondt, A.A. (1998). Risk factors associated with mycoplasmal conjunctivitis in house finches. *Journal of Wildlife Diseases*, **34**:281–88.

Hawley, D.M., Sydenstricker, K.V., Kollias, G.V., and Dhondt, A.A. (2005). Genetic diversity predicts pathogen resistance and cell-mediated immunocompetence in house finches. *Biology Letters*, **1**:326–29.

Haydon, D.T., Randall, D.A., Matthews, L., Knobel, D.L., Tallents, L.A., Gravenor, M.B., et al. (2006). Low-coverage vaccination strategies for the conservation of endangered species. *Nature*, **443**:692–95.

Hess, S.C., Jeffrey, J.J., Ball, D.L., and Babich, L. (2006). Efficacy of feral pig removals at Hakalau Forest National Wildlife Refuge, Hawai'i. *Transactions of the Western Section of the Wildlife Society*, **42**:53–67.

Hines, A.M., Ezenwa, V.O., Cross, P., and Rogerson, J.D. (2007). Effects of supplemental feeding on gastrointestinal parasite infection in elk (*Cervus elaphus*): Preliminary observations. *Veterinary Parasitology*, **148**:350–55.

Hobbelen, P.H.F., Samuel, M.D., LaPointe, D.A., and Atkinson, C.T. (2012). Modeling future conservation of Hawaiian honeycreepers by mosquito management and translocation of disease-tolerant Amakihi. *PLoS ONE*, **7**:e49594

Hochachka, W.M. and Dhondt, A.A. (2000). Density-dependent decline of host abundance resulting from a new infectious disease. *Proceedings of the National Academy of Sciences*, **97**:5303–06.

Hohenlohe, P.A., McCallum, H.I., Jones, M.E., Lawrance, M.F., Hamede, R.K., and Storfer, A. (2019). Conserving adaptive potential: Lessons from Tasmanian devils and their transmissible cancer. *Conservation Genetics*, **20**:81–87.

Jones, M.E., Paetkau, D., Geffen, E.L., and Moritz, C. (2004). Genetic diversity and population structure of Tasmanian devils, the largest marsupial carnivore. *Molecular Ecology*, **13**:2197–209.

Joseph, M.B., Mihaljevic, J.R., Arellano, A.L., Kueneman, J.G., Preston, D.L., Cross, P.C., et al. (2013). Taming wildlife disease: Bridging the gap between science and management. *Journal of Applied Ecology*, **50**:702–12.

Katahira, L.K., Finnegan, P., and Stone, C.P. (1993). Eradicating feral pigs in montane mesic habitat at Hawaii Volcanoes National Park. *Wildlife Society Bulletin*, **21**:269–74.

Kau, A.L., Ahern, P.P., Griffin, N.W., Goodman, A.L., and Gordon, J.I. (2011). Human nutrition, the gut microbiome and the immune system. *Nature*, **474**:327.

Kearns, P.J., Fischer, S., Fernandez-Beaskoetxea, S., Gabor, C.R., Bosch, J., and Bowen, J.L. (2017). Fight fungi with fungi: Antifungal properties of the amphibian mycobiome. *Frontiers in Microbiology*, **8**:12.

Kessler, M.K., Becker, D.J., Peel, A.J., Justice, N.V., Lunn, T., Crowley, D.E., et al. (2018). Changing resource landscapes and spillover of henipaviruses. *Annals of the New York Academy of Sciences*, **1429**:78–99.

Kilpatrick, A. M. 2006. Facilitating the evolution of resistance to avian malaria in Hawaiian birds. Biological Conservation **128**:475-485.

Knutie, S.A., McNew, S.M., Bartlow, A.W., Vargas, D.A., and Clayton, D.H. (2014). Darwin's finches combat introduced nest parasites with fumigated cotton. *Current Biology*, **24**:R355–56.

Lachish, S., Miller, K.J., Storfer, A., Goldizen, A.W., and Jones, M.E. (2011). Evidence that disease-induced population decline changes genetic structure and alters dispersal patterns in the Tasmanian devil. *Heredity*, **106**:172–82.

Lloyd-Smith, J.O., Schreiber, S.J., Kopp, P.E., and Getz, W.M. (2005). Superspreading and the effect of individual variation on disease emergence. *Nature*, **438**: 355–59.

Margres, M.J., Jones, M.E., Epstein, B., Kerlin, D.H., Comte, S., and Fox, S. (2018a). Large-effect loci affect survival in Tasmanian devils (*Sarcophilus harrisii*) infected with a transmissible cancer. *Molecular Ecology*, **27**: 4189–99.

Margres, M.J., Ruiz-Aravena, M., Hamede, R., Jones, M.E., Lawrance, M.F., Hendricks, S.A., et al. (2018b). The genomic basis of tumor regression in Tasmanian devils (*Sarcophilus harrisii*). *Genome Biology and Evolution*, **10**:3012–25.

Martin, A.M., Burridge, C.P., Ingram, J., Fraser, T.A., and Carver, S. (2018a). Invasive pathogen drives host population collapse: Effects of a travelling wave of sarcoptic mange on bare-nosed wombats. *Journal of Applied Ecology*, **55**:331–41.

Martin, A.M., Fraser, T.A., Lesku, J.A., Simpson, K., Roberts, G.L., Garvey, J., et al. (2018b). The cascading pathogenic consequences of *Sarcoptes scabiei* infection that manifest in host disease. *Royal Society Open Science*, **5**:180018.

Martin, A.M., Ricardo, H., Tompros, A., Fraser, T. A., Polkinghorne, A., and Carver, S. (2019). Burrows with resources have greater visitation and may enhance mange transmission among wombats. *Australian Mammalogy*, 41(2):287–90.

Matchett, M.R., Biggins, D.E., Carlson, V., Powell, B., and Rocke, T. (2010). Enzootic plague reduces black-footed ferret (*Mustela nigripes*) survival in Montana. *Vector-Borne and Zoonotic Diseases*, **10**:27–35.

McCallum, H., Fenton, A., Hudson, P.J., Lee, B., Levick, B., Norman, R., et al. (2017). Breaking beta: Deconstructing the parasite transmission function. *Philosophical Transactions of the Royal Society of London B: Biological Sciences*, **372**:p.20160084.

McCallum, H., Kerlin, D.H., Ellis, W., and Carrick, F. (2018). Assessing the significance of endemic disease in conservation-koalas, chlamydia, and koala retrovirus as a case study. *Conservation Letters*, **11**: e12425.

Medzhitov, R., Schneider, D.S., and Soares, M.P. (2012). Disease tolerance as a defense strategy. *Science*, **335**: 936–41.

Miller, M.W., Runge, J.P., Holland, A.A., and Eckert, M.D. (2020). Hunting pressure modulates prion infection risk in mule deer herds. *Journal of Wildlife Diseases*, **56**(4): 781–90.

Nyari, S., Booth, R., Quigley, B.L., Waugh, C.A., and Timms, P. (2019). Therapeutic effect of a *Chlamydia pecorum* recombinant major outer membrane protein vaccine on ocular disease in koalas (*Phascolarctos cinereus*). *PLoS ONE*, **14**(1):e0210245.

Old, J.M., Sengupta, C., Narayan, E., and Wolfenden, J. (2018). Sarcoptic mange in wombats—A review and future research directions. *Transboundary and Emerging Diseases*, **65**:399–407.

Polkinghorne, A., Hanger, J., and Timms, P. (2013). Recent advances in understanding the biology, epidemiology and control of chlamydial infections in koalas. *Veterinary Microbiology*, **165**:214–23.

Prentice, J.C., Fox, N.J., Hutchings, M.R., White, P.C., Davidson, R.S., and Marion, G. (2019). When to kill a cull: Factors affecting the success of culling wildlife for disease control. *Journal of the Royal Society Interface*, **16**(152):20180901.

Råberg, L., Graham, A.L., and Read, A.F. (2009). Decomposing health: Tolerance and resistance to parasites in animals. *Philosophical Transactions of the Royal Society of London B: Biological Sciences*, **364**: 37–49.

Read, A.F., Baigent, S. J., Powers, C., Kgosana, L.B., Blackwell, L., Smith, L.P., et al. (2015). Imperfect vaccination can enhance the transmission of highly virulent pathogens. *PLoS Biol*, **13**(7):e1002198.

Rhodes, J.R., Ng, C.F., De Villiers, D.L., Preece, H.J., McAlpine, C.A., and Possingham, H.P. (2011). Using integrated population modelling to quantify the implications of multiple threatening processes for a rapidly declining population. *Biological Conservation*, **144**:1081–88.

Roelke-Parker, M.E., Munson, L., Packer, C., Kock, R., Cleaveland, S., Carpenter, M., et al. (1996). A canine distemper virus epidemic in Serengeti lions (*Panthera leo*). *Nature*, **379**:441–45.

Rowe, M.L., Whiteley, P.L., and Carver, S. (2019). The treatment of sarcoptic mange in wildlife: A systematic review. *Parasites & Vectors*, **12**:1–14.

Ryan, S.J. and Walsh, P.D. (2011). Consequences of non-intervention for infectious disease in African great apes. *PLoS ONE*, **6**:e29030.

Scheele, B.C., Hunter, D.A., Grogan, L.F., Berger, L., Kolby, J.E., McFadden, M.S., et al. (2014). Interventions for reducing extinction risk in chytridiomycosis-threatened amphibians. *Conservation Biology*, **28**:1195–205.

Sousa, W.P. (1993). Interspecific antagonism and species coexistence in a diverse guild of larval trematode parasites. *Ecological Monographs*, **63**:103–28.

Stone, C. and Loope, L. (1987). Reducing negative effects of introduced animals on native biotas in Hawaii: What is being done, what needs doing, and the role of National Parks. *Environmental Conservation*, **14**: 245–58.

Vial, F., Cleaveland, S., Rasmussen, G., and Haydon, D.T. (2006). Development of vaccination strategies for the management of rabies in African wild dogs. *Biological Conservation*, **131**:180–92.

Viana, M., Cleaveland, S., Matthiopoulos, J., Halliday, J., Packer, C., Craft, M.E., et al. (2015). Dynamics of a morbillivirus at the domestic–wildlife interface: Canine distemper virus in domestic dogs and lions. *Proceedings of the National Academy of Sciences*, **112**: 1464–69.

Voyles, J., Woodhams, D.C., Saenz, V., Byrne, A.Q., Perez, R., Rios-Sotelo, G., et al. (2018). Shifts in disease dynamics in a tropical amphibian assemblage are not due to pathogen attenuation. *Science*, **359**:1517–19.

Vredenburg, V.T., Briggs, C.J., and Harris, R.N. (2011). Host pathogen dynamics of amphibian chytridiomycosis: the role of the skin microbiome in health and disease. In: Olson, L., Choffnes, E., Relman, D., and Pray, L. (eds.), *Fungal Diseases: An Emerging Threat to Human, Animal, and Plant Health*, pp. 342–55. National Academy Press, Washington, DC.

Waugh, C., Khan, S. A., Carver, S., Hanger, J., Loader, J., Polkinghorne, A., et al. (2016). A prototype recombinant-protein based *Chlamydia pecorum* vaccine results in reduced chlamydial burden and less clinical disease in free-ranging koalas (*Phascolarctos cinereus*). *PloS ONE*, **11**(1):e0146934.

Waugh, C.A. and Timms, P. (2020). A proposed roadmap for the control of infections in wildlife using *Chlamydia* vaccine development in koalas *Phascolarctos cinereus* as a template. *Wildlife Biology*, **2020**(1)https://doi.org/10.2981/wlb.00627

Weigler, B.J., Girjes, A.A., White, N.A., Kunst, N.D., Carrick, F.N., and Lavin, M.F. (1988). Aspects of the epidemiology of *Chlamydia psittaci*-infection in a population of koalas (*Phascolarctos cinereus*) in southeastern Queensland, Australia. *Journal of Wildlife Diseases*, **24**:282–91.

White, N.A. and Timms, P. (1994). *Chlamydia psittaci* in a koala (*Phascolarctos cinereus*) population in southeast Queensland. *Australian Wildlife Research*, **21**: 41–47.

Wiehn, J. and Korpimäki, E. (1998). Resource levels, reproduction and resistance to haematozoan infections. *Proceedings of the Royal Society of London B: Biological Sciences*, **265**:1197–201.

Woodhams, D.C., Bletz, M., Kueneman, J., and McKenzie, V. (2016). Managing amphibian disease with skin microbiota. *Trends in Microbiology*, **24**:161–64.

Woodhams, D.C., Bosch, J., Briggs, C.J., Cashins, S., Davis, L.R., Lauer, A., et al. (2011). Mitigating amphibian disease: Strategies to maintain wild populations and control chytridiomycosis. *Frontiers in Zoology*, **8**:8.

Woodworth, B.L., Atkinson, C.T., LaPointe, D.A., Hart, P.J., Spiegel, C.S., Tweed, E.J., et al. (2005). Host population persistence in the face of introduced vector-borne diseases: Hawaii amakihi and avian malaria. *Proceedings of the National Academy of Sciences*, **102**:1531–36.

CHAPTER 15

Infectious Diseases as Biological Control Agents

The introduction of a few rabbits could do little harm and might provide a touch of home, in addition to a spot of hunting.

Farmer responsible for the introduction and release of the first 24 rabbits into Australia

15.1 Introduction

Much of this book's focus has been on the control and eradication of infectious agents that may have deleterious conservation consequences for the communities in which they occur. However, the reverse possibility is also becoming increasingly relevant for conservation: the purposeful introduction of infectious diseases, typically as biological control agents against invasive species. *Biological control* or *biocontrol* is the use of natural enemies to suppress or eliminate unwanted pests or invasive taxa. It has been widely used in agriculture and, to a lesser extent, for control of vectors of disease in humans or livestock, and for broader environmental benefits (Vincent et al. 2007). Biological control can be classified into three broad types: classical biological control, in which the agent is released and then maintains itself in the environment; inundative biological control, in which large numbers of the agent are released, but they are not able to maintain themselves in the environment; and conservation biological control, which entails maintaining communities of existing natural enemies in the environment. Agents of biological control may be predators, parasitoids, macroparasites (such as

nematodes), or a variety of microparasites, including fungi, protozoa, bacteria, and viruses. Targets of biological control are usually weeds or invertebrates such as insects or mites. Less commonly, and within the scope of this book, biological control using parasites or microparasites has been used or has been proposed as a means of controlling or eliminating invasive vertebrate species.

Some of these cases are summarized in Table 15.1. The two best-known cases where pathogens were successfully used to control invasive vertebrate species are for rabbits and cats, although at the time of writing, a proposal to release a herpesvirus to control European carp (*Cyprinus carpio*) in Australia is under active consideration. Because each of these campaigns had important implications for conservation of local biodiversity, and because they provide useful insights for similar future efforts, they are worth discussing in more detail.

15.2 Myxomatosis as a biological control for rabbit populations

European rabbits (*Oryctolagus cuniculus*) were introduced into south-eastern Australia in 1859, and subsequently spread throughout most of the continent,

Infectious Disease Ecology and Conservation. Johannes Foufopoulos, Gary A. Wobeser and Hamish McCallum, Oxford University Press.
© Johannes Foufopoulos, Gary A. Wobeser and Hamish McCallum (2022). DOI: 10.1093/oso/9780199583508.003.0015

Table 15.1 Use of infectious pathogens for biological control of vertebrate populations.

Agent	Target	Location	Status	Conservation benefit or risk	References
Deployed in the field					
Rabbit myxoma virus (myxomatosis, Poxviridae)	Rabbit (*Oryctolagus cuniculus*)	Australia, UK, Europe	Introduced Australia 1950–1951; France 1951; UK 1953 (from France); continental Europe 1953–1960s (from France)	Large reductions in rabbit populations, but recovery following coevolution. Benefits to native herbivores and endangered flora. Reduction in food supply for Iberian lynx. See text	Ross and Tittensor (1986), Fenner and Fantini (1999), Lees and Bell (2008)
Rabbit hemorrhagic disease virus (RHDV, Caliciviridae)	Rabbit (*Oryctolagus cuniculus*)	Australia, New Zealand	Introduced to Australia 1995	Sustained reductions but no eradication of rabbit populations	Bruce et al. (2004), Bruce and Twigg (2005), Cooke and Fenner (2002)
Rabbit hemorrhagic disease virus (RHDV, Caliciviridae)	Rabbit (*Oryctolagus cuniculus*)	Macquarie Island, Australia	Virus released in 2011 with initial population reductions of 85–95%	RHDV together with poison baiting and hunting resulted in rabbit eradication in 2014	Cooke et al (2017)
Feline panleukopenia virus (FPV, Parvoviridae)	Domestic cat (*Felis catus*)	Marion Island, South Africa, sub-Antarctic Indian Ocean	Introduced 1977	Substantial initial reduction of cat numbers. After shooting, trapping, and poisoning, cats eliminated in 1982	Van Rensburg et al. (1987), Bester et al. (2002)
Proposed/under investigation					
Feline immunodeficiency virus (FIV, Retroviridae)	Domestic cat (*Felis catus*)	Oceanic islands	Proposed in 1999, but never implemented	-	Oliveira and Hilker (2010), Cleaveland et al. (1999), Courchamp and Sugihara (1999)
Unspecified viruses	Cane toad (*Rhinella marina*)	Australia	No suitable wild agents found in South America. Genetically modified organisms investigated, but none used	-	Shanmuganathan et al. (2010)
Hemogregarine-like blood parasites (Hemogregarinidae, Apicomplexa)	Brown tree snake (*Boiga irregularis*)	Guam	Unlikely to be effective	-	Caudell et al. (2002)

continued

Table 15.1 *Continued*

Agent	Target	Location	Status	Conservation benefit or risk	References
Virally vectored immunocontra-ception	Brushtail possum (*Trichosurus vulpecula*)	New Zealand	No agent yet developed		McLeod et al. (2007)
Genetically modified (immuno-contraceptive) *Parastrongyloides trichosuri* (Nemotoda)	Brushtail possum (*Trichosurus vulpecula*)	New Zealand	Under laboratory investigation	Substantial risk to Australian marsupial fauna	Ji (2009)
Genetically modified (immuno-contraceptive) murine cytomegalovirus (MCMV, Herpesviridae)	House mouse (*Mus musculus*)	Australia	Genetically modified agent suppresses reproduction, but transmission rate was too low to be effective as control agent. No longer under investigation	Potential risk to endangered endemic Australian rodents	Singleton et al. (1994), McLeod et al. (2007)
Genetically modified (immuno-contraceptive) myxomatosis (Poxviridae)	Rabbit (*Oryctolagus cuniculus*)	Australia	Some reduction in fertility achieved in laboratory. Issues with immunity and reproduc-tive compensation mean agent unlikely to be successful. No longer under active investigation		Twigg et al. (2000), McLeod et al. (2007)
Genetically modified (immunocon-traceptive) virus	Red fox (*Vulpes vulpes*)	Australia	No suitable agent found		McLeod et al. (2007)
Aleutian disease virus (ADV, Parvoviridae), mink enteri-tis virus (MEV, Parvoviridae), canine distemper virus (CDV, Paramyxoviridae)	Stoats (*Mustela erminea*)	New Zealand	Concerns with host specificity		McDonald and Lariviere (2001)
Cyprinid herpesvirus 3 (CyHV-3, Alloherpesviridae)	European carp (*Cyprinus carpio*)	Australia	Release delayed until at least 2022		Brown and Gilligan (2014), McColl et al. (2014), McColl et al. (2016)

(a)

(b)

Figure 15.1 (a) The results of a successful day rabbiting in Murrungowar, East Gippsland, ca. 1915. Despite high annual harvests during hunting campaigns, Australian rabbit population control was not effective until the introduction of myxomatosis. (Source: Museums Victoria Collections,) (b) Rabbit infected with myxomatosis. (Source: Dutch National Archives.)

with the exception of the far north (see also Chapter 3, Section 3.2). Their explosive population growth, beyond leading to direct competition with the local sheep industry for resources, had a devastating effect on the continent's vegetation cover and by extension on many native wildlife species. In 1951, after many other methods of control had failed (Figure 15.1a), the myxoma poxvirus was introduced into Australia as a biological rabbit-control agent. The Australian virologist Frank Fenner was central in this introduction and described its history and background in a seminal book (Fenner and Fantini 1999), on which much of the discussion that follows is based. The book itself provides important details for anyone interested in the use of pathogens as biological control agents.

Myxoma is a double-stranded DNA virus. Its natural host is the South American rabbit species *Sylvilagus brasiliensis*, in which it causes relatively minor clinical disease. It is transmitted through direct contact between hosts or vectored by biting arthropods, predominantly mosquitoes and fleas. When European rabbits were first exposed to the virus, there was almost 100% mortality. After a series of trials in the late 1940s, the virus

escaped from the trial rabbit populations and rapidly spread through south-eastern Australia, by natural transmission, supplemented by some inoculations. Initially, rabbit populations declined by more than 99% (Figure 15.1b). However, evolution in both the rabbit host and the myxoma virus (see also Chapter 3, Section 3.2) have reduced the effectiveness of the virus as a rabbit population control tool through time, with rabbits evolving resistance, and less pathogenic strains of myxomatosis being selected for (Fenner and Fantini 1999). Myxomatosis was also introduced in France in 1951, with onward spread to the remainder of Europe, including the UK. Whether this onward spread was "natural" or unofficially assisted by humans remains unclear. In both the UK and Australia, strains of intermediate virulence eventually replaced the original highly pathogenic myxoma virus, but avirulent strains have not dominated (Fenner and Fantini 1999). As a result, impact of the epizootic started declining and rabbit populations across Australia started rebounding again (see Figure 15.2). To further facilitate disease transmission between rabbits, and hence increase the impact on the host population, at least two species of rabbit fleas were introduced to Australia in the 1960s and 1970s.

15.3 Rabbit hemorrhagic disease to control rabbit populations

Rising rabbit populations prompted Australian authorities to consider other pathogens, including rabbit hemorrhagic disease virus (RHDV—also known as rabbit calicivirus), as possible additional biocontrol agents. This virus, appearing first in farmed rabbits in China in 1984, became the focus of intense research in Australia starting in the 1990s (Kovaliski 1998). While the pathogen was being investigated at a remote research facility on Wardang Island off the coast of South Australia for its suitability to reduce rabbit populations, it escaped onto the Australian mainland, across more than 4 km of open water. This transport of the virus, most likely by bushflies carried by wind across the straits, illustrates the difficulty in containing viral pathogens of wildlife. Once on the mainland, the virus spread rapidly through the rabbit populations, infecting and killing millions of rabbits during the first weeks (Figure 15.2). Since then, its effectiveness as a control agent for rabbits in Australia has been variable. While it has substantially reduced rabbit numbers in arid areas,

in wetter, more fertile regions, its effectiveness has been mixed, and rabbit populations have recovered in some areas (Mutze et al. 2010). Several hypotheses have been advanced to account for this variable impact of the virus on rabbit populations (Mutze et al. 2014). One possibility is coevolution of virus and host, as exemplified earlier by myxomatosis. Hence, there has been both evidence for evolution in RHDV, as well as evidence for evolution of resistance among rabbits (Mutze et al. 2014; Elsworth et al. 2012). Another possibility is that mortality is lower in juveniles than in adults, and surviving juveniles are then immune to further infections (Mutze et al. 2014). Maternal antibodies protecting juveniles may also play a role (Wells et al. 2015). For example, the illegal release of RHDV in New Zealand in 1997 did not achieve the desired rabbit population control, in part because of the timing of the introduction. The unauthorized release of the virus, possibly by local farmers, occurred soon after the end of the rabbit reproductive season, which meant that numerous young rabbits of the year—which are largely resistant to the disease—survived the initial epizootic and formed the founding stock of a partially immune population that expanded

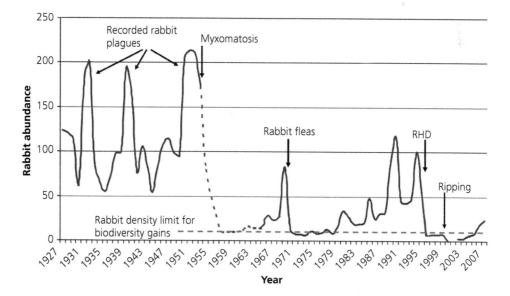

Figure 15.2 Estimated changes in rabbit abundance at select sheep stations in NE South Australia, showcasing the impacts of various biocontrol efforts. These include the introduction of myxomatosis, followed by the release of various flea species facilitating the transmission of the pathogen, as well as the introduction of rabbit hemorrhagic disease (RHD). (Source: Saunders et al. 2010.)

again in subsequent seasons. In addition, there is some evidence that preexisting strains of non-pathogenic caliciviruses that may circulate throughout rabbit populations in some regions provide cross-immunity to RHDV (Strive et al. 2013). Nonetheless, under the right circumstances, the virus has a great deal of potential in reducing rabbit populations, as its release on Macquarie Island (South Pacific Ocean) in 2011 showed. There, release of RHDV was used in conjunction with poison baiting and targeted hunting to drive the extirpation of rabbits from the island in 2014 (Cooke et al. 2017). The variable results of different RHDV-release programs illustrate how the impact of a viral biological control agent depends on a complex interplay between the environment, natural selection, immunity, as well as other pathogens circulating in the system.

In the past few years the epidemiological landscape of Australian rabbit populations has become significantly more complex. An exotic type of rabbit hemorrhagic disease virus (RHDV-2), first detected in 2014, has been spreading through Australian rabbit populations, and leading to substantial declines, while at the same time partially replacing the original RHDV virus (Mahar et al. 2018). Furthermore, another strain (RHDV-K5), which was intentionally released as a biocontrol agent in 2017, seems to have had little impact on resident rabbit populations (Ramsey et al. 2020). The concurrent existence of multiple viruses circulating is leading to potentially complicated host–parasite dynamics. While in some cases different viral strains appear to suppress or outcompete each other, in other cases, they have asymmetric or synergistic effects in their impact on the rabbit populations (Barnett et al. 2018; Wells et al. 2018; Ramsey et al. 2020). This suggests that any releases of pathogen biocontrol agents needs to consider a comprehensive range of factors that may potentially affect the expected outcome.

15.3.1 Rabbit biocontrol—Ecological consequences

Although use of pathogens for rabbit biocontrol typically results in significant benefits to local ecosystems, the ecological consequences can sometimes be quite unexpected. On the Australian mainland, collapses of rabbit populations were followed by partial recoveries of native vegetation and wildlife, generating spectacular conservation benefits in some places. Nonetheless, not all ecological changes following rabbit control have been beneficial for local biodiversity. For example, myxomatosis and rabbit hemorrhagic disease were introduced and subsequently led to substantial declines in the rabbit populations on the Iberian Peninsula over the past 40 years. Rabbits are the primary prey item for a number of endangered species in the area, including the Iberian lynx (*Lynx pardinus*) and the Spanish imperial eagle (*Aquila adalberti*). Following the introduction of RHDV, both of these predator populations declined (Lees and Bell 2008; Moreno et al 2008). On the British Isles, rabbits have detrimental effects on a number of rare plant species on islands, but on the UK mainland, rabbits maintain short swards, which are important in sustaining populations of a number of rare plant species and some animal species (e.g., the stone curlew *Burhinus oedicnemus*, the red-billed chough *Pyrrhocorax pyrrhocorax*, and the woodlark *Lullula arborea*) (Lees and Bell 2008). Following the introduction of myxomatosis and the concomitant decline of rabbit populations, overgrown vegetation resulted in the extirpation of the large blue butterfly (*Maculinea arion*) from the UK—the species has been reintroduced since (Sumption and Flowerdew 1985). Lastly, on Macquarie Island, rabbit eradication appears to have allowed both a rapid rebound of the island's plant endemics but also the spread of an invasive plant species that was previously kept under check by rabbit herbivory.

15.4 Feline panleukopenia virus against cats

The only other documented use of a pathogen as a successful biocontrol agent against an invasive vertebrate has been the introduction of feline panleukopenia virus (FPV) to control feral cat (*Felis catus*) populations on Marion Island (Van Rensburg et al. 1987; Bester et al. 2002). On this sub-Antarctic Indian Ocean island, known for its important seabird colonies, feral cats were responsible for a catastrophic decline in several endangered seabird populations. While the introduction of the pathogen itself did not lead to the eradication of the feral cat population on the island, it did reduce the population in 5 years to an estimated 18% of the predisease population, making it possible for the subsequent intense hunting effort to eliminate the remaining few individuals in 1982 (Figure 15.3a). Because the virus had particularly severe impacts on very young cats, it resulted in a near halving of the reproductive output of the cat population and a general shift toward older individuals (Figure 15.3b, c). Interestingly, during the last stages of the eradication campaign, hunting had reduced cat density to the point where FPV transmission had essentially ceased. The Marion Island cat eradication campaign indicates not only that pathogen releases are best considered as one among several eradication tools, but also that careful thought needs to be given to possible interactions between different control methods.

15.5 Herpesvirus as a control agent for European carp

A proposal to introduce a herpesvirus as a control agent for European carp (*Cyprinus carpio*) in Australia provides an excellent case study of the issues involved in using pathogens as vertebrate biocontrol agents. European carp have been present in Australia since the nineteenth century, but became a major environmental problem when, in the 1970s, an aquaculture-adapted variety escaped into Australia's major river system, the Murray-Darling (McColl 2016). Carp now constitute more than 90% of the fish biomass in some parts of the system, exceeding 3,100 kg/ha and destroying aquatic plants, as well as having major effects on water turbidity (Koehn 2004).

Cyprinid herpesvirus-3, also known as CyHV-3, was first recorded in 1998 and has killed up to 70% of carp in some wild populations in Japan (McColl et al. 2014). As CyHV-3 appears to be specific to carp and there are no native cyprinids in Australia, it has been explored as a potential biological control agent for carp populations (McColl et al. 2014). Multiple important issues need to be evaluated before release of an agent can be considered—at its simplest, the agent must be shown to be both safe and

effective. For CyHV-3, extensive testing was undertaken on 13 species of native teleost fishes, one introduced teleost of importance to sport fishing (*Oncorhynchus mykiss*), two amphibian species, two reptiles, a freshwater crustacean, chickens, and laboratory mice. Despite some mortalities in nontarget species, and low levels of positive polymerase chain reaction (PCR) detections of virus in some nontargets, McColl (2016) concluded that the risk of spillover infections into other species was overall very low In addition to direct impacts on nontarget species, there is also the possibility of epidemic mortalities in carp leading to the accumulation of decaying fish bodies, eutrophication, declines in dissolved oxygen, etc., resulting in major impacts on the overall ecosystem and possibly human health (Lighten and van Oosterhout 2017). Concerns continue to be raised about the efficacy of the pathogen as a biocontrol agent, in terms of the possibility of existing resistance-conferring genetic polymorphisms in carp, the ability of carp populations to quickly recover due to high fecundity rates, and of the likelihood of high water temperatures blocking virus transmission (Marshall et al. 2018; Becker et al 2019; Mintram et al. 2020).

Other criticisms stem from concerns as to whether CyHV-3 can retain its high virulence in carp in the long run. The long-term effectiveness of a biocontrol agent is closely tied to its virulence, as the Australian myxomatosis introduction emphasized. In particular, if there are evolutionary trade-offs between transmissibility and virulence, lower-virulence virus strains may eventually prevail. McColl et al. (2016) suggest that close contact between carp is necessary for transmission of CyHV-3 to occur and also that there may be latent infections in survivors, with recrudescence of the infection during periods of stress. If this is the case, transmission may be optimized by the evolution of low-virulence strains and the initial high virulence of the virus in carp may eventually decline with time. While many of these concerns have been acknowledged and partially addressed, there is general recognition that CyHV-3 releases will alone not be sufficient to suppress carp populations in the long term and that other measures will be needed (McColl et al. 2017, 2018; Becker et al. 2019). Because of these concerns, at the time of writing of this

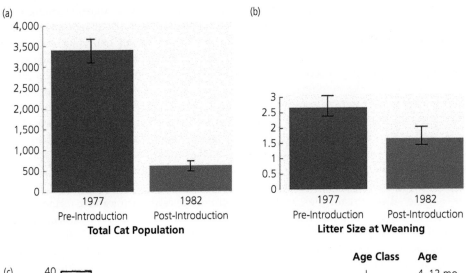

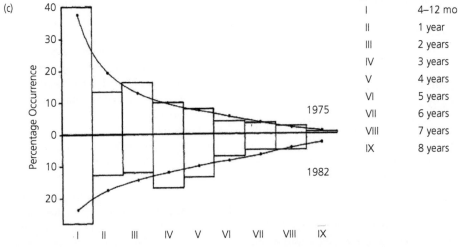

Figure 15.3 Changes in feral cat population characteristics on Marion Island, before (1977), and after (1982), the release of FPV. The introduction of the pathogen resulted in a reduction of the total island cat population to an estimated 18% of the original population size (a). Infection-associated litter size decrease further depressed cat populations (b). Because FPV disproportionally impacts young animals, the introduction resulted in a shift in the proportions of the different age classes in the cat population, with a reduction in young animals (Age Class I) being particularly noticeable (c). While the introduction of FPL failed to complete the eradication of the cat population, it did succeed to decrease it to a level where other approaches were able to eradicate the population (Source: Van Rensburg et al. 1987.)

chapter, the proposal to release the virus remains controversial (Kopf et al. 2019), but a detailed plan is in place (McColl and Barwick 2018).

15.6 Potential biological control of cane toads

The cane toad *Rhinella marina* (formerly *Bufo marinus*) was introduced into Australia in 1935 as an unsuccessful control agent for cane beetles in sugarcane. It continues to spread across northern Australia and has had severe impacts on Australian biodiversity (Shine 2010). There have been a number of attempts to search for, or even engineer appropriate biological control agents (Shanmuganathan et al. 2010). To date, none of these have been successful. While Australia has a large diversity of anurans, there are no native members of the family *Bufonidae*, so a pathogen specific to the family would have a low probability of infecting endemic Australian frogs.

There are a number of viruses and other parasites that do cause mortality in cane toads, but all so far investigated are generalist anuran pathogens. Attempts to search for an appropriate parasite or pathogen within the native range of the cane toad in northern South America have been unsuccessful, and no suitable agent has been found to date (Shanmuganathan et al. 2010). By analogy with myxomatosis, which was originally a relatively benign virus of the Brazilian cottontail (*Sylvilagus brasiliensis*), not the European rabbit (*Oryctolagus cuniculus*), it may be more fruitful to search for a virus among the cane toad's relatives in South America, outside its native range, rather than in the cane toad itself.

15.7 Virally vectored immunocontraception

Immunocontraception is the administration of agents that stimulate an immune response against an animal's own reproductive processes. Immunocontraceptive agents typically trigger an immune response against the production of gametes, or prevent proper gamete function, or inhibit successful implantation of an embryo. By using a genetically modified microorganism, the aim of biocontrol efforts is to infect an invasive species with an agent that produces immunocontraceptive antigens and thus undermines its reproductive output. At present there is an extensive research program in both Australia and New Zealand to develop genetically modified parasites and pathogens that deliver immunocontraceptive agents to control a range of introduced mammals. The program in Australia originally targeted foxes and rabbits. However, research into a vector-delivered immunocontraceptive for foxes was abandoned relatively early in the absence of a suitable species-specific vector (McLeod et al. 2007). Development of an immunocontraceptive vectored by a low-pathogenicity strain of myxoma continued for some years (McLeod et al. 2007). While there was some success in developing an immunocontraceptive agent in the laboratory, field research showed that it was necessary to sterilize in excess of 80% of all female rabbits in order to have a noticeable effect on rabbit population densities, because compensatory increases in juvenile survival reduced the impact of lowered fertility (Williams et al. 2007). Development of an immunocontraceptive for house mice (*Mus musculus*) vectored by murine cytomegalovirus (MCMV) proved to be more technically feasible. However, modeling suggested that transmission of the virus did not occur at sufficiently high rates for the pathogen to be effective as a control agent and, furthermore, as Australia has a large number of endangered native murids, there remain concerns about the host specificity of the vector (McLeod et al. 2007).

In New Zealand, there have been attempts to develop an immunocontraceptive for brushtail possums using the nematode *Parastrongyloides trichosuri* as a vector (Ji 2009). However, there are very real concerns that such an agent might be transferred to Australia, where it might threaten populations of a variety of native marsupials (Gilna et al. 2005).

If a genetically modified version of a pathogen that is already present in the target host is being considered as a control agent, as was the case with the proposal to insert immunocontraceptive genes into a myxoma strain, then the genetically modified pathogen must be able to outcompete the wild-type pathogen. If the pathogen is engineered to increase mortality, it will almost certainly have lower fitness than the wild type, as the wild strain is likely to have evolved toward an optimum R_0 (May and Anderson 1983) and any increase in pathogenicity is likely to lead to a decreased R_0. This will not necessarily be the case for a pathogen modified to reduce host fecundity. However, if there is any density-dependent component to transmission, an effective biocontrol agent will reduce the local population density of its host and therefore will have a locally decreased R_0 relative to the wild type (Hood et al. 2000). This means it is likely to be at a selective disadvantage.

15.8 Synthesis and conclusions

The use of pathogens for biological control of invasive vertebrates, while often suggested, remains a challenging endeavor. On the one hand, pathogens have, when successfully controlling invasive populations, the potential to generate tremendous economic and ecological benefits. Furthermore, these advantages may be sustained over the long term if the introduced pathogen leads to the extirpation of the target host, or it becomes a stable member of the local ecosystem. While complete eradication of a target species may be the ultimate desired outcome, this is neither an easy nor actually a necessary goal for pathogen biocontrol releases. Even partial impacts, leading to a suppression of the target species population, whether through increased mortality or through attenuated recruitment, are often sufficiently beneficial to warrant pathogen release.

On the other hand, identification of an appropriate biocontrol agent is a notoriously difficult endeavor. At a minimum, the pathogen needs to be specifically infectious to the species that needs to be controlled. It should be of high virulence, as well as environmentally stable and persistently infectious under a range of ambient conditions and host densities. If genetically modified agents are being considered, then any release would need to be consistent with the Cartagena Protocol on Biosafety (Secretariat of the Convention on Biological Diversity 2000). Release of pathogens as biocontrol agents can have many unexpected effects, both direct and indirect, which need to be considered ahead of time. Direct effects include collateral infection and mortality in nontarget taxa, including endangered species and pets. Such impacts may even occur in very different ecosystems if a pathogen spreads secondarily beyond the original deployment region. Effects on the focal species may be nonlinear and depend in complex ways on host behavior, population density, as well as social and demographic population structures. Furthermore, host impacts may depend on locally prevailing environmental conditions, as well as the presence of additional stressors that act synergistically with the disease. As revealed by the Australian rabbit control efforts, the presence of related microorganisms that circulate "silently" throughout the target population and conferring cross-immunity may severely undermine the success of a release. Many of these responses may even shift over the years as both the target species and biocontrol agent evolve in a coevolutionary "cat-and-mouse" game.

Beyond these direct impacts, additional *in*direct effects may occur, especially if the pathogen successfully depopulates the focal species. As exemplified by the carp example (see Section 15.5), large-scale disease mortalities may have intense, but transient impacts on the ecosystem, which need to be considered and managed. Second-order effects of the decline of rabbits in Britain included the disappearance of species that were dependent on the open vegetation created by rabbits. Similarly, following the eradication of rabbits in south-western Europe, endangered native predators feeding on rabbits declined severely, as they ran out of food (Lees and Bell 2008).

15.8.1 Logistics of biocontrol

Past experience has shown that the logistics of the pathogen release can be critically important to the success of a biocontrol project. Prevailing environmental conditions at the time of release or the presence of partially immune juveniles can both affect whether an epizootic takes off. Where a biocontrol agent release is used in conjunction with other population control methods, as was the case in the eradication campaigns of cats on Marion Island, it is important to understand how the different eradication methods may affect each other (see Section 15.4).

15.8.2 Required prerelease knowledge

Because once released, pathogens intended for biological control cannot be recalled, it is important to evaluate adverse ecological effects and contingencies ahead of time. At the heart of the information needed is the type and rate of transmission in the focal species under varying densities and different conditions (e.g., different temperature and moisture regimes; McColl et al. 2014). Related, and also of obvious importance, are the types of impacts on the focal species (e.g., effects on fecundity or on mortality). Also important is to understand whether the relationship between biocontrol agent mortality and other types of mortality, such as predation or food availability, is additive, compensatory, or interactive. Last, but not least, it is essential to investigate interactions with other pathogens that might be circulating in the population, both in terms of interactive impacts on the host (Wells et al. 2018), and in terms of cross-immunity and competition that may undermine the effectiveness of the pathogen released (Mahar et al. 2018). Risk of infection for other nontarget hosts needs to be evaluated based on phylogenetic proximity to the biocontrol target species, as well as on the type of pathogen (e.g., RNA viruses are more likely than DNA viruses to successfully "jump" host species because of the higher mutation rate of the former; Longdon et al. 2014). Beyond the pathogen itself, managers need to understand the trickle-down effects of the decline of the target population on other local species and on the greater ecosystem in general. All of this information is best obtained using a combination of approaches that include computer modeling exercises, in vitro and in vivo laboratory studies, as well as ecological field experiments.

15.8.3 Postrelease activities

Following the release of an agent, it is critical to monitor and assess the effects of the introduction, both in the biocontrol target population (including the other pathogens circulating) and the broader species community. These effects will likely change over time and will likely be most pronounced in the first years following the release (McColl et al. 2014). Past experience has also shown that impacts on the target population will be strongest if the biocontrol agent release is accompanied by a sustained array of other activities that will further suppress the population. For example, in the case of Australian rabbits, recent RHDV releases have had little long-term effect on rabbit populations unless they were followed by other suppression activities such as mechanical control (removal of breeding harbor via warren ripping) or chemical control, yet such activities get routinely forgotten (Elsworth 2019).

In summary, biocontrol research is expensive and represents a high-investment, high-payout enterprise. Past experience has shown that diseases as biocontrol agents are most successful if they are released in a well-circumscribed region, and on a host that does not have close evolutionary relatives in the area. For a variety of reasons, it is difficult for pathogens to achieve complete extinction of their hosts (deCastro and Bolker 2005). As a result, they are best used to rapidly reduce target species populations rather than achieving complete host extinction, unless they are released in a limited size area and used in conjunction with other control methods. Pathogen biocontrol efforts are most successful if they are preceded by careful planning and if ongoing evaluation and adaptive management techniques are used. Finally, both regulatory oversight and societal acceptance are necessary ingredients to ensure a successful outcome.

References

Barnett, L.K., Prowse, T.A., Peacock, D.E., Mutze, G.J., Sinclair, R.G., Kovaliski, J., et al. (2018). Previous exposure to myxoma virus reduces survival of European rabbits during outbreaks of rabbit haemorrhagic disease. *Journal of Applied Ecology*, **55**:2954–62.

Becker, J.A., Ward, M.P., and Hick, P.M. (2019). An epidemiologic model of koi herpesvirus (KHV) biocontrol for carp in Australia. *Australian Zoologist*, **40**:25–35.

Bester, M.N., Bloomer, J.P., Van Aarde, R.J., Erasmus, B.H., Van Rensburg, P.J., Skinner, J.D., et al. (2002). A review of the successful eradication of feral cats from sub-Antarctic Marion Island, Southern Indian Ocean. *South African Journal of Wildlife Research*, **32**:65–73.

Brown, P. and Gilligan, D. (2014). Optimising an integrated pest-management strategy for a spatially structured population of common carp (*Cyprinus carpio*) using meta-population modelling. *Marine and Freshwater Research*, **65**:538–50.

Bruce, J.S. and Twigg, L.E. (2005). The reintroduction, and subsequent impact, of rabbit haemorrhagic disease virus in a population of wild rabbits in south-western Australia. *Wildlife Research*, **32**:139–50.

Bruce, J.S., Twigg, L.E., and Gray, G.S. (2004). The epidemiology of rabbit haemorrhagic disease, and its impact on rabbit populations, in south-western Australia. *Wildlife Research*, **31**:31–49.

Caudell, J.N., Whittier, J., and Conover, M.R. (2002). The effects of haemogregarine-like parasites on brown tree snakes (*Boiga irregularis*) and slatey-grey snakes (*Stegonotus cucullatus*) in Queensland, Australia. *International Biodeterioration & Biodegradation*, **49**:113–19.

Cleaveland, S., Thirgood, S., and Laurenson, K. (1999). Pathogens as allies in island conservation? *Trends in Ecology & Evolution*, **14**:83.

Cooke, B., Springer, K., Capucci, L., and Mutze, G. (2017). Rabbit haemorrhagic disease: Macquarie Island rabbit eradication adds to knowledge on both pest control and epidemiology. *Wildlife Research*, **44**:93–96.

Cooke, B.D. and Fenner, F. (2002). Rabbit haemorrhagic disease and the biological control of wild rabbits, *Oryctolagus cuniculus*, in Australia and New Zealand. *Wildlife Research*, **29** (6):689–706.

Courchamp, F. and Sugihara, G. (1999). Modeling the biological control of an alien predator to protect island species from extinction. *Ecological Applications*, 9:112–23.

De Castro, F. and Bolker, B. (2005). Mechanisms of disease-induced extinction. *Ecology Letters*, **8**:117–26.

Elsworth, P.G. (2019). Reorganising the rabbit control toolbox: Do we need to reach for virus first? *Proceedings of the Pest Animal and Weed Symposium*, **2019**:139–44.

Elsworth, P.G., Kovaliski, J. and Cooke, B.D., 2012. Rabbit haemorrhagic disease: are Australian rabbits (*Oryctolagus cuniculus*) evolving resistance to infection with Czech CAPM 351 RHDV?. *Epidemiology & Infection*, 140: 1972-81.

Fenner, F. and Fantini, B. (1999). *Biological Control of Vertebrate Pests: The History of Myxomatosis; An Experiment in Evolution*. CAB International, Wallingford.

Gilna, B., Lindenmayer, D.B., and Viggers, K.L. (2005). Dangers of New Zealand possum biocontrol research to endogenous Australian fauna. *Conservation Biology*, **19**:2030–32.

Hood, G.M., Chesson, P., and Pech, R.P. (2000). Biological control using sterilizing viruses: Host suppression and competition between viruses in non-spatial models. *Journal of Applied Ecology*, **37**:914–25.

Ji, W. (2009). A review of the potential of fertility control to manage brushtail possums in New Zealand. *Human–Wildlife Conflicts*, **3**:20–29.

Koehn, J.D. (2004). Carp (*Cyprinus carpio*) as a powerful invader in Australian waterways. *Freshwater Biology*, **49**:882–94.

Kopf, R.K., Boutier, M., Finlayson, C.M., Hodges, K., Humphries, P., King, A., et al. (2019). Biocontrol in Australia: can a carp herpesvirus (CyHV-3) deliver safe and effective ecological restoration? *Biological Invasions*, **21**:1857–70.

Kovaliski, J. (1998). Monitoring the spread of rabbit hemorrhagic disease virus as a new biological agent for control of wild European rabbits in Australia. *Journal of Wildlife Diseases*, **34**:421–28.

Lees, A.C. and Bell, D.J. (2008). A conservation paradox for the 21st century: The European wild rabbit *Oryctolagus cuniculus*, an invasive alien and an endangered native species. *Mammal Review*, **38**:304–20.

Lighten, J. and Van Oosterhout, C. (2017). Biocontrol of common carp in Australia poses risks to biosecurity. *Nature Ecology & Evolution*, **1**:0087.

Longdon, B., Brockhurst, M.A., Russell, C.A., Welch, J.J., and Jiggins, F.M. (2014). The evolution and genetics of virus host shifts. *PLoS Pathogens*, **10**:p.e1004395.

Mahar, J.E., Hall, R.N., Peacock, D., Kovaliski, J., Piper, M., Mourant, R., et al. (2018). Rabbit hemorrhagic disease virus 2 (RHDV2; GI. 2) is replacing endemic strains of RHDV in the Australian landscape within 18 months of its arrival. *Journal of Virology*, **92**:e01374–17.

Marshall, J., Davison, A.J., Kopf, R.K., Boutier, M., Stevenson, P., and Vanderplasschen, A. (2018). Biocontrol of invasive carp: Risks abound. *Science*, **359**:877.

May, R.M. and Anderson, R.M. (1983). Epidemiology and genetics in the coevolution of parasites and hosts. *Proceedings of the Royal Society of London B: Biological Sciences*, **219**:281–313.

McColl, K. (2016). *Final Report: Phase 3 of the Carp Herpesvirus Project (CyHV-3)*. PestSmart Toolkit publication. Invasive Animals Cooperative Research Centre, Canberra.

McColl, K.A. and Barwick, M. (2018). Re: The release of the Koi herpesvirus. *Australian Veterinary Journal*, **96**:N24–25.

McColl, K.A., Cooke, B.D., and Sunarto, A. (2014). Viral biocontrol of invasive vertebrates: Lessons from the past applied to cyprinid herpesvirus-3 and carp (*Cyprinus carpio*) control in Australia. *Biological Control*, **72**:109–17.

McColl, K.A., Sheppard, A.W., and Barwick, M. (2017). Safe and effective biocontrol of common carp. *Nature Ecology & Evolution*, **1**:1–1.

McColl, K.A., Sunarto, A., and Holmes, E.C. (2016). Cyprinid herpesvirus 3 and its evolutionary future as a biological control agent for carp in Australia. *Virology Journal*, 13: 1-4.

McColl, K.A., Sunarto, A., and Neave, M.J. (2018). Biocontrol of carp: More than just a herpesvirus. *Frontiers in Microbiology*, **9**:2288.

McDonald, R.A. and Lariviere, S. (2001). Diseases and pathogens of *Mustela* spp., with special reference to the biological control of introduced stoat *Mustela erminea* populations in New Zealand. *Journal of the Royal Society of New Zealand*, **31**:721–44.

McLeod, S.R., Saunders, G., Twigg, L.E., Arthur, A.D., Ramsey, D., and Hinds, L.A. (2007). Prospects for the future: Is there a role for virally vectored immunocontraception in vertebrate pest management? *Wildlife Research*, **34**:555–66.

Mintram, K.S., Van Oosterhout, C., and Lighten, J. (2020). Genetic variation in resistance and high fecundity impede viral biocontrol of invasive fish. *Journal of Applied Ecology*, **58**:148–57.

Moreno, S., Beltrán, J.F., Cotilla, I., Kuffner, B., Laffite, R., Jordán, G., et al. (2008). Long-term decline of the European wild rabbit (*Oryctolagus cuniculus*) in southwestern Spain. *Wildlife Research*, **34**:652–58.

Mutze, G., Kovaliski, J., Butler, K., Capucci, L., and McPhee, S. (2010). The effect of rabbit population control programmes on the impact of rabbit haemorrhagic disease in south-eastern Australia. *Journal of Applied Ecology*, **47**:1137–46.

Mutze, G.J., Sinclair, R.G., Peacock, D.E., Capucci, L., and Kovaliski, J. (2014). Is increased juvenile infection the key to recovery of wild rabbit populations from the impact of rabbit haemorrhagic disease? *European Journal of Wildlife Research*, **60**:489–99.

Oliveira, N.M. and Hilker, F.M. (2010). Modelling disease introduction as biological control of invasive predators to preserve endangered prey. *Bulletin of Mathematical Biology*, **72**:444–68.

Ramsey, D.S., Cox, T., Strive, T., Forsyth, D.M., Stuart, I., Hall, R., et al. (2020). Emerging RHDV2 suppresses the impact of endemic and novel strains of RHDV on wild rabbit populations. *Journal of Applied Ecology*, **57**:630–41.

Ross, J. and Tittensor, A.M. (1986). The establishment and spread of myxomatosis and its effect on rabbit populations. *Philosophical Transactions of the Royal Society of London B: Biological Sciences*, **314**:599–606.

Saunders, G., Cooke, B., McColl, K., Shine, R., and Peacock, T. (2010). Modern approaches for the biological control of vertebrate pests: An Australian perspective. *Biological Control*, **52**:288–95.

Secretariat of the Convention on Biological Diversity. (2000). *Cartagena Protocol on Biosafety to the Convention on Biological Diversity: Text and Annexes*. Secretariat of the Convention on Biological Diversity, Montreal.

Shanmuganathan, T., Pallister, J., Doody, S., McCallum, H., Robinson, T., Sheppard, A., et al. (2010). Biological control of the cane toad in Australia: A review. *Animal Conservation*, **13** (Suppl. 1):16–23.

Shine, R. (2010). The ecological impact of invasive cane toads (*Bufo marinus*) in Australia. *Quarterly Review of Biology*, **85**:253–91.

Singleton, G.R. (1994). The prospects and associated challenges for the biological control of rodents. *Proceedings of the Vertebrate Pest Conference*, **16**:301–07.

Strive, T., Elsworth, P., Liu, J.N., Wright, J.D., Kovaliski, J., and Capucci, L. (2013). The non-pathogenic Australian rabbit calicivirus RCV-A1 provides temporal and partial cross protection to lethal rabbit haemorrhagic disease virus infection which is not dependent on antibody titres. *Veterinary Research*, 44 (1):51.

Sumption, K.J. and Flowerdew, J.R. (1985). The ecological effects of the decline in rabbits (*Oryctolagus cuniculus* L.) due to myxomatosis. *Mammal Review*, 15 (4): 151–86.

Twigg, L.E., Lowe, T.J., Martin, G.R., Wheeler, A.G., Gray, G.S., Griffin, S.L., et al. (2000). Effects of surgically imposed sterility on free-ranging rabbit populations. *Journal of Applied Ecology*, **37**:16–39.

Van Rensburg, P.J., Skinner, J.D., and Van Aarde, R.J. (1987). Effects of feline panleucopaenia on the population characteristics of feral cats on Marion Island. *Journal of Applied Ecology*, **24**:63–73.

Vincent, C., Goettel, M.S., and G. Lazarovits. (2007). *Biological Control: A Global Perspective: Case Studies from Around the World*. CAB International, Wallingford.

Wells, K., Brook, B.W., Lacy, R.C., Mutze, G.J., Peacock, D.E., Sinclair, R.G., et al. (2015). Timing and severity of immunizing diseases in rabbits is controlled by seasonal matching of host and pathogen dynamics. *Journal of the Royal Society Interface*, 12, 20141184 .

Wells, K., Fordham, D.A., Brook, B.W., Cassey, P., Cox, T., O'Hara, R.B., et al. (2018). Disentangling synergistic disease dynamics: Implications for the viral biocontrol of rabbits. *Journal of Animal Ecology*, **87**:1418–28.

Williams, C.K., Davey, C.C., Moore, R.J., Hinds, L.A., Silvers, L.E., Kerr, P.J., et al. (2007). Population responses to sterility imposed on female European rabbits. *Journal of Applied Ecology*, **44**:291–301.

CHAPTER 16

Ethical and Public Outreach Considerations

16.1 Introduction

Wildlife diseases have the potential to cause severe impacts to wildlife and domestic animal populations and are often precipitated by human activities (Dobson and Foufopoulos 2001). Because zoonotic wildlife pathogens can have profound implications for human well-being and health, they need to be managed with corresponding seriousness (Figure 16.1). Wildlife disease management is a complex field that involves invasive activities such as culling of wildlife or extensive environmental manipulations, which people may find objectionable for a variety of practical and ethical reasons. Beyond the strictly scientific dimensions of disease management, two additional aspects need to be considered carefully, as they can determine the success or failure of a disease management campaign. First, explicit consideration of the ethical dimensions of a campaign is crucial because they frequently shape public opinion and as a result determine support and "buy-in" of the general public to a disease management program. Related to this, effective communication and public outreach are critically important because they not only facilitate the establishment of consensus, but also result in a more informed and hopefully supportive general public. These two aspects of a wildlife disease management effort will be the focus of this chapter and will be discussed individually in separate subsections.

16.2 Ethics in wildlife disease management

Wildlife management, whether in the context of infection or not, is frequently invasive in nature and has the potential to raise many ethical concerns (Table 16.1). As a result, public attitudes toward wildlife disease management can vary widely between full embracement and support on one side, and open hostility on the other. Underlying and informing these differing public attitudes are typically powerful systems of ethical beliefs that can diverge widely across space, time, and socioeconomic background. They provide the context against which any disease management campaign needs to be evaluated.

A variety of philosophies exist regarding the responsibilities that humankind has toward the environment, and how environmental resources are best managed. Two of the more prominent schools of thought are conservation (sometimes referred to as shepherdism) and preservationism (also referred to as biocentrism). These philosophies differ strongly in their prescriptions of how humans could, and should, interact with nature, and their proponents are therefore engaged in heated debates about the exact nature of environmental policy. For example, Isle Royale National Park is a federally designated wilderness area located in Lake Superior, near the US–Canadian border. It is home

Infectious Disease Ecology and Conservation. Johannes Foufopoulos, Gary A. Wobeser and Hamish McCallum, Oxford University Press.
© Johannes Foufopoulos, Gary A. Wobeser and Hamish McCallum (2022). DOI: 10.1093/oso/9780199583508.003.0016

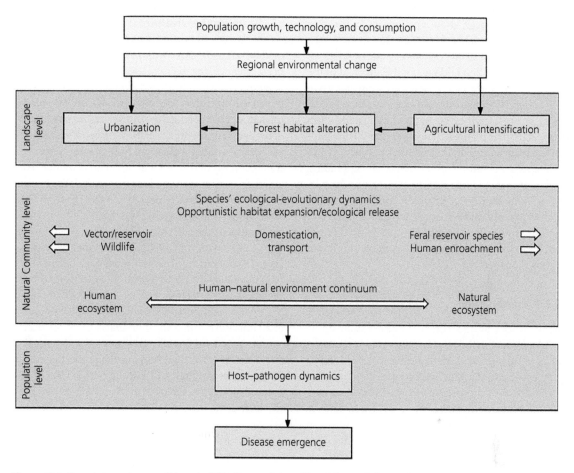

Figure 16.1 Nexus between human activity and wildlife diseases. (Adapted from Wilcox and Ellis 2006.)

Table 16.1 Ethical issues linked to types of wildlife management and conservation practices (adapted from Minteer and Collins 2008).

Practice type	Ethical issues (examples)
Sampling techniques (e.g., animal tissue)	Harm to sentient creatures; increased extinction risk imposed on endangered populations
Wildlife population control (e.g., culling, biological control, sterilization, translocation)	Harm to sentient creatures; evaluating lethal vs. nonlethal control techniques
Choosing conservation targets	Justifying differential valuation of species; value conflicts between ecosystem integrity and animal welfare
Release/reintroduction of captive-born or confiscated animals into the wild	Harm to sentient creatures; issues of human agency/intervention in reintroductions; introduction of diseases
Treating wildlife disease	Interference in the dynamics of wild populations; risks of unforeseen consequences of treatment; positive duties to promote wildlife health through treatment interventions

to a small and declining wolf (*Canis lupus*) population that has been showing signs of inbreeding depression (Hedrick et al. 2014; Robinson et al. 2019). To alleviate the negative consequences of genetic deterioration and ensure the survival of the resident wolf population—and following extensive debate—the National Park Service introduced additional wolves from other populations (Räikkönen et al. 2009; US National Park Service 2018). Supporters of this intervention would be considered to exhibit a conservation mindset. In contrast, holding the belief that it is important to let nature "run its course" and therefore not to intervene, would reflect a preservation mindset. In the sections that follow, we will further examine both schools of thought and, in turn, explore them in the context of wildlife disease management.

16.2.1 Conservation

The principle at the core of conservationism is the idea of facilitating a nondestructive interaction between humans and the natural world, essentially by assuming that the human role in the natural world can be a nonnegative one. While humans need to extract and utilize resources from the natural world, conservationists argue that this can be done in a sustainable manner. The idea that humans can engage in nonnegative interactions with the environment has been an important point of contention between conservationists and preservationists (Vantassel 2008). In the case of wildlife disease management, the debate in practice often boils down to two questions: First, should there be action taken to manage an epizootic? The answer to this is determined by how we weigh individual host suffering against the various economic costs resulting from an epizootic. Second, is it acceptable to impose limitations on, and changes to human behaviors (e.g., by putting up fences and policies to dissuade certain interactions with wildlife), or is it only acceptable to alter the conditions for animals (e.g., through relocation, culling, fences, or other control methods)? This answer depends on the willingness to accept human inconvenience.

Ethicists define moral behaviors as behaviors that lessen or avoid causing pain (defined broadly) to organisms capable of experiencing pain and weighing the value of human pain against the value of pain in other creatures (Lockwood 1988). This principle has important implications for wildlife disease control. For example, wildlife managers must balance the pain caused by disease control actions to wildlife host and even arthropod vector populations against the pain caused by *not* enacting a policy. Choosing the appropriate management activities may involve complex trade-offs between individual animal well-being (which may suffer the repercussions of disease management activities like culling), versus the well-being of populations and even whole species (which would benefit from the eradication of a pathogen).

Farmers in the US state of Michigan were faced with this situation in 1998 when bovine tuberculosis (bTB) spread from white-tailed deer (*Odocoileus virginianus*) to domestic cattle. In order to prevent the spread of infection, farmers often had to eliminate large numbers of their livestock (Carstensen et al. 2011). Although the euthanasia practices used caused pain to the cattle, this method was judged to be less painful than leaving the cattle to possibly become infected and succumb to bTB. Other control strategies for bTB used culling to reduce the overall density of white-tailed deer, as well as selectively removing infected animals. Control measures included extended hunting seasons, reduced restrictions on hunting, and implementing a feeding ban (Cool et al. 1997)—all of which surely caused some degree of pain to the deer. However, the pain resulting from these practices was judged by policy makers to be less than the pain that would be incurred by deer, cattle, and even humans, if deer populations were allowed to surge unabated, providing conditions for bTB to flourish.

The previous example illustrates a typical conservationist approach to the management of bTB in white-tailed deer and cattle. As with any wildlife management and disease control practices, the proposed control measures had varying levels of public support. The greater the chances of human suffering, the higher the public support for more invasive control strategies (Reiter et al. 1999; Peterson et al. 2006). This is not surprising: the interest in self-preservation is a potent motivator, and it powerfully shapes the public perception of what constitutes ethical measures in wildlife control.

16.2.2 Preservationism

The ethics of preservationism are derived from bio-centrism, a philosophy that places all living things on equal ethical footing and does not value human pain above the pain of any other living being (Taylor 1986). Preservationism also advocates for minimizing human impacts to the environment and would typically assert that the best disease management is no management at all. Some authors (e.g., Katz 1997, also see Hettinger 2002) go as far as arguing that humans are not capable of protecting the natural world without succumbing to anthropocentric prejudices, because wishing to do so satisfies a human desire, and therefore, humans have no business trying to restore ecosystems through human means (Katz 1997). While such views may fall outside of mainstream opinions, the majority of preservationists merely wish for natural systems to be left alone without human disturbance. In the case of the wildlife disease management, this would imply letting epizootics run their course, with whatever implications there might be for both wildlife and human well-being.

16.2.3 A special case: Global eradication of pathogens or parasites

Disease management ethics debates center mostly on the rights of various groups of vertebrate hosts and stakeholders, while the ethics of parasite elimination are rarely discussed. The moral issue of parasite extirpation typically arises only in the special situation where transmission has been interrupted to the point where the parasite is in the process of going extinct. The global eradication of a pathogen from its natural hosts has at this point occurred twice, once for a human microparasite (smallpox virus; Pennington 2003), and once for an animal microparasite (rinderpest virus; Normile 2008, 2010). While the intentional extirpation of a parasitic organism is still a highly unlikely event, this is a situation that, given the rapid expansion of human activities, is likely to occur with increasing frequency in the future (see also Chapter 1, Section 1.5). Eradicating a wild species, even if it is a pathogen normally deemed harmful, is a step that should not be taken lightly since it involves a complex web of ethical but also logistic and ecological ramifications. Thus, beyond the ethical question on humanity's moral right to cause the willful disappearance of a wild species, there are various practical implications of this organism's extinction on natural food webs and on human activities. Traditionally this debate has broken down into two opposing camps of "destructionists" versus "preservationists." Nonetheless, advances in synthetic biology are rendering this debate increasingly moot. The ability to increasingly re-create pathogens that have gone extinct will remove the element of extinction irrevocability from these difficult pathogen elimination debates (Kupferschmidt 2017).

16.2.4 Implications for wildlife disease management

Managing wildlife diseases typically involves weighing the interests of humans against animal welfare and conservation concerns. Invasive and expensive processes such as culling, vaccination, and individual animal patient treatment raise substantial ethics considerations, and there is currently no universal ethical basis on which to inform these complex decisions (Minteer and Collins 2008). Although there are well-documented methods to control disease in domestic animal populations, the feasibility of control options in wildlife is restricted (see Chapter 13). Sufficient treatment of infected populations is in general near impossible because of the scale of such an effort, which involves at a minimum capture, detection and diagnosis, successful treatment, and finally release of a substantial fraction of a wild species population. Although this approach may alleviate suffering in individual animals, it is unlikely to slow down significantly the spread of an outbreak (McCallum and Hocking 2005).

Vaccination has been shown to be effective as a measure for controlling some diseases, but it raises ethical questions. Success of large-scale rabies vaccination programs of mesopredators in North America and Northern Europe has been typically attributed to the fact that the vaccine can be orally delivered to a single target species (Cross et al. 2007). However, since the rabies vaccine virus is a recombinant self-replicating entity, and since delivery methods often use baiting systems, there are nontrivial ethical concerns over potential hazards to nontarget species (Rupprecht et al. 2002).

Because wildlife infectious diseases typically need to be addressed at the population level, treatment and vaccinations are frequently cost-prohibitive control measures. This oftentimes leaves culling as the most feasible and economical strategy, although it is also the most contested and controversial option. In 1992, the European Union adopted a nonvaccination policy to combat major outbreaks of foot and mouth disease, avian influenza, and classical swine fever in domesticated animal populations (Cohen et al. 2007). Under this policy, millions of infected and healthy animals were culled in efforts to eradicate the diseases, causing trauma to the people involved and raising questions about animal welfare and the ethics and morality of large-scale culling. In addition, the general public was confronted with news coverage of improper handling and slaughter of animals, as well as the distress and anger of farmers and animal handlers. This situation galvanized the European Union into reevaluating the nonvaccination policy and to discuss alternative methods that would have broader general public support.

16.2.5 Ethics of wildlife disease management: Some practical suggestions

The effort to make ethics more relevant in environmental policy and practices has gained notable momentum since the 1990s (Light and Katz 1996). Although currently no universal tools of ethical analysis exist with respect to everyday wildlife management decisions, the increased attention to this aspect of environmental management has led to the emerging field of ecological ethics. Minteer and Collins (2008) proposed an ecological ethics model where ethics fits within the conceptual scheme of applied practices. In this model, ecological ethics is situated between the fields of ecology and conservation biology on the one hand and normative ethical theory on the other (see Figure 16.2).

Ecological ethics therefore both influences, and is shaped, by ecological research and conservation practices, in addition to relevant professional codes of conduct that convey professional and scientific values, and which promote responsible conduct and accountability. Ecological ethics rests on established normative ethical theories and aims to provide researchers and conservation managers with a practical tool for identifying, clarifying, and coordinating values and positions in difficult research and management situations.

Current practical applications involve engaging in thoughtful deliberations of arguments representing different ethical approaches to wildlife disease. In practice, ethical considerations follow an ethical discourse that involves:

1. Cataloging all possible reasons for and against a particular management decision. This step involves ensuring that all stakeholder concerns have been identified.
2. Developing and assessing all arguments, which involves treating each identified reason as a premise, and identifying any missing or invalid premises. This step can be further broken down into three strategies: (i) hold the conclusion fixed and determine if the premises leading to that conclusion are valid; (ii) correct invalid premises and determine if the conclusion is still supported; or (iii) correct invalid premises and revise conclusion to keep overall argument valid.
3. Synthesis of all arguments analyzed (Decker et al. 2012).

This ethical discourse process is especially practicable when integrated into stakeholder involvement strategies such as workshops (further discussed in Section 16.3.2).

This section highlights the difficult ethical questions wildlife managers must face as part of their day-to-day jobs. Ethical considerations are imperative to the wildlife disease control management process, especially in light of changing public views on animal worth and welfare. Where society once only considered animals for their economic value, the past century has experienced a monumental shift toward a more respectful and protective view. Several high-profile cases related to conservation actions and research have been halted altogether due to media coverage from animal rights groups and public outrage (Dalton 2005; Jones et al. 2012). This holds important implications for wildlife management and speaks to the increasing significance of public relations and outreach.

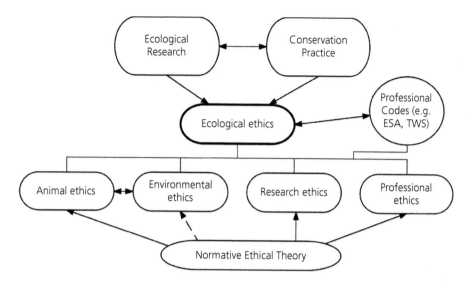

Figure 16.2 Ecological ethics model ESA = Ecological Society of America; TWS = The Wildlife Society. (Adapted and redrawn from Minteer and Collins 2008).

16.3 Wildlife disease management and public outreach

Because of the diverse and often emotionally charged opinions that are held about wildlife and its value, it is critically important to consider the public's reaction to a wildlife management plan, especially if it involves controversial actions such as culling or removal of diseased animals. Indeed in some cases, public acceptance has the ability to shape management decisions altogether. Wildlife managers are therefore tasked with the difficult job of selecting a plan of action to handle problematic wildlife situations by integrating multiple, often contradictory considerations. These include weighing public attitudes against scientific considerations all in the context of the relevant policy circumstances when determining a particular management action (Figure 16.3).

16.3.1 Communicating information about wildlife-associated disease: Interacting with the community

While transparency in wildlife disease management is generally recommended, it becomes particularly important if a wildlife management action may be deemed controversial in the public realm. Citizens who are engaged in a thoughtful way, and are educated about the long-term costs and benefits

of wildlife disease management regarding both the affected species and the ecosystem in general, are also more likely to understand and support the management practice in question.

16.3.1.1 Standard practices for internal collaboration

Past experience suggests that success both in the realm of disease control and effective communication to the general public depends on the establishment of effective interdisciplinary collaborations and communication. Management of wildlife epizootics frequently necessitates the involvement of multidisciplinary teams of specialists, including veterinarians, wildlife ecologists, and—in the case of zoonotic pathogens—public health professionals (Rodriguez and Peterson 2020). Underscoring the importance of open communication channels, Decker et al. (2011) highlight partnerships between three specific groups whose collaboration is critical for effective disease outbreak management (Figure 16.4):

1. Wildlife veterinarians and wildlife management professionals.
2. Wildlife management professionals and public health professionals.
3. Wildlife professionals and stakeholders (i.e., the public).

Figure 16.3 Multiple, often overlapping factors influence wildlife disease management. (Adapted from Decker et al. 2012.)

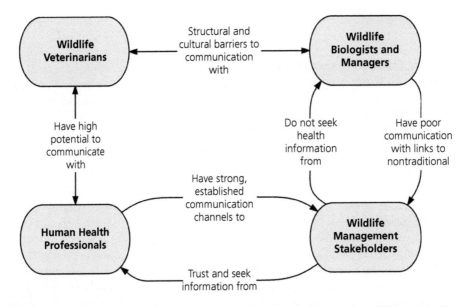

Figure 16.4 Schematic representation of communication relationships among key players during a typical wildlife-associated disease risk communication effort. (Adapted from Decker et al. 2011.)

A particularly sensitive situation arises when a zoonotic pathogen spills into a domesticated animal or even human population (Box 16.1). In these circumstances, the media tends to turn toward the human health professional community when a zoonotic disease is first detected, even though this community may not have the most experience with the particular disease. For example, during the 2020 COVID pandemic, even in the early spillover stages, communication to the general public was taken over by medical professionals despite the fact that they were less familiar with the zoonotic origins of the virus. Collaborating with the public health professional community prior to the rush of news accompanying the emergence of a disease is therefore essential for wildlife professionals aiming to communicate accurate messages to the public. Perceived risk rather than actual risk plays a larger role in shaping public opinion, and media outlets can contribute to the social amplification of risk events by affecting cognitions, affect, attitudes, and behavioral intentions (Heberlein and Stedman 2009). Risk communication therefore needs to be conducted with great care, and messages originating from the professional community should be noninflammatory to avoid inflating risk perceptions.

As standard practice, all professionals should speak with a unified voice and avoid contradicting each other. Beyond this coordination, different wildlife professionals will likely focus on different tasks: for example, wildlife veterinarians tend to focus on offering medical expertise while wildlife managers and biologists often have more connections with stakeholders. Medical training and knowledge can also allow wildlife veterinarians to serve as a liaison between wildlife managers or biologists and public health professionals, a collaboration that can help communicate a "balanced message about zoonotic disease to the public: one that provides for appropriate cautions but does not inadvertently demonize wildlife" (Decker et al. 2011).

A review of wildlife infectious disease prevention and control systems proposes modeling a national/international control structure based upon the American Committee on Arthropod-borne Viruses system (Murphy et al. 1995). Such a system would focus first on organizing the professional community involved with wildlife infectious diseases and encouraging cooperation and collaboration. Any epidemiology program within this wildlife-focused system should integrate data on (i) the role of disease agents; (ii) the factors affecting host susceptibility; and (iii) modes of transmission, to help shape overall management actions. The authors further propose the creation of a national or international database that would provide information on prevention and control.

Ultimately, it is the strong collaboration between the different wildlife, public health, and veterinary professionals that underlies both the sound wildlife health management decision-making process, and the successful communication with stakeholders and the general public.

Box 16.1 The comparative management of bovine tuberculosis (bTB) in two US states

The detection of bTB in wildlife populations requires immediate and often aggressive responses to prevent widespread transmission to domestic livestock and risk to human health. Public acceptance can significantly influence the types of control measures implemented and the overall success of disease management efforts. Carstensen et al. (2011) provide a case study comparing the management strategies of the Minnesota DNR and Michigan DNR on two separate outbreaks of bTB.

Figure 1 Winter feeding of deer (here red deer in Denmark) often leads to crowding, which in turn can promote disease transmission. (*Photo:* Jens Cederskjold, under Creative Commons Attribution 3.0.)

bTB was first detected in domestic cattle and deer in north-western Minnesota in 2005. The discovery prompted aggressive disease deer management actions that included a combination of liberalized hunting, targeted culls, and bans on recreational winter feeding. The last activity can lead to aggregation of deer during winter and facilitate the spread of the disease in the affected area (Figure 1). The DNR's management strategy reduced the disease's incidence to an undetectable level, and was successful largely due to support from cattle ranchers, hunters, politicians, and the general public.

In contrast, bTB was first discovered in Michigan deer in 1975 but control measures were not put into place until 1995. By then, the disease had spread to domestic cattle populations and culture-positive deer had been found as far as 116 miles from the outbreak area. Although liberalized hunting and bans on baiting and feeding have reduced the disease prevalence, the lack of widespread

---

Box 16.1 *Continued*

support from hunters, farmers, and the general public for more aggressive disease control measures (e.g., culling) led to a prolonged outbreak period.

16.3.1.2 Standard practices for communicating risk and management actions to stakeholders

Engaging the public in a wildlife removal or control management plan should ideally strike a balance between being able to implement effective action, and achieving public support and understanding. Early and continuous involvement of the affected community in wildlife management decisions should serve as a standard practice, particularly when it involves controversial actions such as a cull (Chess and Purcell 1999; Reed 2008). If stakeholders are initially neglected, they may enter the process later to challenge management decisions, often through litigation that can suspend implementation and lead to economic losses (Decker et al. 2012). For example, when chronic wasting disease (CWD) was discovered in deer in the US state of Wisconsin in 2002, the Wisconsin Department of Natural Resources (DNR) responded by making hasty decisions and neglecting stakeholder concerns (Heberlein 2004). By 2004, the Wisconsin DNR had failed to achieve their goal of minimizing "the negative impact of chronic wasting diseases on cervid populations, the state's economy, hunters, landowners, and other people dependent upon healthy, wild, and farmed populations of deer and elk" (Bartelt et al. 2003). Deer hunt licensing dropped drastically, revenues to the agency decreased, and deer densities in the eradication

zone remained high. Additionally, landowners who had opposed the management decision signed a petition not to partake in the proposed cull and wildlife recreational viewers persuaded the legislature to end the statewide feeding ban. Had the Wisconsin DNR implemented a long-term management strategy that involved stakeholders from the beginning, they might have been able to effectively eradicate CWD with fewer negative human impacts.

Public engagement is most effective when outreach strategies match the geographical, temporal, and social scale of the issue at hand, as public acceptance and viewpoints vary regionally. In order to facilitate effective communication with stakeholders, wildlife managers must be able to discern their various interests, beliefs, and characteristics. Stakeholder analysis is often used to systematically assess sectors of the public relevant to the decision at hand. It is a process that (i) identifies potential sociocultural issues that may emerge as a result of a management action; (ii) identifies individuals or groups that are potentially affected by a management action; and (iii) characterizes these individuals and groups for involvement (Reed 2008). When it comes to its positions on a controversial issue, the public can be characterized into three types: (i) a nonaware public; (ii) a latently aware public; and (iii) an active public (Grunig 1983). Active individuals are typically stakeholders who are directly affected by a management decision and are primary in the decision-making process; latently aware individuals are those who disseminate information about management actions; and nonaware public individuals are those who are potentially impacted by management decisions, but are unaware of them (Table 16.2). By identifying these various stakeholder types,

---

Table 16.2 Communication strategies for types of publics (adapted from Decker et al. 2012).

Type of public	Strategy	Media type	Messages	Audiences
Nonaware	Education	Media, publications	Define problem	General public
Latently aware	Education, recruitment, conversion activities	Mixed	Focus on perceived constraints	Targeted audiences
Active	Meetings, consultations, demonstration projects, reward activities	Interpersonal, small groups, organizational	Involving, focus on data, solutions	Known audiences

wildlife managers can assess the level of problem recognition, constraint recognition, and level of involvement, and coordinate public outreach and strategize accordingly.

16.3.2 Stakeholder engagement methods

Once public demographics and interests have been identified, several stakeholder engagement methods can be employed (Jacobson et al. 2020):

1. Dissemination of information through mass media outlets like social media, mailing lists, and press releases.
2. Setting up of open houses and public meetings as interactive outreach opportunities.
3. Solicitation of comments from stakeholders through public comment periods.
4. Distribution of surveys by telephone, mail, in-person interviews, or over the internet.
5. Use of focus groups for information-gathering sessions.
6. Establishment of workshops, task forces, and advisory committees.
7. Organization of large-group planning and decision-making processes such as conferences.
8. Negotiation and ratification of agreements between the management agency and stakeholders.

Information dissemination methods involve distributing print and electronic media (i.e., factsheets, leaflets) using multiple outlets, including at visitor and educational centers, agency websites, mailing lists, and through social media (see Box 16.2). Typically, these activities are best conducted by a public outreach and communications specialist, especially if the information to be communicated involves a potentially contentious management action, or is using mass media outlets.

Box 16.2   Social media as a public outreach platform

Widespread availability and access to the internet now makes it possible for managers to disseminate more information than ever to a wider audience. Social media consists of a variety of free online communication tools that have the ability to share data across platforms, enabling widespread circulation of information. Tools for web-based outreach are always evolving so that any list is likely to be out of date, but some commonly used tools include:

Social networking sites are highly effectual means to spread information, photos, articles, and links, and to build a loyal following. Managers can use these platforms to communicate news, upcoming events, and publications. Popular social networking sites include Twitter, Facebook, Instagram, and LinkedIn.

Video and photo sharing sites like YouTube and Flickr make it easy to upload multimedia content and are useful for spreading educational messages and promoting agency actions.

Blogs serve as an interactive method to share expertise and information and solicit comments from the public. They are open to the public, and readers often leave comments as a way to interact with both agencies and other readers. Blogs are usually hosted on an agency's website and can be promoted via social media.

In addition to its ability to reach a wide audience in a short amount of time, social media can be used to monitor public interest around issues by tracking number of visitors, number of followers, viewer demographics, and user comments. Many government agencies, including the US Fish and Wildlife Service, have taken advantage of social media as a public outreach platform (Figure 16.6).

Open houses and public meetings serve as informal settings that allow for face-to-face interactions between concerned stakeholders and the public officials or managers involved in the matter (McComas 2001). These events are used both to provide information and collect public feedback on a proposed management action, and are often held in a large room that allows attendees to move among displays and learn about the components of a proposed management action. Handouts and factsheets are typically provided as supplemental material, along with relevant contact information.

Public comment periods are used to formally gather comments from the public on proposed management actions and decisions (Eisner 2012; Jacobson et al. 2020). The key to effective public comment periods is to ensure that interested stakeholders are aware of when comments will be

Figure 16.6 Links to social media platforms on the US FWS homepage.

accepted, for how long they will be accepted, and how to submit them. For formal comment periods, managers can post public notices in regionally appropriate media outlets, such as newspapers or community bulletins. Comments are typically sent by mail, e-mail, or over the internet. Prior to the initiation of a public comment period, there should be a well-organized plan in place to receive, catalogue, and respond to all comments.

Surveys are useful for gauging all perspectives around a proposed management action, and are systematically administered in-person, over the telephone, by mail, or over the internet. Given the voluntary nature of surveys, it is important to recognize that there may be voluntary response bias, which tends to bias results toward those who have strong opinions or motivations. Nonetheless, the responses gathered often reveal important information about values and beliefs that can help wildlife managers predict how the public will respond to certain management decisions. For example, a study conducted by the US Department of Agriculture's Wildlife Services program found that respondents overwhelmingly preferred nonlethal methods of control to lethal ones (Reiter et al. 1999). However, the same survey found that citizens consider human health to be of utmost importance in selecting management methods, followed by animal suffering, plan effectiveness, and environmental impacts. The results suggest that citizens want a role in wildlife management policy formation when it involves hunting, trapping, or controlling wildlife, but otherwise respect wildlife professionals' judgment in specific management situations.

A more recent survey evaluated the opinions of residents in the US state of Michigan on lethal wildlife management in the state and found that a majority of respondents would be supportive of lethal methods being used in the case of diseased wildlife (Koval and Mertig 2004). This group was

also overwhelmingly in favor of these methods when it meant that the action would ensure the species' survival, preserve ecological health, and help manage population levels. In contrast to the previously cited study, while respondents did not prefer lethal methods generally, they were overwhelmingly in support of them when they were used to manage a diseased wildlife population.

Focus groups typically consist of a few representatives from each stakeholder group. These are useful for gathering more in-depth knowledge on stakeholder concerns, but do not aim to educate the stakeholder on the overarching issue at hand (Rowe and Frewer 2000). In contrast to focus groups, workshops and task forces involve a larger subset of stakeholder representatives that are established to perform a specific task, such as developing a list of management alternatives or recommendations. Stakeholders on an advisory committee serve as informants to managers on a continuous basis, usually as part of a particular management program.

Conferences stimulate direct interactions between many different stakeholders and provide an effective avenue for exchanging knowledge and concerns. Because of the logistics of organizing and running a conference, it is often beneficial to outsource the planning to a professional event service. These services can be easily found online or through recommendations from colleagues and venues. A trained facilitator is integral to this engagement method, as stakeholders are likely to reject a policy if they are not part of the negotiating process or do not understand how they will be affected. Trained facilitators should be perceived as impartial, open to diverse perspectives, approachable, capable of maintaining positive group dynamics and open dialogue, and experienced in handling dominating or offensive individuals (Reed 2008).

Negotiated agreements are written agreements between multiple agencies and stakeholders that define each party's responsibilities when management actions are to be implemented in a co-managerial manner (Stricker et al. 2020). A memorandum of understanding is a common type of negotiated agreement and is often used when wildlife management actions fall under multiple jurisdictions, as with the case of the conservation and management of the greater sage-grouse (*Centrocercus urophasianus*) sagebrush habitats throughout the western United States and Canada (Western Association of Fish and Wildlife Agencies et al. 2008). Project-based agreements are another type of negotiated agreement and are typically instituted between agencies and the public. The Partners for Fish and Wildlife Program, administered by the US Fish and Wildlife Service (2003), is an example of a project-based agreement. The program provides technical and financial assistance to private landowners who set aside a portion of their property to help meet the habitat needs of endangered, threatened, or candidate species. Negotiated agreements are formed on a case-by-case basis and are a practical way to directly involve the local community with wildlife management.

Successfully engaging stakeholders also involves continuous evaluation of the efficacy of engagement methods. Possible obstacles to public participation include perceived inability among stakeholders to influence issues, a difficulty bridging the gap between technical material and interested laypersons, lack of time to participate, and lack of involvement due to historical, social, or cultural reasons (US EPA 2001). The ability to identify and target these root causes is an important step toward having a representative and diverse set of stakeholders, which will in turn ensure the establishment of effective and long-lasting policies.

The examples listed highlight the importance of communicating disease risk and management activities to the public. Past experience has shown that integrating stakeholder knowledge and concerns in the decision-making process tends to improve governance and accountability, and leads to increases in overall effectiveness and durability of a management action (Beierle 2002; Wagenet and Pfeffer 2007). In summary, by taking a cue from risk management research, organizations need to focus on communicating facts about the wildlife disease, as well as information about the level of risk to animal and human health, and the decisions, actions, or policies that need to be taken (Gore et al. 2006).

References

Bartelt, J., Pardee, J., and Thiede, K. (2003). *Environmental Impact Statement on Rules to Eradicate Chronic Wasting Disease in Wisconsin's Free-Ranging White-tailed Deer Herd*. Wisconsin Department of Natural Resources, Madison, WI.

Beierle, T.C. (2002). The quality of stakeholder-based decisions. *Risk Analysis*, **22**:739–49.

Carstensen, M., O'Brien, D.J., and Schmitt, S.M. (2011). Public acceptance as a determinant of management strategies for bovine tuberculosis in free-ranging U.S. wildlife. *Veterinary Microbiology*, **151**:200–04.

Chess, C. and Purcell, K. (1999). Public participation and the environment—do we know what works? *Environmental Science and Technology*, **33**:2685–92.

Cohen, N.E., Van Asseldonk, M., and Stassen, E.N. (2007). Social-ethical issues concerning control strategies of animal diseases in the European Union: A survey. *Agricultural and Human Values*, **24**:499–510.

Cool, K.L., Haveman, J., and Wyant, D. (1997). *Recommendations for Elimination of Bovine Tuberculosis in Free-Ranging White-Tailed Deer in Michigan*. Department of Fisheries and Wildlife, Michigan State University, East Lansing, MI. 58.

Cross, M.L., Buddle, B.M. , and Aldwell, F.E. (2007). The potential of oral vaccines for disease control in wildlife species. *The Veterinary Journal*, **174**:472–80.

Dalton, R. (2005). Animal-rights group sues over "disturbing" work on sea lions. *Nature*, **436**:315.

Decker, D.J., Riley, S.J. , and Siemer, W.F. (2012). *Human Dimensions of Wildlife Management*, 2nd edn. Johns Hopkins University Press, Baltimore, MD.

Decker, D.J., Siemer, W.F., Wild, M.A., Castle, K.T., Wong, D., et al. (2011). Communicating about zoonotic disease: Strategic considerations for wildlife professionals. *Wildlife Society Bulletin*, **35**:112–19.

Dobson, A. and Foufopoulos, J. (2001). Emerging infectious pathogens of wildlife. *Philosophical Transactions of the Royal Society of London B: Biological Sciences*, **356**:1001–12.

Eisner, N. (2012). *Rulemaking Requirements*. US Department of Transportation, Washington, DC. 48. https://www.transportation.gov/regulations/rulemaking-requirements-2012 (accessed July 15, 2018).

Gore, M.L., Knuth, B.A., Curtis, P.D., and Shanahan, J.E. (2006). Stakeholder perceptions of risk associated with human-black bear conflicts in New York's Adirondack Park campground: Implications for theory and practice. *Wildlife Society Bulletin*, **34**:36–43.

Grunig, J.E. (1983). *Communication Behaviors and Attitudes of Environment Publics: Two Studies*, Monograph 81. Association for Education in Journalism and Mass Communication, Columbia, SC.

Heberlein, T.A. (2004). "Fire in the Sistine Chapel": How Wisconsin responded to chronic wasting disease. *Human Dimensions of Wildlife*, **9** (3):165–79.

Heberlein, T.A. and Stedman, R.C. (2009). Socially amplified risk: Attitude and behavior change in response to CWD in Wisconsin deer. *Human Dimensions of Wildlife*, **14**(5):326–40.

Hedrick, P.W., Peterson, R.O., Vucetich, L.M., Adams, J.R. and Vucetich, J.A. (2014). Genetic rescue in Isle Royale wolves: Genetic analysis and the collapse of the population. *Conservation Genetics*, **15** (5):1111–21.

Hettinger, N. (2002). The problem of finding a positive role for humans in the natural world. *Ethics and the Environment*, **7** (1):109–23.

Jacobson, S.K., Brown, H.O., and Lowe, B.S. (2020). Communications and outreach. In: Silvy, N.J. (ed.), *The Wildlife Techniques Manual, Vol. 2 Management*, 8th edn. John Hopkins University Press, Baltimore, MD.

Jones, M., Hamede, R., and McCallum, H. (2012). The devil is in the detail: Conservation biology, animal philosophies and the role of animal ethics committees. In: Banks, P., Lunney, D., and Dickman, C. (eds.), pp. 79–88, *Science Under Siege: Zoology Under Threat*. Royal Zoological Society of New South Wales, Mosman.

Katz, E. (1997). *Nature as Subject: Human Obligation and Natural Community*. Rowman & Littlefield Publishers, Lanham, MD.

Koval, M.H. and Mertig, A.G. (2004). Attitudes of the Michigan public and wildlife agency personnel toward lethal wildlife management. *Wildlife Society Bulletin*, **32** (1):232–43.

Kupferschmidt, K. (2017). How Canadian researchers reconstituted an extinct poxvirus for $100,000 using mail-order DNA. *Science*, July 6. doi:10.1126/science.aan7069.

Light, A. and Katz, E. (1996). *Environmental Pragmatism*. Routledge, Abingdon.

Lockwood, J.A. (1988). Not to harm a fly: Our ethical obligations to insects. *Between the Species: An Online Journal for the Study of Philosophy and Animals*, **4** (3):204–09.

McCallum, H. and Hocking, B.A. (2005). Reflecting on ethical and legal issues in wildlife disease. *Bioethics*, **19** (4):336–47.

McComas, K.A. (2001). Theory and practice of public meetings. *Communication Theory*, **11** (1):36–55.

Minteer, B.A. and Collins, J.P. (2008). From environmental to ecological ethics: Toward a practical ethics for ecologists and conservationists. *Scientific Engineering Ethics*, **14**:483–501.

Murphy, F.A., Fauquet, C.M., Bishop, D.H.L., Ghabrial, S.A., Jarvis, A., Martelli, G.P., et al. (eds.). (1995). *Virus Taxonomy: The Classification and Nomenclature of Viruses: Sixth Report of the International Committee on Taxonomy of Viruses*. Springer-Verlag, Vienna.

Normile, D. (2008). Driven to extinction. *Science*, **319**:1606–09.

Normile, D. (2010). Rinderpest, deadly for cattle, joins smallpox as a vanquished disease. *Science*, **330**:435.

Pennington, H. (2003). Smallpox and bioterrorism. *Bulletin of the World Health Organization*, **81**:762–67.

Peterson, M.N., Mertig, A.G., and Liu, J. (2006). Effects of zoonotic disease attributes on public attitudes toward wildlife management. *The Journal of Wildlife Management*, **70** (6):1746–52.

Räikkönen, J., Vucetich, J.A., Peterson, R.O., Nelson, M.P. (2009). Congenital bone deformities and the inbred wolves (*Canis lupus*) of Isle Royale. *Biological Conservation*, **142**:1025–31.

Reed, M.S. (2008). Stakeholder participation for environmental management: A literature review. *Biological Conservation*, **141**:2417–31.

Reiter, D.K., Brunson, M.W. , and Schmidt, R.H. (1999). Public attitudes toward wildlife damage management and policy. *Wildlife Society Bulletin*, **27** (3):746–58.

Robinson, J.A., Räikkönen, J., Vucetich, L.M., Vucetich, J.A., Peterson, R.O., Lohmueller, K.E., and Wayne, R.K. (2019). Genomic signatures of extensive inbreeding in Isle Royale wolves, a population on the threshold of extinction. *Science Advances*, **5**:p.eaau0757.

Rodriguez, S.L. and Peterson M.N. (2020). Human dimensions of wildlife management. In: Silvy, N.J. (ed.), *The Wildlife Techniques Manual, Vol. 2 Management*, 8th edn. pp. 39–58 John Hopkins University Press, Baltimore, MD.

Rowe, G. and Frewer, L. (2000). Public participation methods: A framework for evaluation in science. *Technology and Human Values*, **29**:88–121.

Rupprecht, C.E., Hanlon, C.A. and Hemachudha, T. (2002). Rabies re-examined. *The Lancet Infectious Diseases*, **2**:327–43.

Stricker, H., Schmidt, P.M., Gilbert, J., Dau, J., Doan-Crider D.L., Hoagland S., et al. (2020). Managing North American indigenous peoples' wildlife resources. In: Silvy, N.J. (ed.), *The Wildlife Techniques Manual, Vol. 2 Management*, 8th edn. pp. 288–304, John Hopkins University Press, Baltimore, MD.

Taylor, P.W. (1986). *Respect for Nature: A Theory of Environmental Ethics*. Princeton University Press, Princeton, NJ.

US Environmental Protection Agency (EPA). (2001). *Stakeholder Involvement & Public Participation at the U.S. Environmental Protection Agency*. Office of Policy, Economics and Innovation, Washington DC.

US Fish and Wildlife Service. (2003). *640 FW 1: Partners for Fish and Wildlife Program*. US Fish and Wildlife Service, Washington, DC.

US National Park Service. (2018). *Record of Decision: Final Environmental Impact Statement to Address the Presence of Wolves at the Isle Royale National Park*. US Department of the Interior, Washington, DC. https://parkplanning.nps.gov/document.cfm?parkID=140&projectID=59316&documentID=88676 (accessed December 30, 2020).

Vantassel, S. (2008). Ethics of wildlife control in humanized landscapes: A response. In: *Proceedings of the Twenty-Third Vertebrate Pest Conference*, pp. 294-300, University of California, San Diego, CA.

Wagenet, L.P. and Pfeffer, M.J. (2007). Organizing citizen engagement for democratic environmental planning. *Society and Natural Resources*, **20**:801–13.

Western Association of Fish and Wildlife Agencies, US Department of Agriculture, US Department of the Interior, and US Department of Agriculture. (2008). *Greater Sage Grouse Management*. US Department of the Interior, Washington, DC.

Wilcox, B.A. and Ellis, B. (2006). Forests and emerging infectious diseases of humans. *Unasylva No. 224*, **57**:11–18.

Index

Explain at the onset what the letters f, t, b behind the page # stand for (Figure, Table, Box)